AF323028

System Engineering Deployment

systems engineering series

series editor
A. Terry Bahill, University of Arizona

System Engineering Deployment

Jeffrey O. Grady
JOG System Engineering, Inc.
San Diego, California

CRC Press
Boca Raton London New York Washington, D.C.

Library of Congress Cataloging-in-Publication Data

Grady, Jeffrey O.
    System engineering deployment / Jeffrey O. Grady.
        p. cm. – (Systems Engineering Series)
    ISBN 0-8493-7839-7 (alk. paper)
    1. Systems engineering. 2. Management. 3. System analysis. 4. Organization.
    I. Title. II. Series.
TA168.G648 1999
620′.001′171–dc21

        99-19953
        CIP

About the author

Jeff Grady has been President of JOG System Engineering, Inc., San Diego, since 1993. JOG is a systems engineering firm focusing on assessment of company current capability coupled with education, leading to planned improvements. He was formerly engineering manager of Systems Development at General Dynamics Space Systems Division, working on space transport and energy systems. His other experience over a 30-year period includes: system engineer with GD Convair on cruise missiles; system engineer, project engineer, and field engineer with Teledyne Ryan Aeronautical on unmanned photo reconnaissance, ELINT, electronic warfare, and target aircraft; and customer training instructor with Librascope on ASROC and SUBROC underwater fire control systems. He served in the Marine Corps for 10 years in the aviation communications field as a radio technician, helicopter squadron electronics chief, and electronics instructor.

Mr. Grady is the author of *System Requirements Analysis* (McGraw-Hill), and *System Integration, System Engineering Planning and Enterprise Identity,* and *System Validation and Verification* (all published by CRC Press). He is a lecturer in the systems engineering certificate programs at the University of California San Diego and the University of California Irvine, serving on the System Engineering Advisory Boards at both institutions. He is a member of SOLE, SOCE, AUVSI, and IEEE, and a charter member of, first elected secretary for, and founding journal editor for the International Council on Systems Engineering (INCOSE).

Mr. Grady holds a bachelor's degree in mathematics from San Diego State University and an M.S. in systems management from the University of Southern California.

Acknowledgments

The reader will encounter the term *shared vision* in Chapter 2. I cannot over-emphasize the importance of this idea in implementing an effective systems approach and very much appreciate the insight into this shared vision notion, which I learned from Mr. Bernard Morais, President of Synergistic Applications, and Dr. Brian Mar, a professor at the University of Washington. They discuss this important concept in a course they have jointly presented for several years.

When the manuscript was essentially complete, I had a conversation over dinner in Harry's Bar in Los Angeles during the 1997 INCOSE Symposium with Bernard Morais about product interfaces, leading to the addition of Chapter 10.

I thank Bernard Morais and Mr. Gordon J. Lawson, director for the systems engineering process, Rocketdyne (Canoga Park) Division of Boeing North American, for reviewing the manuscript. Mr. William E. Greenlee, system engineering manager at Lockheed Martin Fairchild Systems, provided valuable service in reviewing selected chapters. Thanks to you all for your critical contributions.

Much of the motivation for this book flowed out of a conversation with Mr. Thomas (Tom) J. Nagle, an engineering director at McDonnell Douglas in Long Beach, CA (now a Boeing division), which took place in early 1997 about the existence of three tiers of system engineering capability: individual knowledge and skill, enterprise process definition, and demonstrated ability to successfully deploy the system approach to programs. We all want the latter but often stop when we think we have spent enough money on the former one or two. Good individual knowledge and a sound process simply form a foundation upon which one can build a successful system engineering capability. None of these three elements is sufficient, and all are necessary. Thanks Tom.

Mr. Marty Watenberg, University of California Irvine Extension marketing, has for years trained people in project management as a member of PMI, vice president of Interstate Electronics, and as a member of the University of California Irvine Extension staff. I appreciate his encouraging me to look for the coincidence of purpose between project management and system engineering. Looking back on my own industry experience, I had to conclude that program managers have a lot more influence in deciding whether or not an enterprise will perform system engineering work well than do those who actually do the work. So the reader will find recognition of the important common interest between these two fields in this book.

Thanks also to Terry Bahill, Systems Engineering Series editor for CRC Press and a professor at the University of Arizona in Tucson, for including this and three other of my book projects in his Systems Engineering series. I also appreciate the opportunity to work again with Bob Stern, a consummate editor, at CRC Press.

Contents

List of illustrations

List of tables

Preface

Over the decade of the 1990s while working on four other book projects, all in publication as I begin this project, I have found repeatedly that all of one's knowledge about anything is of fleeting value. Any number of times I have vainly convinced myself that I had come to a nearly complete understanding about ways to organize work and people for the purpose of developing systems to satisfy complex needs. These moments have been short-lived as new knowledge forced rethinking and recombination of old ideas. As our sphere of knowledge expands in a particular direction, we are forced to reorganize our current understandings of related matters to encourage mutual consistency between them as we add new knowledge. The result is a continuous period of intellectual turmoil with a trailing turbulent wake of broken knowledge. At any snapshot in time we may pluck a framework from the bow wave and attempt to describe it only to find that it is in the wake by the time the ink is dry.

One of several central elements behind all four of my previous books has been a detailed set of process diagrams with an attempt to fit all of the work discussed into a common process mold. It is not the same set of process diagrams in all books because of this phenomenon that affects the inquisitive of continuously increasing knowledge. One of the most profound changes in my thoughts about the system development process came through exposure to U.S. Air Force efforts to define an integrated view of the overall management process for programs that resulted in their development of integrated master plans and schedules (IMP and IMS). My last three books have included comment about this process, in each case inching toward a more comprehensive understanding on my part of the optimum marriage between program peculiar planning and an enterprise's generic identity. It is intended that this book will complete that journey for me.

This book is about the transform between an enterprise identity and an individual program. It provides a progressive prescription for making that transform, especially as it applies to system engineering. Therefore, the book focuses on the deployment of a quality systems capability into programs. In order for this to work, it is necessary that the enterprise have an identity maintained and improved over time based on lessons learned from programs derived using metrics. Therefore, we discuss ways to create the generic infrastructure controlled and owned by a functional department structure, the program structures that use the deployed capabilities, and the deployment of the former into the latter. Another necessity is a knowledgeable workforce skilled in the techniques of blending individual knowledge into a common vision. So this book covers ways to connect the in-house process definition (what to do) with how-to knowledge sources to encourage the development of a skilled workforce.

In writing this book, the content of which is driven by the need for specialization, I have finally come to realize that specialization is at the root of many of the problems enterprises have in making the system development process work. We currently educate system engineers, program managers, and functional managers independently and mostly through the school of hard knocks. The reality is that these three casts of characters must act together in predictable ways about which they can be educated together. This conclusion has driven me to realize that the right target is not improving system engineering, as important as that is, but improving the system development process in which these three professions must interact. That is, an enterprise is a system with an architecture of people and organizations and interfaces formed through good or bad human communications.

Jeffrey Grady
January 1999

chapter one

Introduction

1.1 Where are we going?

It has been said that the most complex undertaking we become involved in is the development of systems to solve complex problems, problems like going to the moon, disposing of radioactive waste, developing an economically viable electric automobile, and satisfying difficult military needs. The author agrees but not for the same reason that many system engineers cite. This work is hard not because the problems are technically difficult or because our technology is limited. It is hard because the related problems have to be solved by human beings. This is a lot like the problem of environmental stress in a system. The noise that impacts everything in a space launch vehicle as it rises off the launch pad would go away if we simply shut down the rocket engines. Many of our development problems would go away if we could simply eliminate humans from the system development solution or if we could modify humans so that they might work together more efficiently.

These solutions are certainly not in our best interest, and the good news is that they are not necessary. A few enterprises have mastered the art of developing systems to satisfy complex problems, but most are still struggling to make their process work even though the systems approach has been understood for decades. This process has been called *system development* (or acquisition), the application of system engineering, the most complex human undertaking, and the engineering of complex systems, as well as being characterized by any number of unflattering phrases. Some organizations have rejected the systems approach as overly complex, placing their trust in TQM (total quality management), QFD (quality function deployment), concurrent engineering, or even the *ad hoc* approach, thinking it will encourage creativity in their personnel. The management of some firms flits between momentarily popular techniques, never developing a sound system development capability and tradition.

This book is based on the premises that the systems approach is the most effective method currently available to develop solutions to complex problems and that it is possible to use the same techniques to design effective systems for the development of product systems. That is, we can apply the systems approach to the design of the systems that we belong to through employment in enterprises that do this kind of work. It seldom occurs that an enterprise will begin as a result of a careful, structured analysis of the enterprise vision. Companies are commonly created by entrepreneurs, who are more often guided by intuitive rather than analytical thinking. The management of surviving enterprises eventually will find that the complexity of the problems they are trying to solve has increased and that the loose working environment that served them well in the beginning is no longer doing so. Therefore, while the techniques covered in this book would work in creating an enterprise, they will find greater application in reengineering existing enterprises.

For those who have not yet convinced themselves that the systems approach has merit, let us consider its foundation. The systems approach is powerful because it recognizes the human condition and provides machinery that is effective in getting the best humans have to offer. It does demand some discipline and a desire to succeed as a body of people rather than individually.

There was a time when it was not easy to find good source material on this process because what was known was located in the corporate memory of enterprises that had succeeded in applying it to programs. Sometimes it was written in manuals but more often it was merely retained in the minds of the people working there. This excuse for not applying the systems approach is no longer valid. The process and all of its techniques have been exposed in any number of books published over the past 5 to 10 years. Some universities offer degrees and certificates in the whole process, or big pieces of it. There is simply no shortage of knowledge about how to do this work well. These views of the process do not always coincide precisely, but there exists a sufficiently large pool of knowledge to describe a profession, that of the system engineer.

This profession is based on a broad yet shallow depth of knowledge, with strength only in some particular discipline through which the engineer enters the field of systems engineering. It is steeped in understanding the interactions between elements in systems and the consequences of these interactions. People in this profession have a broad interest in systems rather than a narrow domain focus. They look at the system development process in the context of three fundamental steps: defining problems (requirements analysis and specifications), solving problems (design by domain specialists, integration, and optimization), and proving the solution relative to the requirements (verification).

The system engineer must apply his/her skills and knowledge to technical systems and related problems. The system engineer also applies many of the same techniques useful in developing good systems to the system of which he/she is a member — the development organization. These systems are composed of organizations, teams, and at the atomic level, individuals. These people commonly have serious problems interacting among themselves, regardless of their best intentions and plans. System engineers must be skilled in product systems, in human organizational systems, and in the bridge between them that is formed by programs the enterprise creates to solve problems brought to them by customers.

It is seldom openly disputed in an enterprise that the systems approach to solving of complex problems is superior to the *ad hoc* approach. Under the *ad hoc* approach, people working on the program are given work to do motivated by the immediate situation as conceived by someone in authority rather than in accordance with a grand plan. Because no one has made an effort to seek out, identify, and prevent problems, those problems often occur and must be addressed before the scheduled work can continue. A program can become so convoluted through firefighting efforts to solve these unforeseen problems that the original plan and schedule, if they existed, are simply unworkable.

So why is the *ad hoc* approach applied in so many enterprises, and why do we perform so poorly in applying the systems approach? Some companies are ill-served by their management for any number of reasons. Others are trying to improve their process, but it is not easy. The author has concluded along with Mr. Thomas J. Nagle of Boeing Military Transports that enterprises applying the systems approach require three layers of capabilities in order to do this work well, and often they focus only on one or two of them.

When a critical mass of personnel in an organization has mastered the theoretical base of knowledge needed to do this work, the organization satisfies the first layer of system engineering sophistication. Just one person who knows it all is not sufficient, except in a very small enterprise. All or most of those charged with defining, coordinating, integrating, and optimizing the work of the specialists must have this knowledge. This includes knowing how to do requirements analysis to define the problems in specifications that the organization will attempt to solve through the application of coordinated specialized engineering skill. It includes being able to integrate across the interfaces spanning system elements under different responsibilities. It includes actively seeking and finding conflicts in the current synthesis and exposing them for attack. It includes the ability to encourage and facilitate communication between the many specialists so that they become the equivalent of one great mind attacking a problem larger than any one of them could master independently. Finally, it includes the knowledge to set up a sound requirements verification process that will produce convincing evidence of design compliance.

This knowledge is available from books, from some people who have mastered it in many companies but are all too often not called upon to help everyone else master it, from an increasing number of institutions of higher learning, and from other training sources such as the one run by the author. There is no excuse for an enterprise not mastering this knowledge. Many engineering and training managers believe the solution is achieved simply by sending people to school, but unfortunately, this knowledge by itself is insufficient to ensure a successful system development capability.

The second layer is achieved by having a written process definition or description that tells generically what work has to be done, in what order, and what departments shall provide the people, practices, and tools to do it. In addition, the responsible departments in each case should have made the connection between the content of this process description and the detailed how-to knowledge one needs to go from *knowing* that particular work must be done to actually *doing* it. This connection need not be in the form of a detailed company manual, essentially a textbook on the process. Alternative sources for the how-to knowledge include an in-house training course, identification of a local university course, reference to a textbook or standard that tells how to do the work, or identification of experts on your staff who have this knowledge. Finally, this layer should include identification of the tools needed to do each piece of work and ways for people to learn to use those tools.

Achievement of these two layers is still not sufficient to ensure success, however. The enterprise must also have the organizational skill and experience to deploy these resources into programs and cause them to break into a symphony of action useful in moving toward program goals. Of these three layers of capability required to perform system engineering well, the latter is the most difficult to achieve. Not only is it necessary to have succeeded in the other two levels, it is also very difficult to manage a group of people toward a common goal in accordance with a predefined plan. It is much easier to gather people together and periodically correct them when they go in the wrong direction. These are the characteristics of the *ad hoc* environment, which has proven to be flawed in so many programs that it cannot be seriously considered by a program manager.

When the goal is very difficult to achieve, when developing systems to solve complex problems, when the demands for communication and interaction are extreme, and when the work must be accomplished within severe time and money constraints, there is simply no better way to proceed than to apply the systems approach.

Other books are good sources of the how-to knowledge to satisfy the first level of competency. In this book we will focus on the latter two layers of competency. The author has tried to provide both the logical exposition of a sound system development process and its deployment to programs for the system engineer, as well as making an appeal to those intuitive people (often in positions of responsibility) who are capable of implementing changes in current methods. It is hoped that the book will increase the pace of the deployment of effective systems capabilities into programs in more companies.

The purposes of this book are:

a. To encourage people in positions of responsibility to move from their current state of denial about the superiority of the systems approach to an attitude of enthusiastic support for this process
b. To show how to make this process work within the context of employing flawed but perfectly normal human beings to do the work
c. To show how an enterprise may connect several fairly recent development insights into a powerful engine of program development, raising the plateau of generic process definition.

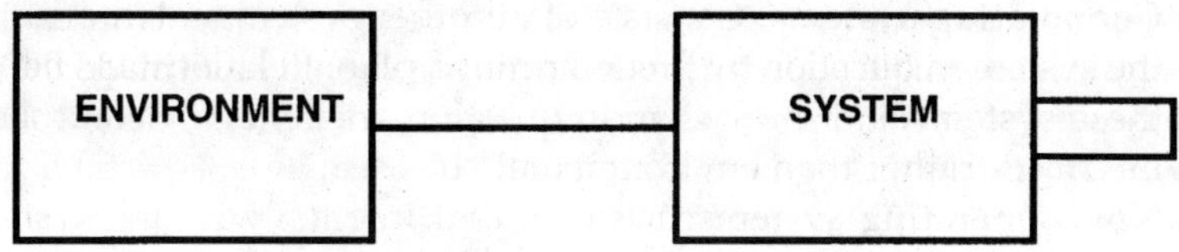

Figure 1.1   The system and its environment.

In a word, this book focuses on the *deployment* of the system engineering process, moving an enterprise from a condition of theoretical acceptance to a condition of implementation excellence. Few enterprises have been successful in deploying an effective systems approach. We seek to understand why this has been commonly the case in the past and what can be done to interrupt this history, encouraging the universal enthusiastic support for the implementation of the systems approach in the development of systems.

1.2   System defined

This book deals with the development of man-made systems, as opposed to natural systems. A man-made system consists of two or more subordinate entities that interact in some fashion to accomplish a process that transforms a set of predetermined inputs into a set of desired outputs, in time achieving a predetermined goal. A system interacts with an external environment, which includes everything in the universe that is not in the system, as shown in Figure 1.1. It is very important to be able to define the boundary between the system and its environment without ambiguity. The system interacts with itself, as indicated by the system-to-system interface line, and with its environment.

Actually, Figure 1.1 illustrates every system that ever was in the past, those that currently exist, and those that will be developed in the future, except one. The one that is excluded is the system we know as the universe itself. That system is all-system or all-environment; there is nothing else. The author made this statement to a physicist who was working on the International Thermonuclear Experimental Reactor and for whom he had great respect. The physicist looked at the author's schematic block diagram of the universe for a few moments and finally said, "Well you might have neglected a worm hole or two." For most of us, it is quite adequate to consider the universe as the collection of many, many galaxies in space. Indeed, for most systems that we must develop, it is not necessary to consider our universe extending much beyond Earth's confines.

The system environment includes several major subsets as follows:

a. **Natural environment:** Consists of space, time, and natural forces and influences. As noted above, this includes everything in the universe not in the system, but we generally do not have to account for relationships that are out of this world, space systems excepted.

b. **Cooperating systems:** Consists of all other systems interacting with the system in question by predetermined plan. Relationships between these systems and the system in question are generally controlled as interfaces rather than environmental stresses.

c. **Noncooperating systems:** Systems that interact with the system in question with no harm intended simply because they share the same space with the system in question. Electromagnetic interference is an example of one of these elements.

d. **Hostile systems:** Systems bent on disrupting achievement of the goals of the system in question through damaging the system or interfering with its delivery of planned services. Hostile elements are sometimes appropriate to commercial systems, as in the potential for a hacker to gain access to data or to introduce a virus into your computer system.

e. **Self-induced effects:** Through normal operation, the system creates effects that interact with natural phenomena to introduce other stresses into the system in question. A rocket engine roars to life on the launch pad. The engine creates acoustic energy that impacts on every part of the launch vehicle rising off the pad. If you could turn off the engine, this acoustic energy would not be a problem.

1.3   Systems beget systems

As pointed out earlier, man-made systems come into being through the creative efforts of men and women organized to work together within another system, known as a program, functioning within an enterprise. In the ideal world, this program and its parent enterprise function efficiently to transform input materials into an output solution in the form of a product system that satisfies the input customer need. The creator system may also be required to perform logistic support and operation services subsequent to delivery of the product system. Our enterprise has all of the characteristics of a system, as we can see in Figure 1.2. It has inputs. It has outputs. It has some internal processes composed of parts that interact and, in so doing, transform the inputs into outputs. Its process is composed of many elements that must work together in some way to achieve an end-use function, in this case to produce useful systems that result in a revenue stream for the enterprise or otherwise satisfy an enterprise vision statement.

The creator system can function with great efficiency in transforming material and the customer's need into an output product system, or not. Clearly, it is desirable for the creator system to be efficient. Enterprises that create systems, whether they are of a military or commercial nature, must function in a very competitive environment, and to remain in business, they must be able to deliver value to their customers. Given two enterprises, the one with the most efficient process, both for generating programs and moving them to conclusion, should prevail in a competitive situation over time. Efficiency in this case can be measured by the enterprise's economic gain relative to the resources expended in making the creative transform. Granted,

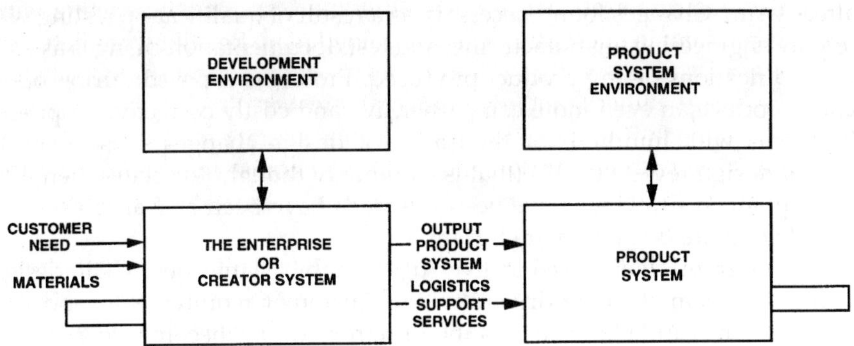

Figure 1.2   Creator and product system relationships.

an enterprise could produce systems of little value to their customers with great efficiency, but they will not survive when other enterprises are producing valuable systems efficiently.

The question we must answer is, "How can we craft an enterprise that can smoothly generate programs that efficiently provide customers with systems they value?" That is the central theme for this book, and it involves two very important activities. First, our enterprise must have a smooth program generator, and second, we must be able to run programs well. The desired outcome does not happen simply because we wish it. We have to work for it, not just once to achieve the desired condition, but for all time. Enterprises are like marriages — they survive if they are nourished, and they wither if they are not. While there are many facets to achieving the desired capability, one of the most important ones is the systems approach for solving difficult problems.

An enterprise that applies an *ad hoc* process runs considerable risk of failing to deliver customer value with efficiency. The reason for this risk is that the enterprise will fail to foresee problems before they become serious and consume resources and reduce economic gain. It is generally agreed that aggregate program cost is inversely proportional to the percentage of resources consumed to good purpose in the early program phases. The expenditure of resources early in a program helps to define the problem to be solved as a prerequisite to solution through design, and exposes problems that can occur later. By mitigating the risks associated with these problems before they materialize, the potential consequences can be avoided.

It is also generally accepted that the earlier we catch a problem, the lower the cost of the cure. Problems avoided through the application of a sound systems approach do not materialize, or materialize with less severe consequences. Problems identified prior to design may cost next to nothing to resolve — the cost of a 15-minute conversation between two engineers. If the cost of labor to our enterprise is $50/hour, this might amount to $50/hour $\times$ 2 $\times$ 0.25 = $25. Problems exposed during the design phase require design (drawing) changes that may reflect on analyses and tests already run. Changes found necessary as a result of qualification testing will require engineering, manufacturing, and test documentation changes as well as modifications of any product produced. Problems exposed during operation produce an even more comprehensive and costly corrective response. Programs with hundreds or thousands of design changes subsequent to critical design review (CDR) (that is commonly thought to occur when 95% of all planned drawings have been released) have been and are in serious trouble and are certainly not efficient.

A program chief engineer may try to alibi a large post CDR design change problem as being driven by new customer requirements, and it is true that this can be a factor. As the design solutions become clear to customers, they may begin to see new opportunities they had not previously visualized. What seem to them to be minor changes could satisfy an altered need. This is generally, however, a weak excuse for failing to do the requirements work well as a prerequisite to the design. If the customer's needs had been fully explored, most of these opportunities would have been exposed. This unearths a very valuable metric for system engineering quality. The total change rate late in the program includes components driven by avoidable and unavoidable changes. It is the avoidable changes that signify poor prior performance of the systems work. The difficulty, of course, is the partitioning of all changes into the correct subset.

Many engineers who advance to a position of management for the development of systems appear to undergo a mental transformation along the way that convinces them that these great truths do not apply to them or their programs. Many managers avoid the systems approach with the rationale that it does not deliver value relative to its cost and it is so rigorous that it adversely affects the creativity of design engineers. These are all symptoms of system engineering poorly applied. It is true that some system engineers believe that the development process revolves around them, or should, while the reality is that this whole process should be focused on providing the creative design engineers with the broadest possible solution space consistent with the system need. System engineering work well done is transparent, and system engineering work badly done is very public, which is another burden the practitioners must bear.

The author has encountered more than a few irritated design and analysis engineering managers in organizations with a pattern of ineffective systems capability. Through ineffective development of system requirements and flowdown of these requirements to the level at which design may safely be accomplished, the system people would sow the seeds of future design failure through poor requirements work, which the system people would be called upon to solve when problems became known. The results were that a system engineer who had performed poorly during the front end received praise later for solving a problem he helped to create. These are valid concerns for design and analysis managers in a company that permits this kind of thing to happen.

Mr. Mack Alford, former Chief Scientist at Ascent Logic and developer of RDD-100, a private consultant when this was being written, has a saying that illustrates the consequences of these attitudes in the software world, but it is immediately understood in the hardware business as well in terms of engineering drawing changes. A software person needs to define the requirements for the software he is developing and he replies, "I don't have time to write down the requirements; I'm too busy debugging." Work on poorly defined problems will grow, especially when defining the problem consists of discovering the boundary conditions uncovered through bad choices exposed too late. All of us not in denial know this to be fact.

It is hard to accept that we are members of a system, because we cannot easily sense its existence from the inside looking out. Besides, we are people, not modules in a machine. This denial is perhaps the basis of our resistance to consciously treating our enterprise as a system. As a result, we miss the important insight that we are in charge of this creator system. We have the option to continue functioning as we are or to consciously work to reengineer our own system. We can choose to optimize our creator system. The choice is ours — to be satisfied with the status quo or to pursue excellence.

Unless our enterprise is much better than average as a creator system, we may also be unwilling to accept that it is a system because we choose not to believe that a system could function so badly. Ah, but there are good systems creating good *and* bad product systems, and there are bad systems creating good *and* bad product systems. Of all of these possibilities, we can easily accept that good systems have a higher probability of creating good systems. Let us accept that notion without proof. As long as we must be elements of a system, would we not prefer to be elements of a good system? We do have a choice. Those who do not have overall responsibility can either work to improve their creator system or migrate to another system (company) that is working better or has a management determined to make it better. If you are running this organization, desertion is not an option; at least you should try to make the enterprise whole first.

Whatever your choice, we must determine what constitutes a good creator system and how we might navigate to that condition. The inputs and outputs of the system are fairly fixed generically so we must work on the internal transform or process. This book offers a generic process model that can be applied to almost any kind of complex problem. It can be applied to software or hardware at the grand system level as well as down through the depths of the system. Many different analysis methodologies have been woven into the process and are available for selection as a function of the product line and preferences of the enterprise. The process can be applied whether you choose to mirror the waterfall, spiral, "V," or "N" development sequence model illustrated in Figure 1.3. It is compatible with rigorously structured development as well as rapid prototyping. It should function on weapons programs as well as commercial endeavors.

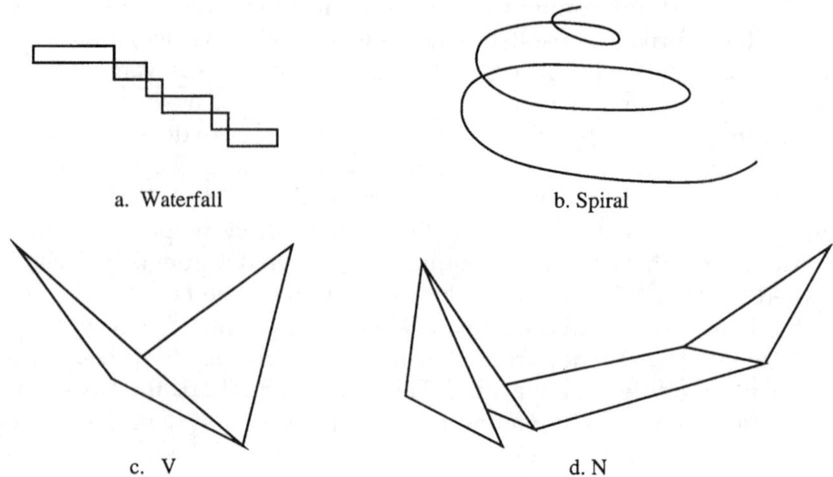

Figure 1.3   Development sequence models.

There are some generalities, however, that are simply too grand. We must agree on the enterprise level. It is true that some problems that man-made systems are designed to solve are so complex that two or more enterprises must work together cooperatively. This book will, however, focus on optimizing the creator system at the single enterprise level. This may be at the division level of a corporation or the corporate level at which contracts can be consummated. Greater aggregates can be constructed by the reader using modules composed of two or more good creator systems with an additional layer of management and technical integration and optimization.

Figure 1.4 offers a generic view of our enterprise as a first expansion of the model shown in Figure 1.2. The business of the enterprise is accomplished through programs, each associated with a particular customer and their needs. Each program has its own management staff focused on achieving customer goals. The author believes that, for enterprises with multiple programs, a matrix is the best structure. This matrix should include a lean functional management organization responsible for providing all programs with qualified personnel, good tools, and proven standard practices. Functional management should also be responsible for maintenance and improvement of resources through continuous process improvement based on lessons learned and benchmarking information from outside the enterprise.

The enterprise has a new business function that maintains contact with outside possibilities in a fashion driven by the customer base. If the enterprise deals with the Department of Defense (DoD), it has a marketing staff that works closely with potential customer acquisition organizations to determine their needs and offer solutions the enterprise can deliver. Customer needs and requests for proposal flow through this function and back to the customer in the form of proposals. Winning proposals result in new program starts. If the customer base is commercial in nature, this function

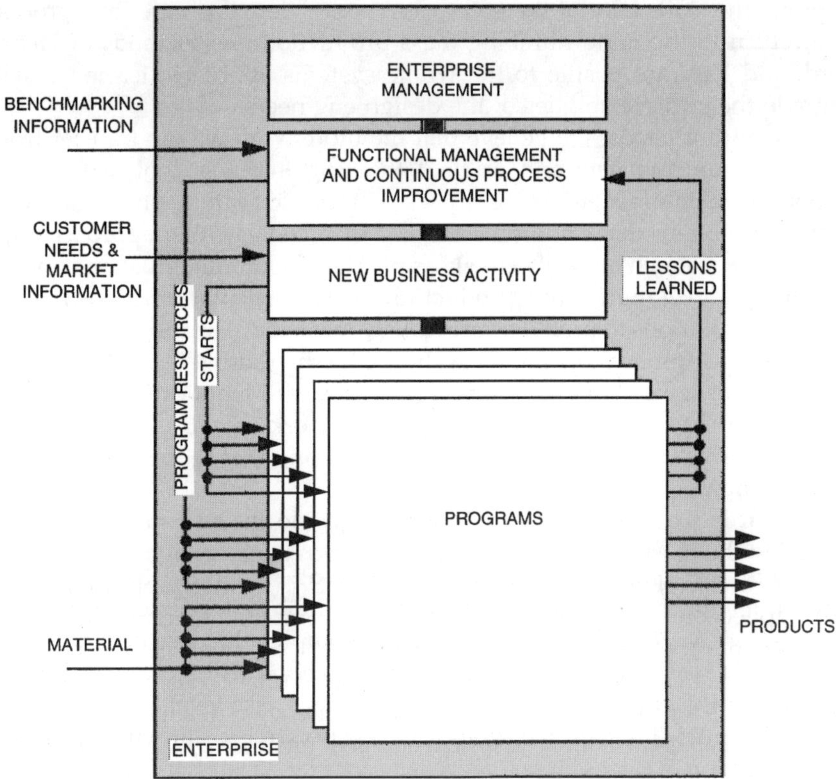

Figure 1.4   The enterprise structure.

may entail a more energetic process of identifying customer needs expressed, not in terms of a request for proposal, but the results of market research, focus groups, and historical precedent.

The single program enterprise is an example of the matrix in which functional and program management merge into one entity. Another example is the author's organizational structure of choice: the matrix with a lean functional staff responsible for the knowledge base, tools, and best practices, combined with physical allocation of personnel and their day-to-day management as a function of the programs.

1.4   Creative vs. canned responses

Programs will come and programs will go for the multiple program enterprise, but the enterprise should continue functioning on all continuing programs in accordance with a standard process incrementally improved over time and tailored perhaps to some extent for the specific program. That is one of the fundamentals encouraged by this book, the adoption of a standard process that can be used as the basis for repetitious work that helps to improve the skills base of the enterprise. Some view this repetitive process as a return to the mind-numbing mass production line methods of Henry Ford, and they are hostile to the whole systems approach, fearing it will degrade the creative abilities of the design engineers.

The author chooses to believe that the more work we can transfer from the pool requiring new and original thinking into the pool of standard responses, the more mental capacity we will turn loose to apply to the really difficult problems that cannot be resolved by a cookie cutter approach. Our intelligence is much more effectively applied to the unique situations uncovered in the development of a product to satisfy particular functions than to recreating solutions to problems previously mastered. The result of applying the systems approach to a program should be the routine disposition of the considerable amount of repetitive work and clear identification of the serious problems to be resolved by our creative workforce. By structuring the work, there will still be plenty of work remaining that requires the unique creative capabilities of the human mind.

It is true that the system engineer can overdo the structured approach, achieving the most worrisome fantasies of design engineering managers. The U.S. Air Force Systems Command Manual series 375 (Forms B) approach, which was applied on many intercontinental ballistic missile programs during the heyday of the application of mainframe computers to system engineering tasks, came close to this result, and the reaction to the resultant constraining feelings was the basis for the concern expressed by a generation of engineers, some of whom are managers today. The system engineer must recognize that the needs for creativity are great in the earliest phases of programs, and this condition changes over time until the program must apply very rigorous procedures to purposefully make it difficult to change the design because of the magnitude of the investment in the decision making process. The system engineer locked in on the mindset appropriate to a later-phase program will do a great disservice to early-phase programs by trying to apply too much structure too early. In this situation, the engineering community is right to rebel.

The challenge for program management and system engineering people is to set up development environments that maximize the opportunities for communication between specialists, minimize the need for communication, and clarify ways for anyone to determine the importance of communicating particular kinds of information. We wish to create an efficient communications system between the people in the program without so loading it down with noise that the important messages fail to get through. Product interfaces are the places where the development of product systems tends to break down, and this is exactly where organizational systems tend to break down as well, at the interfaces between teams, between people. These interfaces are completed through human communication, which is essential in the development of solutions to complex problems. In fact, this need for effective human communications is at the very foundation of the systems approach.

1.5 *The current plateau*

The content of this book applies to what the author characterizes as a plateau of current capability. It may be possible in the future to make significant advances in development technology by adapting computer technology more directly to our intelligence by a means that is not understood today. The author of *How Brains Think*, William H. Calvin, paints a picture of extremely complex processes working at the lower levels of our minds, above the molecular level but still very low in the hierarchical chain. At some point, it may be possible to understand how our mind really works at the micro as well as macro level and network our mental capabilities at a more detailed level than is possible through communication implemented by spoken and written language and pictures. One can imagine walk-around computers (currently available) interfaced with our mind in a more intimate fashion than we can currently imagine and interfaced together with mainframe, mini, and desktop computers via radio ports into these enterprise networks. In this environment, it may be possible to implement an effective system development process in ways that are beyond our current insight.

The author has chosen to accept that there is a current reality within which we will have to work for some time. That reality can be characterized by several constructs. Enterprises are formed to solve problems defined by their customers or themselves in return for some form of compensation. An enterprise may have more than one problem-solving process, called programs, at work at one time. These enterprises are comprised of people, and the enterprise associates these people with the programs in-house based on the knowledge needs of the programs and the availability of personnel.

All persons have minds that are capable of understanding ideas communicated to them in speech, written, or pictorial form, applying their knowledge, experience, and skill to transform generic knowledge in some specialized field into useful knowledge particular to a specific problem and its solution. These people are all specialized, and therefore limited, in terms of their depth and breadth of knowledge. The more relevant the knowledge brought to bear on a problem, the faster it will be solved and, generally, the solution derived will be improved.

Programs are structured to associate people having the right knowledge mix to problems focused on the things that must be created, encourage sound thinking about the problem, and effective communication between the several people involved in solving the problem, to create the effect of one mind with all of the knowledge needed to solve the problem. Today, our best approach for encouraging the desired result is to organize the right specialists into product-oriented physically collocated teams and organize a time-sharing arrangement so specialists have time to apply their detailed knowledge to the problem in isolation mixed with the synergism of the team through meetings.

1.6   New vs. modified

This book has probably been unconsciously constructed based on programs involved in the creation of a new product system rather than on the more general situation where we may have to create a really new system or modify one that has already been created. The author believes that the techniques in this book can be applied to any complex problem-solving process where the problem is to create a truly new capability where none exists or where there is a current reality we wish to improve. In each case, a program will be required that first works to understand the problem and then acts to create design solutions (hardware, software, and procedures), and finally proves that the design solution satisfies the predefined requirements.

The book will not point out in every case how the ideas included can be applied in both cases, but it is hoped that the reader will be able to apply the content by extension to the product system modification situation. Generally, in a modification program we should focus our development insight through the precedented architecture overlaid by the change in system functionality. In an unprecedented application, we would focus first on the needed functionality and, through an application of structured analysis, determine what shall accomplish that functionality.

The author's system development thinking has been shaped by long adherence to the ideas of Sullivan, the late 19th century architect who maintained that a structure's form should be driven by its function, leading to the phrase "form follows function." The author has discovered, however, that there is a wide range of structured analysis environments, not all of which are function driven. While one could apply an object-oriented approach to the reengineering of the enterprise, the author has chosen to apply Sullivan's ideas. That is, the author has consciously selected the functional analysis approach as the model for understanding how to design or redesign the enterprise.

1.7   The book

In Chapters 2 and 3, we will explore the human component and its relationship to the knowledge (K) needed by the enterprise based on its vision statement, related customer and product bases, and the organizational structures appropriate to the functional axis of our matrix. In Chapter 4, we will further dissect the enterprise vision statement to create an effective generic process for accomplishing the work of the enterprise as illustrated in Figure 1.5. In Chapter 5, we will draw a map leading from the process steps and functional department charters to their associated knowledge base responsibilities. This map defines the link between functional departments and the work that must be done on programs, and this linkage is called up later in Chapters 11 through 13. In Chapter 6, each of the process task blocks exposed in Chapter 5 is described in terms of the work that must be done within the context of a hierarchical structure of functional department manuals.

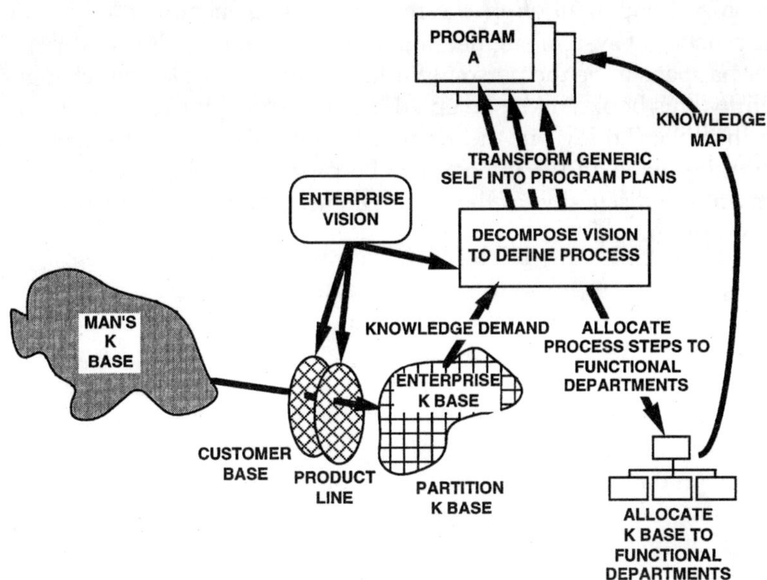

Figure 1.5   Program resource generation.

Chapter 7 tells how to structure program organizations around the product that will be developed, drawing resources from the functional departments and linked clearly and simply into program cost, schedule, and technical responsibilities, interface responsibilities, and work definition. These programs must encourage excellent communications between the teams and individuals from which they are built. Chapter 8 encourages that result. The functional department resources generated up to this point must be applied to a program, and in Chapter 9 we offer a structured process for defining the things in a product system around which the planning will coalesce. Chapter 10 encourages the clear identification of the interfaces between the elements in the product system and draws an analogy to similar interface patterns among the humans who staff the teams.

Chapter 11 introduces a powerful engine of program generation channeling our generic resources into program tasks, focused on customer need, into an infrastructure that offers relative management ease. Chapters 12 and 13 expand upon Chapter 11, offering two implementation approaches. Chapter 14 provides insight into estimating the program job both generically and programmatically. Chapters 15 and 16 build the infrastructure for access to the how-to information people need in order to implement the enterprise process in terms of a training program and references to other material telling how to do the tasks called for in the generic process diagram mapped to the functional organizational structure. Chapter 17 offers avenues to pursue in metrics to measure program performance of system engineering work and to assess the quality of performance of this work on programs.

Finally, Chapter 18 offers the reader ways to encourage improvements in the practice of system engineering without crippling his/her career. If the reader happens to be the person with the power to implement change, then hopefully, this book will be accepted as a cookbook for improvements. The intent has been to lay out the content in the sequence that improvements can best be made and is intended to be prescriptive in nature. Chapter 19 closes out the discussion of the deployment of system engineering capabilities into programs.

chapter two

The human foundation for system engineering

2.1   The expanding knowledge base and its effects

Throughout all prior time and for much of the first half of the 20th century, a single engineer could generally accomplish the complete design task for an entity as illustrated in Figure 2.1a for two reasons. First, the problems posed for engineers could be solved within the context of a single engineering discipline, such as electrical or mechanical engineering. Second, the expanse of knowledge required to solve these problems could be mastered by a single engineer with no help from specialists focused narrowly on particular facets of the engineering process.

Beginning before World War II during the late 1930s and continuing immediately thereafter during the early cold war years of the late 1940s and the 1950s, the military problems posed for engineers became increasingly complex and often would not yield to a solution solely within the context of a single engineering discipline. It was also found that it was not possible to depend on a single engineer in any one field to master the complete range of knowledge due to an expansion of the knowledge base for science, engineering, and weapon systems. As a result, it became necessary to apply the specialization paradigm, narrowing the scope of the traditional engineering disciplines and adding others, such as reliability. Engineers from several disciplines had to work together as suggested in Figure 2.1b, each supported by a cast of specialists.

The intent was to cover the complete range of knowledge needed to solve the problems encountered in the development of increasingly complex systems. In too many organizations managed through a matrix, the mistake was made to place the responsibility for program work in the hands of functional department management rather than people reporting to the program manager. Specialty engineering integration during development became a very difficult problem, with all of the people physically collocated by functional departments and budget partitioned in a crazy quilt relationship

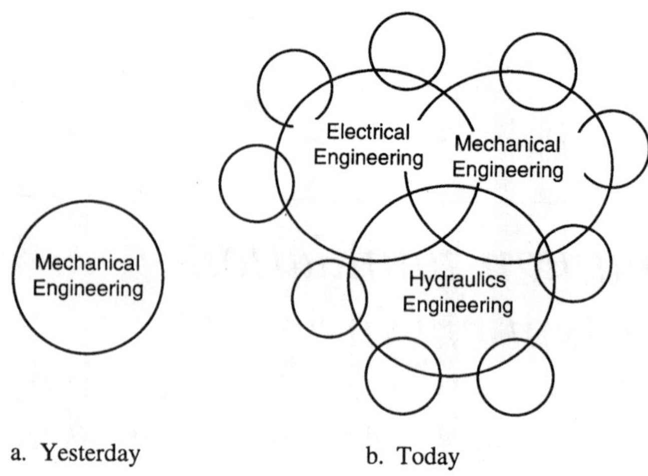

Figure 2.1   The world has changed.

between the items in the system (identified by a work breakdown structure) and the functional departments contributing to item development.

The difference between the two situations illustrated in Figure 2.1 is significant in the emergence of the systems approach. It is at the very foundation of the system engineering process driven by the explosive expansion of man's knowledge base beyond the early specialization framework applied to the engineering field, consisting of civil, mechanical, electrical, and chemical engineering, forcing further specialization of the people who do this work.

2.2   Knowledge dimensions

Specialization has more than a single dimension of depth. We should recognize that a specialist cannot achieve depth in their field and also master great breadth of knowledge. When we specialize, we are limited in other fields as suggested in Figure 2.2. Any development program will require access to particular knowledge but not all knowledge. The nature of the program that the several specialists in Figure 2.2 will be working on determines the kinds of specialists required. A project involving flight in the atmosphere will require an aerodynamicist, for example, whereas a computer system operating in a fixed facility probably will not. These several specialists might have some difficulty communicating except for a shared knowledge in, at least, mathematics and a language (checkerboard pattern shared by all of the specialists) permitting them to discuss their common problem.

All specialists do not, however, specialize in depth. A system engineer (cross-hatched knowledge space in Figure 2.2) specializes in breadth of knowledge in order to understand and integrate the products of the specialists. It is not possible for any person, even a system engineer, to master both

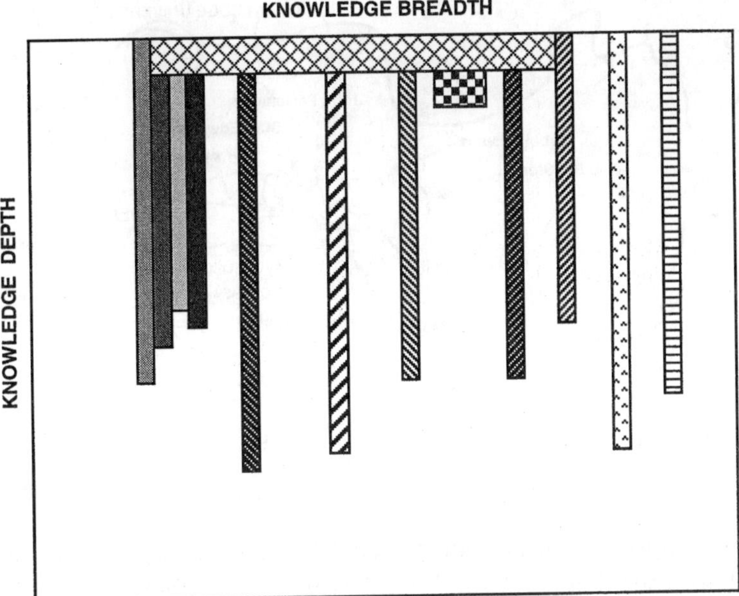

Figure 2.2 Breadth and depth of knowledge.

depth and breadth so there will be specialized fields about which the system engineer is not well informed. To the extent that the system engineer is thin from a knowledge perspective, problems not detectable by the specialists may become embedded in the evolving design. This can especially occur at the product interfaces.

Every system engineer has to come from a past experience that probably involves an engineering discipline providing a knowledge strength. The author has known system engineers with a background in the normal engineering disciplines (mechanical, electrical, chemical, and industrial), as well as textiles, physics, mathematics, economics, and journalism. No matter the past experience, the system engineer must accumulate sufficient knowledge about the many disciplines involved in a system development to communicate ideas in those disciplines. In today's world, involving an intense appeal to software, the system engineer, many of whom come from a hardware background, must extend their knowledge base into software modeling to the extent necessary to enable intelligent communication.

So how does the system engineer master both depth and breadth sufficiently to provide a useful service on a program? Figure 2.3 offers one route. You first take an inventory of your knowledge and compare it with the knowledge space required for common programs upon which you must work. Build a list of the areas not well understood and identify someone in those fields whose knowledge you trust. You and the knowledge base formed by your personal support team create a whole system engineer with a broader knowledge mastery than any one real person can realize. It is not

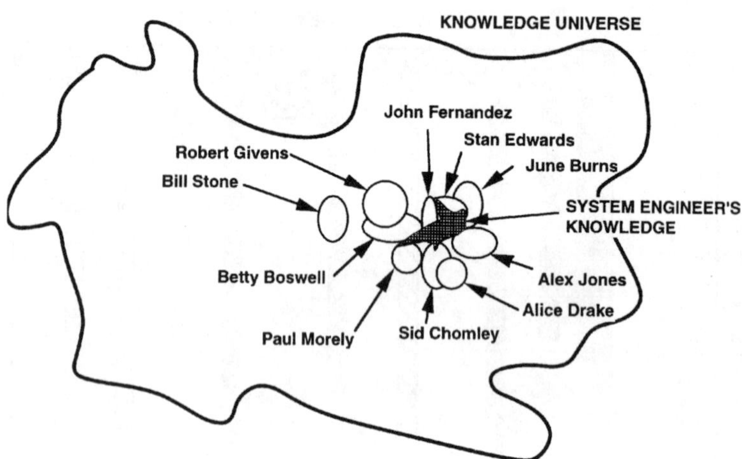

Figure 2.3   The whole person.

necessary to inform all of your personal team members of their selection, but it is helpful to write this list down and as people leave your team you can replace them with another person of similar skills. This is an example of what many people call networking, of course, but networking for a very specific purpose, to create a whole system engineer.

Systems to solve complex problems cannot be created by individuals because of their demands for knowledge beyond what any one person can master, despite the pooled knowledge technique discussed above. Also, it becomes necessary for an enterprise to assign more than one person of the many disciplines involved in these developments in order to get things accomplished within the customer's schedule. The knowledge base for an enterprise, therefore, can best be represented by the model in Figure 2.2 with an added third dimension of depth reflecting the number of specialists in the several fields as shown in Figure 2.4. A development enterprise may have one or more specialists in every field required by their product line and customer base but may not be able to bring to bear some of these specialists on all programs because of a depth problem. All of the specialists may be fully involved in other programs, and one program could suffer as a result.

2.3   Knowledge bond

A man-made system is a collection of things that interact to satisfy a pre-defined need, as noted earlier. The customer need for a system defines a problem that must be solved. Man-made systems are created by system development enterprises. It is generally accepted that possession of more knowledge relevant to a problem by such an enterprise is better than less in solving problems. Efficient and effective problem-solving organizations are rewarded in a competitive marketplace, encouraging system developers to

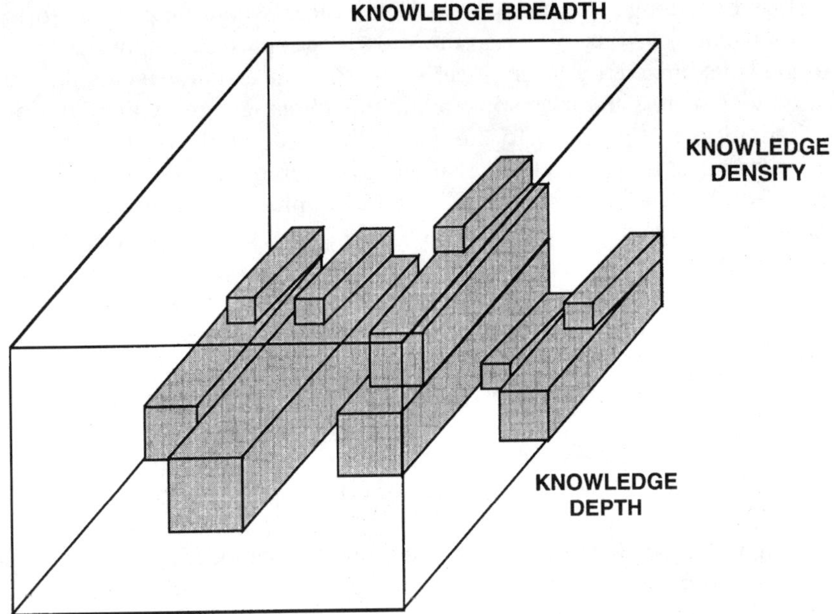

Figure 2.4   Organizational knowledge dimensions.

master the knowledge related to particular kinds of systems on which the organization chooses to focus.

There has been, over the past 10,000 years, an explosive growth in knowledge, while human knowledge capacity has remained fairly constant. A very long time ago, man's knowledge base exceeded the individual knowledge limit. Man has solved this knowledge deficiency problem when confronting problems requiring access to a larger knowledge base than any one person can master by applying the concept of specialization. Under specialization, individuals focus on subsets of knowledge such that no one among them has mastered the whole knowledge set needed to solve the problem, but collectively some number of people do possess the needed knowledge.

While specialization solves the problem of knowledge coverage, it partitions or isolates the aggregate of all the knowledge needed to solve a complex problem into the minds of many different people. The collective or aggregate knowledge of the several cooperating specialists can only be applied efficiently in solving complex problems if it can be shared. Humans have developed the ability to communicate by verbal and visual means in various media permitting the sharing of knowledge. It is intended that the specialized knowledge of individuals working together will be shared through some form of human communication. The generic specialized knowledge is less important in this communication than the system-peculiar consequences of that knowledge applied to solve the problem represented by the system.

This knowledge-sharing process is enhanced when the people working together share a vision and when the knowledge space corresponding to this vision is maximized. In the engineering field, some essential elements of this shared vision are a common spoken and written language, elementary logic, mathematics, and physics. To the extent that the people working together on a common problem can efficiently share problem-related knowledge in near real time, the fruits of their labor will be enhanced in terms of reduced risk of failure, less cost, shorter solution time frame, and greater effectiveness of the solution.

The process of encouraging the intense cooperation of specialists toward solution of a common problem, once referred to as specialty engineering integration, is now often called concurrent engineering or development. Clearly, this process depends on good communication between the specialized participants in order to expand the shared vision of the system under development. Good communication can be encouraged through several steps that recognize the fundamentals of human psychology.

a. It is known that the quality and magnitude of communication between people is increased as an inverse function of the distance between them. So people who must communicate frequently should be placed close together, especially when the concepts they must communicate are complex.

b. Knowledge may be shared in a wide range of patterns as suggested in Figure 2.5, showing communication paths between eight people A through H. Figure 2.5b calls for everyone to communicate through a single person. Figure 2.5c encourages communication between anyone and everyone in each case. Figure 2.5a provides for communication in any pattern we may choose assuming that any number of communication connections may be enabled for any one communication action. The problem with the 2.5c configuration is that many of the communication actions are of no interest to some people, so we are detracting from the focus of those people who need not participate in particular communication actions when we force them to do so. The problem with the 2.5b configuration is that the synergism of the group may be lost in that direct communication between individuals is discouraged. In order to make the 2.5a configuration work, we must adopt the goal of maximizing the opportunity for communications while minimizing the need to do so. The magnitude of the shared vision determines the need. The greater the shared vision, the less the need to communicate, but the shared vision is achieved through communication.

c. We need to place close together the people who have the most difficult communication task. All of the specialists of a particular kind, like reliability, working on the development of a system have a simpler communication problem between themselves than all of the specialists of different kinds working to develop a particular item in the

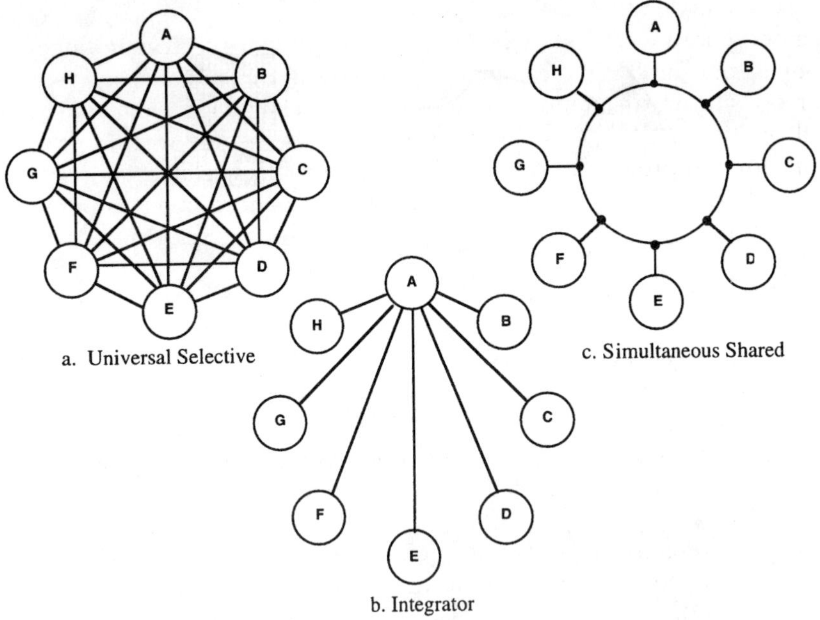

Figure 2.5   Some human organizational communication patterns.

system. So we should organize by cross-functional teams in accordance with the system architecture and collocate physically as a function of those teams.

d. The knowledge retained in the minds of the people working on a development and the direct human communication network via voice and visual images by various media should be supplemented by machine-stored information connected via machine networking to encourage sharing ongoing work products in near real time. The information on the machine media should be organized by team and discipline coinciding with the organization of the product and the people working on the product development. The people must be encouraged to place their in-process work product in the machine storage media and update it in keeping with current conclusions.

2.4   Knowledge and its exchange

Programs represent strings of knowledge generation, accumulation, and sharing in time. Knowledge can exist in one of four forms or nodes: (1) that which is stored in a human mind, (2) that which is stored in a digital computer, (3) physically stored information in the form of paper documents, computer storage media (tape, disk, etc.), and (4) the special case of knowledge resulting from synergism of the group. Figure 2.6 illustrates some sample situations in the transfer or exchange of knowledge between these nodes.

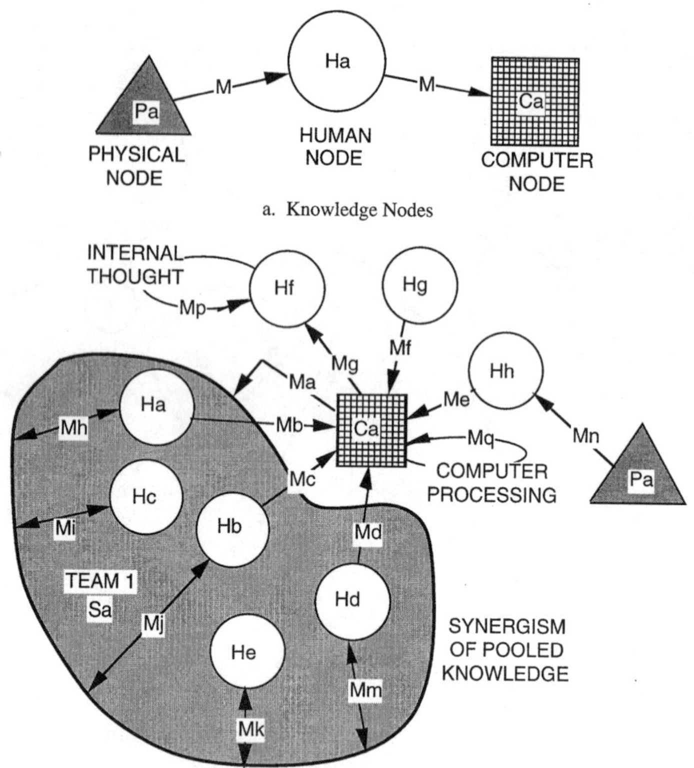

a. Knowledge Nodes

b. Representative Knowledge Transfer Situation

Figure 2.6   Kinds of knowledge and transfers.

Figure 2.6a identifies three of these nodes coupled together through messages. Physical node Pa may be a paper document being read by human Ha who enters data into a computer Ca based on the new information derived from the paper document possibly modified through internal thought by Ha.

In Figure 2.6b we have a more complicated situation in which there are five humans inputting information into a computer (Hg via message Mf, Hh via Me, Ha via Mb, Hb via Mc, and Hd via Md). Human Mg is deriving information from the computer via message Mg and thinking about it via Mp to create new knowledge. Human Hh is deriving knowledge from a paper document Pa via message Mn. Data being entered is being processed to create new knowledge via message Mq. Several members of Team 1 are meeting in a group deriving information from the computer via message Ma and interacting as a group, resulting in synergism Sa not possible by any other means.

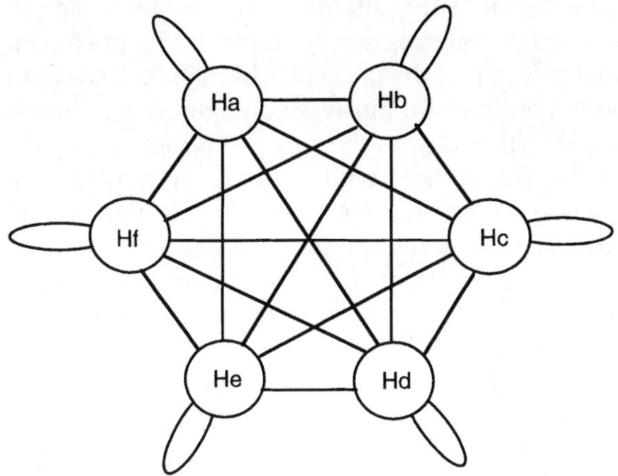

Figure 2.7   Communication possibilities.

2.5   *Individual vs. group work*

The reader can think of many other combinations of these communications building blocks interconnected by messages in various ways. One might feel that maximizing communications might be the right action in all cases. However, if everyone assigned to a program maximizes his/her communications with everyone else on a program, the result could be to deprive everyone of any opportunity of doing any real work. For example, if we had a team of six engineers working on a common problem, as shown in Figure 2.7, we would find that there are hundreds of communications situations.

One of these situations is everyone talking and no one listening. Another six situations cover one person talking and five others listening, as in meetings. Another 15 situations cover different combinations of two people simultaneously talking and four people listening. Yet another situation is all six people internally communicating, more often called thinking, signified in Figure 2.7 by the circular lines with both ends on the same human node. Benefits derived from this situation must be balanced against the goal of maximizing communications on a program.

Individual specialists need time to think and apply their specialized skill and knowledge to the problem at hand, and continuous forced intense interaction between team members can have the effect of suppressing the specialized work that will result in useful information products, thereby reducing the rate of progress of the program. On any one program we are challenged to maximize the opportunities for messages between all of the human nodes while minimizing the need for them so as to provide time for the development of knowledge and related messages by the people representing the nodes reflected in Figure 2.7.

All of these communication situations will not be accomplished with the same effectiveness in a real program environment. Some may be suppressed by the emotional makeup of the particular humans selected to staff the team, while others are overdone. An introverted engineer may not be assertive in communicating information he/she has produced or in gaining access to information needed from others, while an extrovert may force his/her correct or incorrect ideas on everyone else while demanding needed information from team mates regardless of schedule realities.

There are several paradigms that could be applied to regulate the communication process within a group of persons. We might conclude that the communication process should not be interfered with in any way, relying on the participant to ask the right questions of the right people at the right times and provide the right answers to the right people at the right times. Alternatively, we may conclude that some form of facilitation or control is required to encourage efficiency in movement toward scheduled events.

2.6   *Communication facilitation and facilitization*

There is no more powerful communication mode than person-to-person conversation. Physical collocation of product-oriented team members from the several different disciplines encourages effective communication using this mode when it is needed either to ask and answer a question or to supply needed but unsolicited information. Formal team meetings should be called periodically to take advantage of the synergism of the group and the informal discussion groups encouraged to handle problems as they arise. The space should have computer projection capability and some wall space, a projection screen, or a nonglare white board surface permitting projected image markup or annotation. Wall space is also needed so the team can place key drawings and data representing the current big picture.

If the team is relatively small and can be located in a single unobstructed, contiguous space, the team can effectively be in a meeting all of the time, with the people involved in the conversation changing over time as it moves from one area of interest to another. A team in this situation can be working in the individual mode, engaging in an all-hands meeting, or something in between, depending on the needs of the moment. The advantage of this arrangement is ease and speed of assembling the right conversation at any time. The disadvantage is that it can be a distracting environment within which the specialist must accomplish his/her detail work.

The other extreme of team space is a fully cubicalized space in which everyone has a measure of privacy within which to work. Obviously, the disadvantage is the difficulty of interaction between members. The team using this arrangement will need a separate or adjoining area within which to meet, and that space should have computer projection capability as mentioned above.

In addition to direct voice communications, teams need to be able to extend their communications via telephone conversation, fax (or its equivalent), video exchange, and computer networks. Eventually, all of these kinds of services will be integrated into a desktop instrument the purpose of which is to seek out and share knowledge between the humans staffing a program development effort. This capability will not expand the individual's capacity for knowledge nor solve the problems created by specialization. Rather, the continuing expansion of knowledge, despite the improvements in communication, will encourage a multilevel grouping of the application of system thinking. The computer will vastly improve communication between people in the workplace. It will integrate more closely the nodes of knowledge residing in the people who staff the teams.

2.7   The enterprise knowledge base

The computer, by whatever name, will become the instrument of shared vision for teams and programs. The work that all of us accomplish should lead to knowledge products composed of documents (drawings, text, database content, and so forth) that provide the results of that work to be shared. In the past, this sharing process has been imperfect, and the only repository has been in the minds of those working on the program. Each specialist has read and retained information thought to be related to his/her work. This information influences subsequent communication between program personnel, each of whom has been impressed with some subset of the aggregate program knowledge. In this situation, the shared program knowledge base, that which everyone knows, might reach 50%.

The work of all of the specialists must be placed in a computer network so that it can be shared by all team and program members. The computer thus becomes a member of the program and its teams. This member happens to be able to master a tremendous range and mass of information as far as remembering it faithfully is concerned. This member, however, can do very little with this information. The computer becomes a valuable though limited team member with a great memory and thus provides the program with the important shared vision function.

Books have, in the past, been the principal shared vision instrument for man's knowledge base, as illustrated in Figure 2.8. Through books we can share in the wisdom of others and develop with others a common perspective (not necessarily a shared belief). The computer is rapidly taking over this role, and it should move into this role for enterprises and programs.

The enterprise knowledge base is some subset of man's knowledge base focused on the enterprise's product line and customer base. This knowledge is retained in the minds of the persons working for the enterprise, but it should also be captured in a documented form available to all. The program knowledge base is initially generated through assignment of personnel, those temporarily not employed on other programs as well as people from other programs. These people carry with them their knowledge of their department's charter and how to do related work.

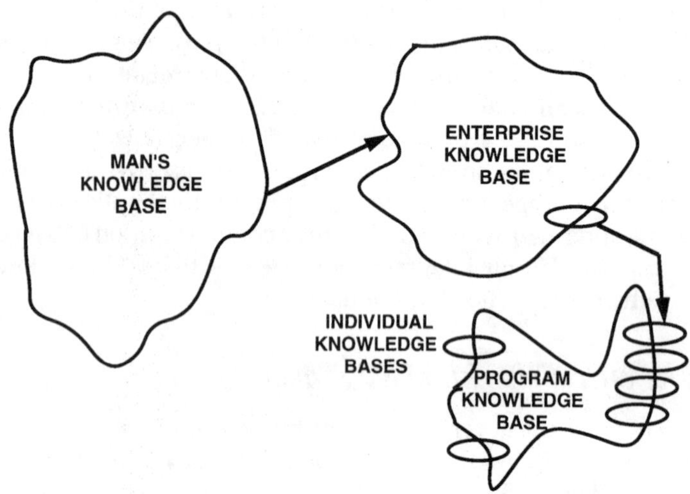

Figure 2.8   A progression of knowledge bases.

This enterprise knowledge base consists of two kinds of information: what-to-do knowledge and how-to-do knowledge. Both kinds should be captured in written form and reside on the computer network. In the case of a specific discipline, this may amount to a checklist of things to do on a program and the sequence in which they should be done to satisfy the what-to-do knowledge. For each item, we also need to know how to do the related work. This may be in the form of an in-house manual, reference to a textbook or standard created by others, or reference to a course available in the company, from a local university, or otherwise accessible.

All of this information should be accessible from everyone's desktop. Specialists who have not yet mastered their specialty can move toward competency faster than they might without this information easily available to them. Those who would manage the work of others can find out what all of these people do at a time of their own choosing. System engineers will become better informed about how the various specialists contribute to the increasing store of knowledge on programs. The whole enterprise and its subsets, called programs, will achieve a shared vision that helps all of the participants focus on achieving the enterprise vision on each program.

In a military unit, the members are bound by a sense of history of that service and unit, a tradition of selfless service, and the knowledge that failures will bring them down in the esteem of their fellows and may lead to punishment as well. They have a shared vision that brings into focus what their purpose is and how to attain it. As a result, we find it possible to cause a group of otherwise sane persons to form a unit and march in cadence in response to their leader's commands. These same people will move into harm's way with energy, knowing full well the personal dangers involved.

An enterprise should seek this same kind of dedicated service in its people but cannot generally appeal to the same forces that shape the ideal Marine infantryman. A common vision shared between management and everyone else, a sense of common justice, a reward system responding to both individual and team efforts, enduring top management support for the current enterprise and methods to continuously improve it (rather than a temporary leap to each new fad), and a means to share the fruits of our specialized knowledge will go a long way toward providing an enterprise with people who will repeatedly march into hell's furnace with a proposal or a program manager. The fine tradition of a beer party from time to time might not hurt.

This bond can be very strong in both directions. As an example, a fine gentleman, Mr. Erich Oemcke was the Engineering Vice President at Teledyne Ryan Aeronautical for a period in the 1980s. He personally led three proposal efforts in a row, all of which the company lost. After the third loss, Erich gathered all of the people in the department and told them that he could not expect his people to follow him on another crusade as selflessly as they had on the previous three, and as a result he was resigning from his position and the company.

chapter three

Enterprise knowledge requirements and the functional organization

3.1   The knowledge transform

Knowledge is what makes it possible to make progress on programs. Generic knowledge in the many specialized disciplines required on programs involved in solving complex problems is applied to the problem and translated into program-specific knowledge. In order to accomplish this, an enterprise must begin the program with the correct generic knowledge mastered by the people assigned to the program, and these people must transform the aggregate of their knowledge into the information needed to solve the problem characterized by their customer to produce a product of value to that customer, as suggested in Figure 3.1.

This transform process is accomplished in performing the planned work on the program. Several later chapters will focus on the related planning problems. In this chapter, we attempt to describe the knowledge the enterprise needs as the input to this process. This enterprise knowledge base is what drives the hiring the enterprise does and the training it should provide for people on the staff. This population of employees should have been selected based on ideas that are traceable from the enterprise vision crafted in response to decisions concerning the intended enterprise customer base and product base. The people in an enterprise are the instruments through which everything happens. We are the physical architecture of the creator system banded together into functional organizations by a common knowledge space.

3.2   Generic knowledge needs

An enterprise that develops systems to solve complex problems requires the services of people specialized in some fields regardless of the product line

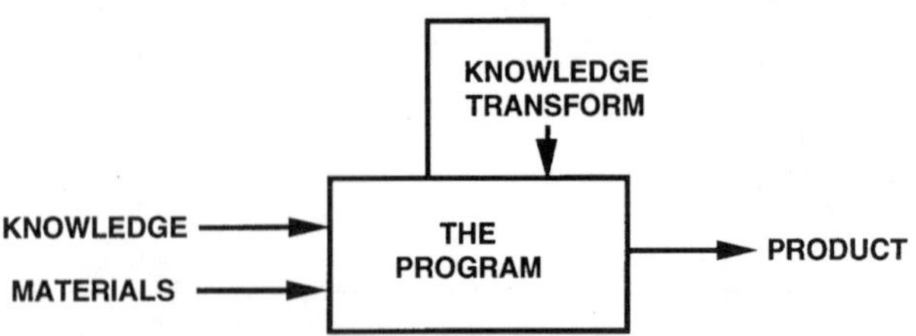

Figure 3.1   Knowledge transform.

and customer base. This includes all of the business-related disciplines: legal, contracts, administration, accounting and finance, human resources or personnel, training and development, security, and transportation.

Given that we assume that the enterprise must produce products based on designs crafted by itself and others, we clearly need an engineering and manufacturing function and the related knowledge spaces. Acquisition of the products produced by others will be achieved through the contracts function.

We will want to ensure that our product is characterized by quality, so we will need that focus in our enterprise. This interest in quality should extend to process as well as product and embrace the system test and evaluation work. All system development activities produce documentation, so we will also need a means to keep track of that documentation and clearly define its meaning relative to our product solution. We need configuration and data management knowledge to control, communicate, retain, and protect this documentation.

The enterprise, like an organism, must provide for its survival through a continuing acquisition of new business and a continuing improvement of its internal capability to accomplish the work acquired through marketing efforts.

3.3   Product line base

The product line and the related technologies determine the kinds of engineering and technical knowledge needed in the enterprise. Generally, the product line shapes the need rather than prescribing it. Most enterprises today involved in system development will need electrical engineers, mechanical engineers, and software engineers. The enterprise will need very different electrical engineers, however, if it deals with electrical power transmission rather than microwave communication instruments. A somewhat different mechanical engineering focus is needed for bridge design than for aircraft structural design. This is not always obvious, however. The gifted designer of the wings of the long-surviving BQM-34 target drone built by Teledyne Ryan Aeronautical since the early 1950s was respectfully known as Bowman the Bridge-Builder.

In addition to the degree of product specialization needed in the basic engineering disciplines, there is also a need for analytical specialists as a function of product. If efficiency of motion of the product relative to Earth's atmosphere is important, then we need aerodynamics. If the fluid is water rather than air, we need people with the same general knowledge but with a different medium, and every aerodynamicist may not be equally knowledgeable for both mediums. On the other hand, an aerodynamicist skilled in Earth's atmosphere would most often be able to adapt to a program dealing with the Martian atmosphere.

3.4   The customer base

The technical disciplines needed in contract work may be slightly influenced by the customer and his preferences. A particular customer may require that certain kinds of work be done that the enterprise might not otherwise perform. For example, the structural integrity required of a naval aircraft is much more demanding than for a land-based aircraft because of the nature of the air field. Corrosion control may also be a much more important discipline in the case of the naval aircraft. These influences may not change the knowledge mix but can change the relative importance of those disciplines.

3.5   Organizing possibilities

Knowledge is necessary in the program development organization, and that knowledge can exist in one of four configurations (individual human, physical media, computer, and shared), as we saw in the previous chapter. There are three fundamental ways to organize an enterprise that is focused on a particular product and customer base, and in the structure we choose there must be an agent for enterprise knowledge acquisition and maintenance.

3.5.1   Projectized enterprises

We could elect to make each project a separate entity, each responsible for its own people, practices, and tools. The advantage in this arrangement is that the project can focus on the customer, ideally leading to a superior product. Clearly, this enterprise will have great difficulty developing a single enterprise-level process around which it can focus continuous improvement. One program or another may excel and bring considerable profit for the enterprise while others do not do so well. It will be relatively difficult to bring all of the programs up to a similar level of excellence because of the differences between them. It may also be inefficient for a program to take advantage of personnel from another program because of organizational, practices, and tools differences between programs. This structure does not provide for a knowledge agent. It simply focuses its knowledge needs on immediate program needs. It acquires and sheds knowledge as a function of the work demand.

3.5.2   *Functionally organized enterprises*

A functional organization is structured around the knowledge that is needed and does have a knowledge agent, functional management. The problem with this kind of organization is that the work is accomplished in the context of functional management. Program work is decomposed in this context and parceled out to functional managers. It is very difficult to coordinate this work in a cross-functional sense because each manager jealously guards his/her own space. The solution to this problem on DoD programs was intended to be the work breakdown structure, which partitions all funds in two orientations: by product entities and by functional responsibilities. A package of work may require contributions by 15 different departments, and if that work were managed with physical collocation of the persons contributing, it would overcome the criticism implied.

3.5.3   *Matrix organized enterprises*

In the matrix structure, there is a pair of axes reaching into a program product organization and the functional departments providing the people to accomplish the work. Obviously, there are two orientations here, and we could pick either one. The matrix strength can be along the functional axis where people are physically collocated by functional department with all of the problems of work coordination of the functionally organized enterprise. Alternatively, the project could be the strong axis, and this is the most effective approach. The people are physically collocated on the product team they serve and are managed primarily from the program axis for however long they are charging time to the program. This supports the cross-functional integrated product development concept supported by the author and others.

At the same time, the matrix provides an agent for knowledge in the functional structure which should provide all programs with coordinated sets of good people, good tools, and good practices. The functional axis has the responsibility to maintain and continually improve the resource base consisting of people, tools, and practices.

We can begin to see the functional organization evolve out of the mist from the generic, product line, and customer base needs discussed earlier. We should collect the knowledge we need into organized chunks because it must be mastered by people who will do the work on our programs and we know that these people will have to be specialists, because of the human knowledge capacity deficit relative to man's knowledge base. True, we could allow our employees to develop individual knowledge skills that appeal to them, but we would realize considerable difficulty in managing the whole enterprise. Assignment of these specialists to program work would require

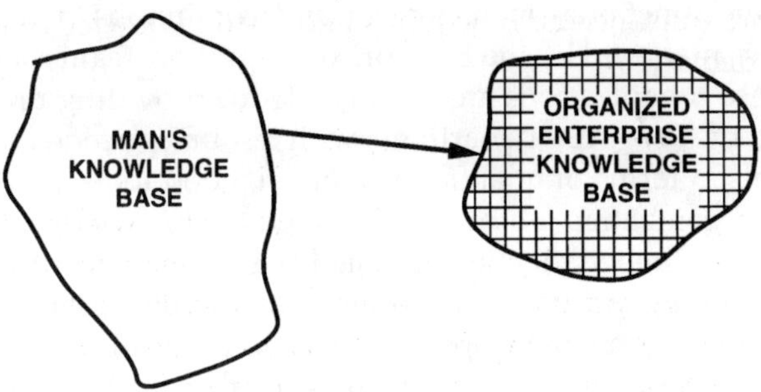

Figure 3.2   Organized knowledge.

matching individual knowledge to program problems rather than known collections of knowledge. Depending on the specific persons available for work on a particular program, we may have to choose a different set of persons for other work. This would be a tremendously complex problem made simple by standardizing the knowledge spaces our enterprise purchases on the labor market.

It is the functional organization structure that offers the enterprise a means to break knowledge into standardized pieces, as suggested in Figure 3.2, such that members of these departments can be treated more or less as interchangeable modules of knowledge, encouraging great flexibility in staffing programs. The way we organize must respect the knowledge needs of our programs coordinated with the availability of knowledge workers. The latter is determined largely by universities and the way they decide to break up all knowledge into departments, but it is also influenced by the industry as a whole and the way they choose to organize for knowledge. While we might conclude that we need one department full of designers with a mix of fluid, mechanical, and electrical design skills, this kind of person may not be available on the labor market. Over time, our employees may fit into the desired pattern, but we probably would not be able to recruit new workers from other companies or from universities in the desired combination of disciplines. So there are two forces shaping our selection of a functional organizational structure. First, there are the knowledge domains we require and our vision of the most efficient way to combine them for management purposes and program application of that knowledge. Second, there exist marketplace constraints driven by university organizational structures that are not always responsive to industry needs.

Table 3.1 begins the process of identifying top-level knowledge spaces within which we will collect the detailed knowledge domains determined by generic needs and the chosen product line. These organizations have been numbered for eventual development into department numbers for use in coordinating tool, practices, and personnel development and work responsibilities. Table 3.1 also correlates these organizations with respect to responsibility for the functionality expressed in Figure 1.4. The organizations marked with an X in the responsibility column cooperate in satisfying or supporting the functional management and continuous process improvement activities managed by the Enterprise Integration Team functional organization. These organizations are responsible for providing programs managed through the Programs department with resources needed to accomplish program goals in terms of qualified personnel, good tools, and sound practices.

In order to associate the more detailed organizational entities with process responsibilities, we will have to develop a finer understanding of the process which will come in Chapter 5. Appendix B includes a detailed functional organization diagram and tabular list providing brief charter statements. There is a leap between Table 3.1 and Table B.1 that needs to be filled in with subsequent material.

3.6 *Program implementation through teams*

In an enterprise employing the matrix structure, every program organization is a temporary entity. True, some programs last for years, but they are generally transient relative to the endurance of the functional organization. Success in implementing the matrix structure is determined by the way the enterprise allocates responsibility to the two axes of the matrix. The biggest mistake companies make in trying to make the systems approach work on programs is in making functional departments responsible for program work. Program work should be the responsibility of program management, using the resources provided by functional departments. The program should organize the people derived from the functional organizations into program teams determined by the architecture of the product. These teams are commonly called integrated product teams (IPT). These teams are responsible for all of the development work associated with the items for which they are responsible.

The IPTs should be established by a system team referred to in this book as a program integration team (PIT) responsible for engineering the complete system. This primarily involves integration work across the team boundaries. The PIT may establish four IPTs based on the top-level architecture. As the work unfolds on these teams, it may become necessary to establish subteams within one or more IPT. These subteams function identically with the first-tier IPT except that the first-tier IPTs act as the integration agent for all of their subteams. This process can continue to expand beneath the first-tier IPTs until a procurement decision terminates the expansion string or the items below the IPT in question are not sufficiently complex to warrant a complete team.

The PIT interacts with all item team tasks in the third dimension (in and out of the page). Imagine the PIT process diagram overlaid in different planes with the item team diagrams with cross-organizational integration occurring in the vertical dimension, as suggested in Figure 3.3.

Table 3.1   Primitive Functional Organization

Dept. Number	Dept. Name	Responsibility
000	EXECUTIVE	Enterprise Management
100	BUSINESS FUNCTIONAL	X
	Human Resources	
	Facilities	
	Contracts	
	Legal	
	Finance	
	Material and Procurement	
	Security	
	Transportation	
	Scheduling	
	Industrial Safety	
200	TECHNICAL FUNCTIONAL	X
210	ENGINEERING	X
	Engineering Administration	
	Mechanical Engineering	
	Electrical Engineering	
	Fluid Engineering	
	Software Engineering	
	System Engineering	
220	MANUFACTURING	X
	Manufacturing Support	
	Material Preparation	
	Manufacturing Engineering	
	Production Facility P	
230	LOGISTICS	X
	Logistics Engineering	
	Logistics Support	
240	QUALITY	X
	Quality Assurance	
	Verification Engineering	
300	ENTERPRISE INTEGRATION TEAM	Process Owner
400	PROGRAMS	All Programs
900	NEW BUSINESS	All New Business Activity

3.7   Building modes

If the reader is going to form a new functional organization for a new enterprise starting tomorrow, the ideal way to proceed would be to first define the functionality of the organization to be formed, because the organization is a system and the best way to develop a system is to follow the pattern of "form follows function." The functionality of this system can be expressed in terms of a process flow diagram, to be developed in Chapter 4. This is a string of blocks connected by directed line segments showing sequence. The blocks represent

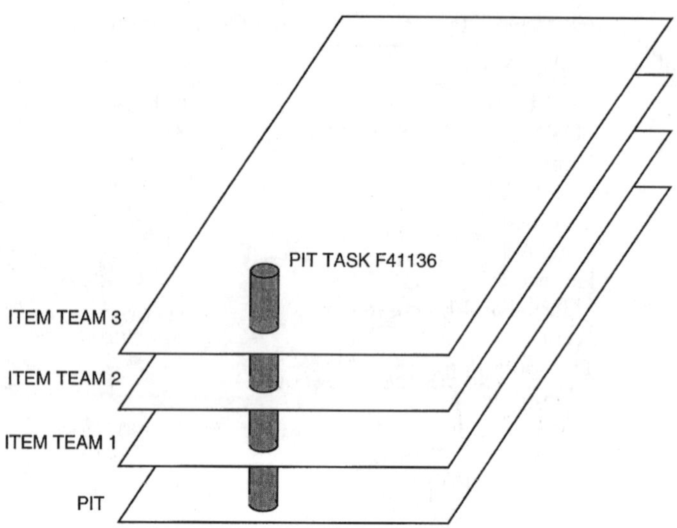

Figure 3.3   Cross-team integration.

tasks that must be accomplished. The knowledge needed to accomplish each
of the tasks on the process diagram should then be mapped to the task blocks
in a fashion identical to the allocation of system functionality to product architecture in the development of a product system.

Concurrently with this allocation process, we need to define the map of
knowledge to functional departments. Since the process blocks have been
identified with respect to the knowledge needed to accomplish related work,
when we allocate (assign) these process steps to functional organizations,
they carry with them the knowledge demand previously attached, as shown
in Figure 3.4.

So we can assemble a three-way relationship between process tasks,
functional departments, and kinds of knowledge in the enterprise knowledge base piecewise by pairs. In this chapter, we are building a beginning
view of the relationship between top-level functionality and a top-level view
of a functional organization (Table 3.1) that would be useful in any development organization. The organizational names suggest particular knowledge spaces completing the top-level trio of ideas. In the next chapter, we
will complete the process definition and in Chapter 6 establish a complete
functional organization based on the allocation of that functionality to the
functional organization. This will still be a generic organization with the
general skills needed in enterprises that seek to solve very complex problems, where we find a pronounced need for the application of the systems
approach. A particular enterprise will probably have to tailor that structure
based on its specific product line and customer base.

Given that our new functional organization has been created as indicated, we now need to staff it. That can be done by simply hiring people

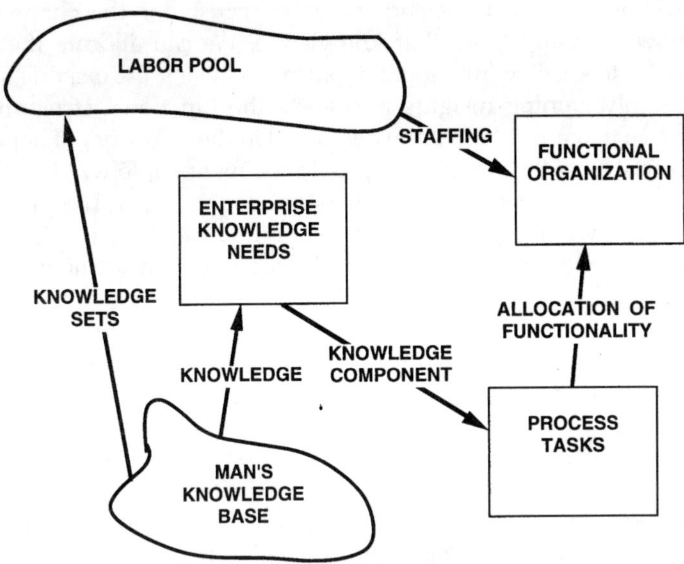

Figure 3.4   Knowledge mapping.

from the labor pool with the combinations of skills and knowledge associated with the functional department charters we have established. In order for this to be as easy as possible, the charters we elect in the functional organizations should be passed through a reality screen to ensure they align with currently available knowledge domains. It may be necessary to refine the knowledge map to bring about coordination between our needs and the real world. The functional organization is now ready to supply all enterprise programs with good people, but must bring about a condition of synergism between its tools, practices, and people through training, assuming the other two components have been acquired as well.

The tools are a matter of selection, and the practices have to be written, describing what has to be done and how to do it coordinated with the process blocks at some level. So we must have a rule of responsibility for acquiring these two components. The recommended rule is to identify all functional departments which must contribute to each process task through the allocation process (there will often be more than one allocation for a given task). Where there is more than one functional department responsible for achieving a task, we must identify a principal department. That principal department should be given the responsibility for tool selection for the task and initial preparation of task practices. Other departments must then learn how to use the preferred tool in accordance with the task description.

The reader is working within an existing reality defined by an enterprise with ongoing work proceeding in accordance with contractual commitments. How would we improve such an organization using these principles? We must first understand the functionality your programs should perform. So we can still perform the front-end analysis suggested in this chapter regardless of how we may apply that knowledge. We can allocate the exposed functionality to specific functional departments using the existing structure, while possibly gaining insights into ways the functional structure can be improved. In that almost all of the personnel in these functional departments will almost always be assigned to programs, few people will be affected in the short term by any changes in functional departments. The problem is in determining to what extent any changes should be applied to programs. The following sequence is suggested to make the initial adjustment to the new structure:

1. Define needed enterprise functionality based on customer base and product line defined in the vision statement.
2. Allocate exposed functionality to functional organizations that will be responsible.
3. Identify the transform between the existing functional structure and the new one. What responsibilities flow from what functional department to what functional department? If necessary, rename any departments as a function of the nature of their final responsibilities.
4. Reassign personnel as necessary to balance the available employee skills and knowledge with the responsibilities of the departments. Apply this new functional structure to all new proposals and programs and bring on board all existing programs as early as possible, but not necessarily right away.
5. Apply the new process description on all new proposals and programs and review existing program contracts, budgets, and schedules to determine to what extent it should be applied (possibly not at all on some programs).

Over time, all surviving programs will come under the new process. This revolutionary step should be viewed as a one-time activity. Thereafter, the enterprise must apply continuous process improvement based on lessons learned from programs to make incremental changes in process, personnel, and tools.

This situation is similar to the reengineering of a product system that remains in service while it is being improved. The product is redesigned and enters a major modification or a series of modifications to achieve new functionality. In that situation, we seek to understand how the initial functionality has changed and allocate the new functionality to the architecture beginning with the initial architecture. When the new functionality can be allocated to an existing item, we do it. When there is nothing in the initial architecture that will accomplish a particular function, we have to consider modifying something that now exists or creating a new item in the architecture. Anything that has not had any functionality allocated to it, we should consider eliminating.

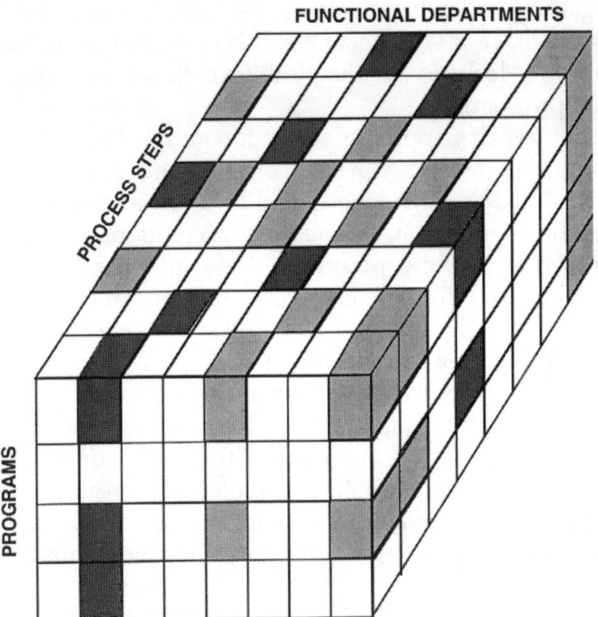

Figure 3.5   Program responsibility map.

Once the resources become available to actually implement the changes in operational use, they have to be phased in so as not to disrupt the ongoing flow of work. For a period of time, the organization will have a dual charter, and we must focus our interest on those areas where process tasks have been allocated differently in the initial and new situations. Each of these conditions should involve a specific decision regarding department responsibility on each program applying that task. Figure 3.5 illustrates the pattern of decisions that must be made. The dark gray intersections reflect the principal department, while the light gray intersections correspond to contributing departments. The white intersections correspond to voids, meaning the indicated functional department has no responsibility for that task.

If a program has not yet applied a task, it can be shifted to the new alignment, provided there is not a cost impact in conflict with the original program plan. If there is, shame on us for improving a task so as to increase its cost of implementation. If the task has already been completed on a program, it can be shifted to the new alignment with respect to that program as well. The problem arises when the task is under way on programs. In each such case, we should consider the impact of applying the new alignment. If it is slight, we should go ahead. If it is more significant, we should delay the new alignment until later when the impact has lessened, perhaps until the task is complete on the program in the extreme.

The process of changing the alignment on a program may involve a transfer of personnel between departments which is of no consequence to a program organization, a change of practices, or a change of tools. The latter two actions can have a significant effect on program work and should be the basis for the decision on the timing of the introduction of the new alignment on programs.

In a real enterprise applying continuous improvement, there will be a condition of process instability in effect at all times, but this instability should be minimized. The rate of change cannot be allowed to be so great that this instability becomes too great. Where great revolutions are to be made, they must be phased in as new programs come into the enterprise. The old processes in this case will fade out as the programs in house at the time the change is made fade out or the tasks in question have been completed.

All of these problems in transition suggest that an enterprise should have a process agent, a process owner, that can regulate the overall improvement process. The author maintains that the functional organization should be responsible for providing programs with good people, good tools, and good practices. However, as in all distributed responsibility situations, we need an integrator and overall owner. That agent, in the author's view, is an entity called an enterprise integration team (EIT), the ultimate process owner for the enterprise.

chapter four

Enterprise generic process definition and functional allocation

4.1 The ultimate enterprise functionality

We commonly do not develop an enterprise in accordance with system principles. An enterprise is most often born small and grows despite the owner's faults into something that satisfies customer needs at a price and with a dependability that customers like. Ideally, we would, in starting an enterprise, identify an ultimate function that the enterprise is intended to satisfy, which the author refers to as its vision statement. Then, applying Sullivan's principle of "form follows function," we could expand the definition of this ultimate functionality through structured decomposition and allocation of the more detailed exposed functionality to organizational entities that must satisfy the allocated functionality. The architecture of enterprises that create product systems is composed of organizations of people who possess knowledge and apply it toward specific goals. The vision statement, expressing its ultimate functionality, is allocated to the complete enterprise just as the customer need, expressing the ultimate functionality of the system the customer requires, is allocated to the top element of the system architecture, the system. Now we must describe the enterprise functionality in some detail.

If we were about to establish an enterprise, this might be a fine way to proceed, but here you are stuck in the middle of an existing enterprise that is not working well. Is there any hope for you and your enterprise short of starting all over? Well, this process can be applied to reengineer an existing enterprise just as the system approach can be applied to develop a new product or reengineer an old one. In so doing, in either case, we have to recognize that there is a current reality on the physical plane that is not going to go away with a snap of the fingers. There is a transition process that is required to move from the existing enterprise organizational architecture and functional responsibilities to some new combination. Just as in reengineering an existing system, we have to determine what architectural elements remain as they are, which ones must be modified, which ones must

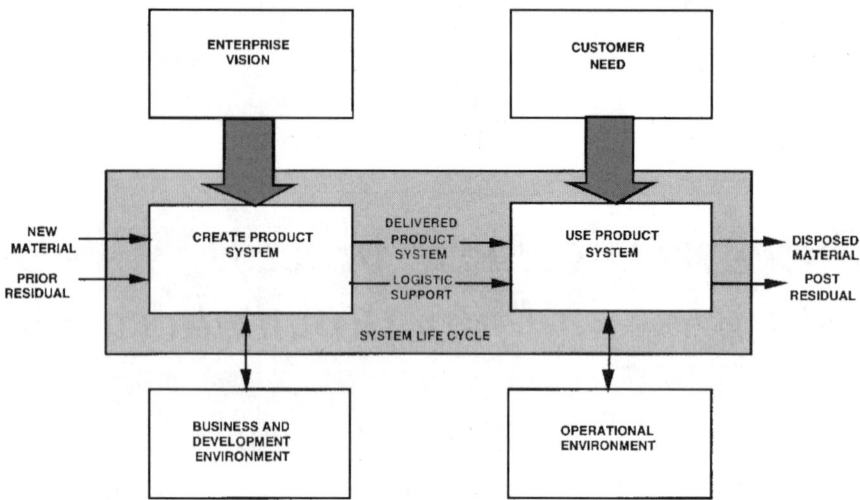

Figure 4.1   The ultimate system functional relations.

be deleted, and what new elements must be added. It may be that the move from the present reality to a new identity cannot be made in one step, leading to a growth path composed of several steps permitting some adjustment in process. At the end of the last chapter we discussed how this transition could be made while continuing to serve the needs of several programs in various stages of completion.

Whatever we conclude the detailed functionality of the enterprise to be, it should have been expanded from an ultimate functionality. The author will use the phrase *enterprise vision* to denote this ultimate functionality. It represents a brief description of what the enterprise must achieve just as a customer need statement is the ultimate system function telling what the system to be produced must achieve. Figure 4.1 illustrates these relationships in their most fundamental form. The enterprise vision is the ultimate function that drives the enterprise process for creating product systems. This enterprise must function within system development and business environments while transforming material and an understanding of customer needs into a product system that satisfies that need over some period of time. Someone or some organization will operate the product system within a predetermined product system environment satisfying their need, possibly drawing logistics support resources from the creator enterprise.

In the end, and the end must come for every man-made system, the residual elements of the system will either be disposed of or recycled into a new system, perhaps one that will continue to satisfy the old one's functionality.

At least some part of the enterprise vision statement should involve an interest in satisfying customer needs through programs focused on those needs. An enterprise may, at any one time be in the process of satisfying

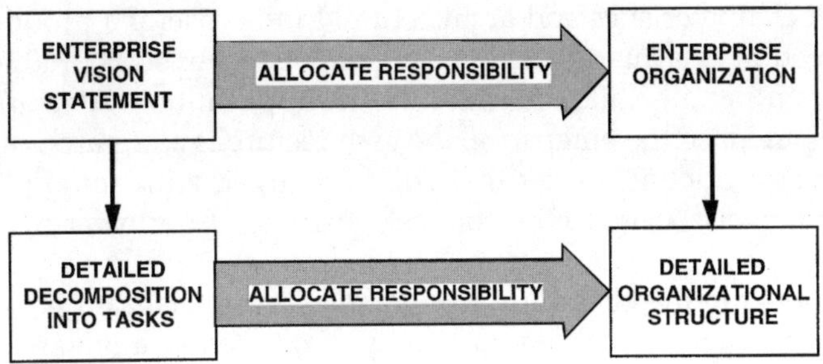

Figure 4.2   Process development and allocation.

more than one customer need, and this is the kind of enterprise this book envisions, a multiple-program enterprise. By definition in this book, the enterprise entity that responds to a customer need is called a *program.*

From the creator's perspective, we can look upon the creation and operation (or use) of the system as functions within the life cycle of the system. Some systems are operated by their creator, and some are operated by others. Space launch vehicles, for example, are often operated by their creator system, whereas fighter or commercial transport aircraft are operated by the customer. A single program may serve multiple customers, as in the case of the Boeing 777 Program building commercial aircraft for American Airlines as well as many other airlines. A single customer may also have interests in more than one program within a particular enterprise as well. In this book we will focus on the generic relationship between one enterprise and one customer through the window offered by a single program. The elements of this single view can be multiplied in any number of ways to appreciate more complex situations in which people find themselves in industry.

4.2   *First-tier functionality expansion*

The create product systems function is a variation on or subset of the enterprise vision statement and will be allocated to an enterprise capable of responding to customer needs, as suggested in Figure 4.2. This is a mindless allocation. By definition, the enterprise is the entity that will satisfy the vision statement, but conscious recognition of this trivial relationship is not a senseless action. It is important to recognize a condition of logical continuity throughout the enterprise structure and this step connects the creative spark of genius giving birth to the enterprise through its vision statement with the detailed description of the enterprise.

Subsequent steps require us to expand on the functionality needed by the enterprise in order to achieve the enterprise vision. These details will take the form of a multilevel functional flow diagram in this chapter. We will then, in later chapters, allocate the identified functionality to organizational entities, thus expanding the detailed definition of the organization. This is exactly like the activity one would apply on a program to understand the system characteristics and architecture. In the case of a product system, the architecture is a hierarchical arrangement of physical product entities arranged in a parent–child structure. In the application of this approach to the development of the enterprise, the architecture is a hierarchical arrangement of organizational entities that will be charged with accomplishing the responsibilities associated with the tasks expressed on lower-tier flow diagrams.

We might inquire that since the organizational diagram is going to be expressed in a hierarchical diagram, why don't we use a hierarchical list of functionality rather than a flow orientation? This approach can be used, but the danger is that it will lead to a bad organizational design. When the flow orientation is different from the hierarchical organizational structure, it encourages a degree of isolation between the efforts to understand needed functionality and defining the organizations responsible. The use of hierarchical structures on both sides tends to encourage a bidirectional coupling rather than the intended unidirectional relationship following the "form follows function" paradigm.

The Create Product System function illustrated in Figure 4.1 is but one of at least four fundamental functions our enterprise will require in order to satisfy the vision statement. Figure 4.3 shows these four fundamental functions. Someone must manage the whole enterprise toward satisfying its vision statement — enterprise management. The enterprise must execute its current business. We choose to group this work into something called programs, and we assume more than one will exist at a time. Each of these programs is responsible for transforming new material and possible prior residuals into a system that satisfies their customer's need within the context of a particular operational environment. Each program requires a cohesive group of people responsible for executing program work directed at satisfying customer needs through complying with contractual commitments as well as planning and scheduling that work. There may be many programs in the works at one time, and programs may contain subprograms sometimes called projects.

The enterprise requires a means for new business acquisition, that is, gaining new programs. This function keeps abreast of potential customer needs and coordinates them with enterprise capabilities, focusing energy on achievable business relationships through customer contact, proposal development, and other marketing activities. As a beginning of this analysis and allocation process, let us allocate responsibility for the four functions exposed in Figure 4.3. Figure 4.4 identifies organizations in our enterprise that will be responsible for the functions exposed in Figure 4.3. Only two programs are identified, and there could be more.

The map should be obvious to the reader. The only allocation that is a little abstract is the allocation of the Resource Maintenance and Continuous Improvement function to the Functional Organization of Figure 4.4. This organization is responsible for acquisition, maintenance, and extension of the enterprise's knowledge base to prepare it for work related to its customer

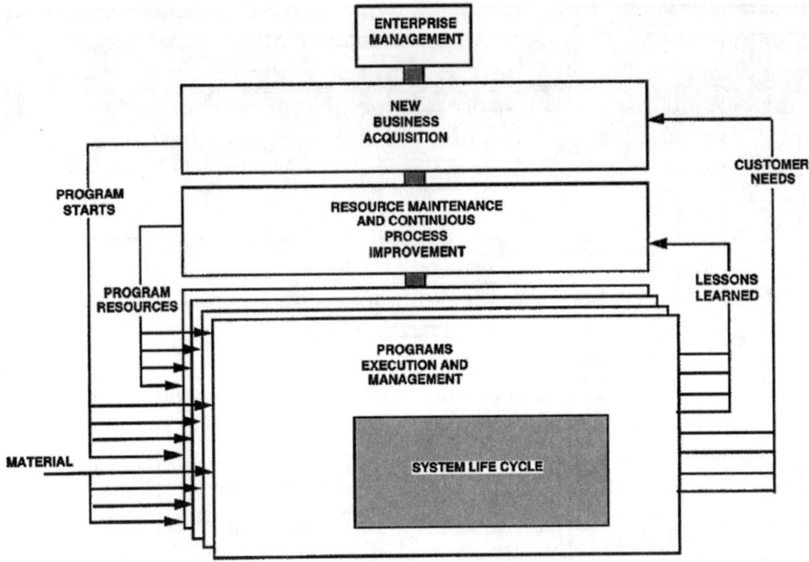

Figure 4.3   The four basic enterprise functions.

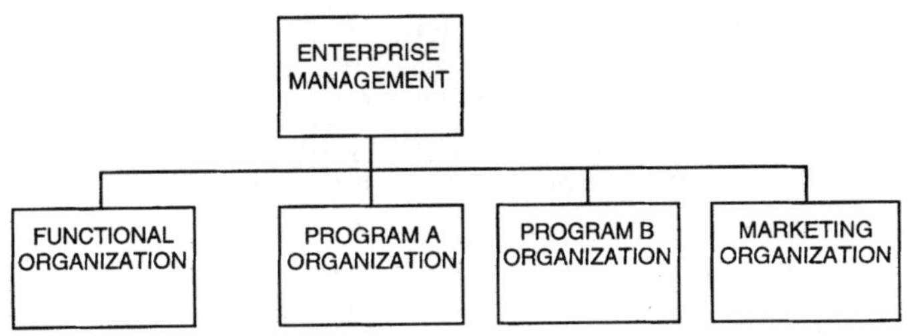

Figure 4.4   Top-level organizational diagram.

base and product line identified in its vision statement. The functional organization will provide programs with the resources they need to succeed. The
marketing organization is, of course, responsible for acquiring new business.

The programs are focused totally on their customers and cannot be
depended on to maintain and improve the enterprise's knowledge base. This
is the responsibility of the functional organization in a matrix organization.
We need both of these focuses in our enterprise: the focus on satisfying
customer needs through program organizations and the focus on maintaining and improving the enterprise and its knowledge base through the functional organization. The functional department structure should provide
every program with people who have the knowledge appropriate to the
program technology and skills related to that department's area of responsibility within the enterprise, best practices improved over time, and good
tools coordinated with company practices and personnel knowledge.

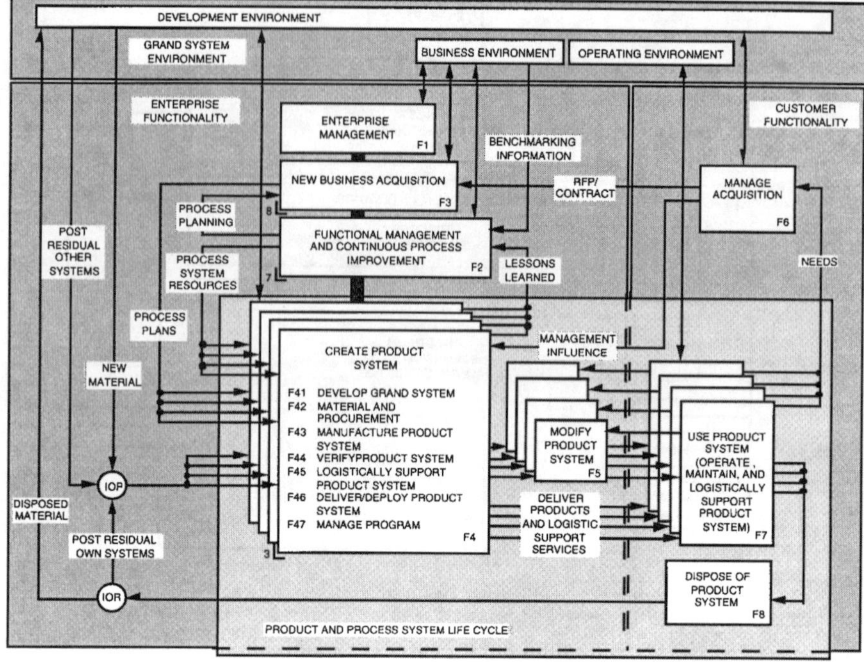

Figure 4.5   Expansion of first-level functional detail.

Results from programs must be consciously reviewed in the context of a continuing search for excellence in program execution. An element of functional management must review the results and draw conclusions about the best improvement actions to take to improve program performance. This may entail improvement of the knowledge base of its people, better practices, or better tools.

Figure 4.5 expands on Figures 4.1 and 4.2 for an enterprise that happens to have four programs in house at the time. Post residual may flow in from the development environment as customer furnished from prior systems past their prime and new material from suppliers based on purchase orders or contractual relationships. Post residual material may also flow in from systems previously created by the enterprise. Other important interfaces also exist between the development environment and the program and between the business environment and the other enterprise functions.

In Figure 4.5 we have identified the lower-tier functionality specifically, and these functions will be addressed in the next section. We have differentiated the normal development cycle from the modification development cycle identified as function F5. These modification processes are actually small programs that should follow the same pattern we define for a full program. Generally, these modification programs will take in product previously delivered and combine it in some way with new product to yield modified product that can accomplish new or expanded functionality or

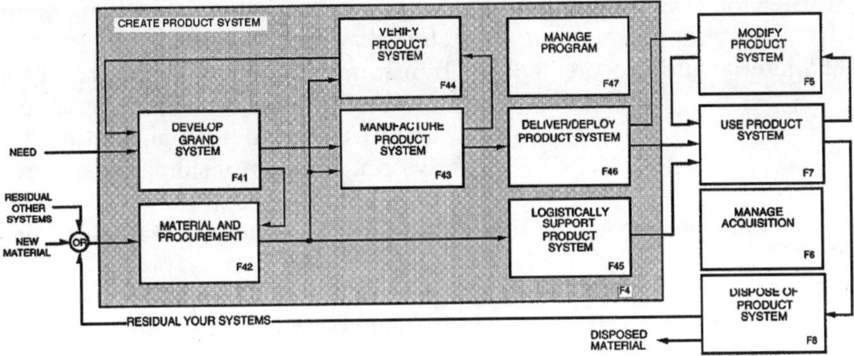

Figure 4.6 Life cycle functional flow diagram.

performance figures. The previous product may actually be brought back to the enterprise physically for modification, or modification kits and instructions may be shipped to the customer locations for installation. The normally complex nature of system development programs becomes evident when more than one of these modifications are under way in different stages of development and implementation while production proceeds.

It remains to expand the create product system function one more level and describe these functions. We can then allocate these functions to organizational elements which will require certain knowledge in order to accomplish work associated with the responsibilities driven by the allocated functionality.

4.3   The marriage of enterprise and system functionality

Figure 4.6 illustrates the major functions of an enterprise that creates product systems using the subfunctions identified in Figure 4.5. Granted, another person might choose to decompose the vision statement differently, but this one does reflect some commonly accepted enterprise functions. These functions are briefly explained below. Only those functions in the product and process life cycle are illustrated. Those listed below are from Figure 4.5.

F1	Enterprise Management	Top-level management of the enterprise.
F2	Functional Management	Organized by focusing on the knowledge base the enterprise needs to create products in its chosen product line and customer base.
F3	New Business Acquisition	New program starts are acquired from potential customers.
F4	Create Product System	The transform from need to delivered product proven to satisfy customer-defined requirements.

F41 Develop Grand System   Transform the customer need into a pre-
                                    ferred system definition.

F42 Material and Procurement   Transform engineering definition of the
                                    product into a stream of material and
                                    services needed to manufacture the
                                    system. Temporary storage of material
                                    is included.

F43 Manufacture Product System   Transform engineering, manufacturing,
                                    material, and quality product defini-
                                    tions into a physical product.

F44 Verify Product System   Evaluate product items and system for
                                    compliance with requirements and
                                    produce evidence of a condition of
                                    compliance.

F45 Logistically Support System   Deliver products and services that en-
                                    sure that operational elements of the
                                    system remain fully operational.

F46 Deploy System   Move the product system from the cre-
                                    ating organization to the operational
                                    organization. This may involve com-
                                    mercial distribution or military deploy-
                                    ment.

F47 Manage Program   Management skill is brought to bear on
                                    all of the other functions encouraging
                                    the people responsible for those func-
                                    tions toward satisfying the goals estab-
                                    lished for them in planning.

F5   Modify System   Change an existing system in planned
                                    ways to improve functionality.

F7   Use System   The system is operated and maintained
                                    throughout its life.

F8   Dispose of System   Transform any hazardous elements into
                                    a nonhazardous state and convert re-
                                    sidual material into other applications
                                    or properly processed sanitary fill.

Each of the functions illustrated in Figure 4.6 can be further decomposed using some method of structured analysis, not necessarily the functional flow diagramming technique applied so far. For example, the procurement and material handling functions associated with F42, Material and Procurement, could be further decomposed using process analysis. The same approach could be applied to the Develop Grand System function, F41, as has been done partially in this chapter and completely in Appendix A for the enterprise used as an example throughout this book. Functional flow diagramming might be used to expand upon our understanding of the customer need, which can be rephrased as Use Product System, F7.

Here is where the enterprise process merges the enterprise vision and the individual program need statement. The vision statement can be decomposed in a generic sense as this chapter will do and that functionality allocated to functional departments which will be called upon to supply program resources to accomplish those functions. The Use Product System function is program specific and must be decomposed as a program task with allocations flowing to product entities. Program process analysis may also develop insights into unique aspects of the process generically represented as part of the enterprise process diagram.

There are any number of structured analysis models that could be used to decompose function F7 for specific programs. If the product system were substantially a software system, one might apply a software-oriented modeling technique such as Hatley–Pirbhai or object-oriented analysis. The supporters of these two modeling techniques would, on the other hand, encourage the use of their model regardless of the nature of the system, whether hardware or software. An enterprising analyst might even apply these techniques to enterprise modeling rather than the flow diagramming approach chosen by the author.

The author differentiates between process flow and functional flow analysis as the former being a sequence of things that must happen that is generally crafted before you know what resources will be available to accomplish those things; the latter is a physical analog of the process with known resources flowing in time through the process diagram. In a functional flow diagram, the directed line segments only indicate a sequence of things that must happen. In a process diagram, the directed line segments can be interpreted both in terms of sequence and a flow of some form of physical material through the process.

The Develop Grand System function will be expanded and identified in terms of functions in this book, but it is understood that there is a one-to-one correspondence between the functional and physical models. This is not usually the case between the Use Product System function (F7) and the physical process defined after one understands the physical system architecture. Thus, we could reidentify all of the develop system functions Fn as processes Pn, where n is a string of base 60 characters. The author may use the words process and function as interchangeable in this regard, but he does not imply that they are synonymous in a more general analytical situation. At the lower tier, these blocks will be referred to as tasks. All of these words describe the same kinds of things — the work that must be done by the enterprise on programs.

4.4   Functionality allocation

The functionality exposed in Figures 4.5 and 4.6 is allocated to the organizations illustrated in Figure 4.4 as shown in Table 4.1. Some of the functions map cleanly and solely to a single organizational entity, but most map to all program organizations for implementation while depending on the functional

Table 4.1 Allocation of Functionality

Function ID	Name	Responsible Organization	Nature of the Responsibility
F1	Enterprise Management	Enterprise Mgmt.	Set, prioritize, and achieve goals
F2	Functional Management and Continuous Process Improvement	Functional Org.	Define process, personnel, and tool needs
F3	New Business Acquisition	Marketing	Define process and acquire new business
F41	Develop System	Program X Org.	Implementation
F42	Material and Procurement	Program X Org.	Implementation
F43	Manufacture System	Manufacturing Org.	Implement with facilitized control
		Program X Org.	Define product and mfg. process
F44	Verify Product System	Functional Org.	Implement with facilitized control
		Program X Org.	Define process
F45	Logistically Support System	Program X Org.	Implementation
F46	Deliver/Deploy System	Program X Org.	Implementation
F47	Manage Program	Program X Org.	Implementation
F5	Modify System	Program X Org.	Implementation
F6	Manage Acquisition	Customer	Manage from customer position
F7	Use Product System	User	User may be program or customer
F8	Dispose of System	Program X Org.	Implementation

organization to provide the programs with the resources needed to accomplish that functionality. The exact manufacturing and test responsibilities may be allocated in many different combinations, but in this example we are assuming that the facilities involve manufacturing facilities and test laboratories that can best be preserved and managed through the functional structure responding to program needs.

We now must identify more finely the needed functionality and associate that functionality with functional organizational entities responsible for the related work in the context of programs responding to customer needs.

4.5 Lower-tier development functionality

To further expose the needed functionality, we have but to study each of the functions listed in Table 4.1 and consider in more detail what things have to happen to encourage success in the function we are expanding. This can continue down to some level of granularity best determined in the organizational plane. When we are confident we understand the full functionality of all of the organizational entities we have created and that we have created all that are necessary to accomplish the vision statement, we may conclude that we have exhaustively examined the needed functionality. Appendix B expands on the functionality expected from our enterprise in some detail. A map is included in Table B.1 between the exposed tasks and responsibilities coordinated with the functional organizational diagram shown in Figure A.1 and the typical program organization diagram illustrated in Figure A.2. Some of the lower-tier functionality is examined in some detail in subordinate paragraphs. The reader may conclude differently than the author about the best way to define one's process but the example included in this book will, hopefully, offer encouragement for crafting a detailed definition that does satisfy the enterprise's process needs as a basis for its functional organizational needs and generic program process.

4.5.1   Develop grand system, F41

Figure 4.7 expands on the Develop Grand System block (function F41) of Figure 4.6 for generic program XXX, and subordinate paragraphs discuss the exposed functionality. This same set of processes may very well be going on in parallel at different points in the life cycle of several programs at the same time. The term *grand system* refers to a need to develop both program product and program process concurrently and optimize both at this grand system level. Refer to Table A.1 for detailed allocation of this functionality to the functional organization. In this chapter, we will give some examples of this allocation process leading to functional department charters (that is, the design of the enterprise) in the next chapter.

Function F41 includes two fundamental tracks that are not entirely clear from the diagrams used in this chapter. These two tracks focus on the two kinds of teams needed to create product systems. The grand system is the responsibility of the Program Integration Team (PIT), which assigns development responsibility for specific system items to Integrated Product Teams (IPT). Some of the tasks in our process diagram must be applied by the PIT at the system level, and subsequently by the several IPTs for lower-tier elements. We will first go through the cycle for the PIT, the integration and optimization agent for the program, and then come back for the IPT cycle.

4.5.1.1   Program integration

The system integration functions, F4113, F4123, and F4133, are allocated to an entity called a Program Integration Team. However, every program activity must have a home in the functional organization which is responsible for providing all programs with personnel, tools, and practices corresponding to that responsibility. Therefore, these functions are allocated to an Enterprise Integration Team (EIT) functional department, yet to be fully characterized. The PIT is initially responsible for system definition (mission analysis, technical program plan development, and system specification preparation), applying one cycle of function F4111, and subsequently for overall system development and integration once it has assigned integrated product team

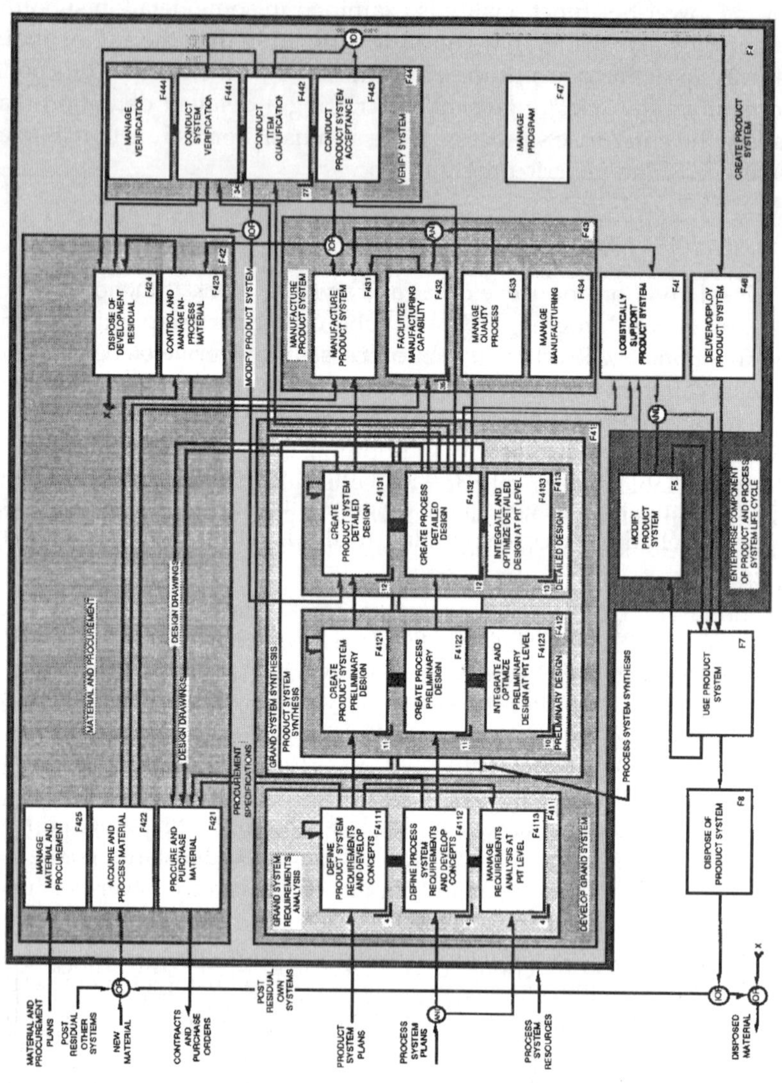

Figure 4.7 Develop grand system.

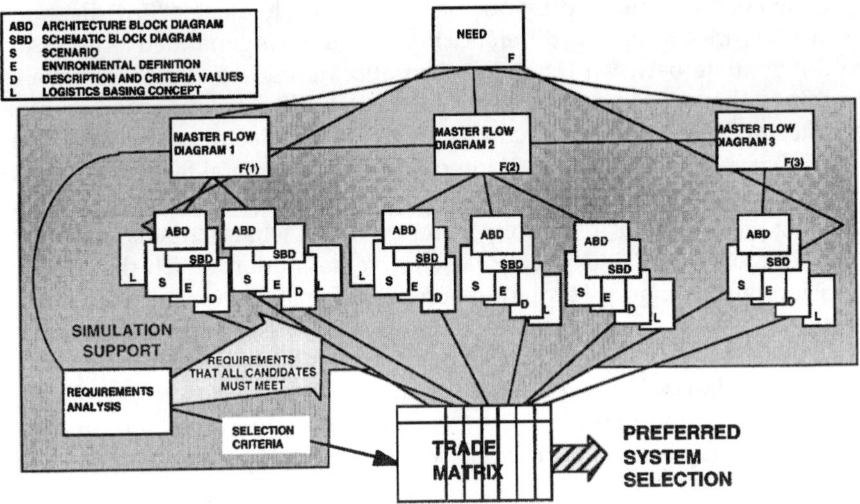

Figure 4.8 Structured selection of the preferred concept.

responsibilities to system elements and staffed the teams. Many functional departments will have to cooperate to accomplish all of the PIT work, as is the case for all of the teams established on a program. But, to see this multiplicity of functional department responsibility clearly, one must look one or two levels deeper corresponding to the level that work is actually accomplished.

4.5.1.1.1 Define product system and develop concepts, F4111.   Mission analysis is a process for translating a customer need into a critical mass of information about a preferred system solution that will satisfy the intended mission. The principal responsible functional department is system engineering. This process may begin with only a customer need conveyed by a potential customer through a request for proposal or deduced by the enterprise based on marketing analysis. In this case, the need acts as the ultimate function for the use system function (F7) and is allocated instantly to the top architecture block named system and expanded to identify lower-tier functionality which, in turn, is allocated to lower-tier elements in the architecture, thus forming the physical view of the system. Also in mission analysis we must seek to understand the one or more ways the customer will use the system, and the results feed the development of the functional view of the system and help visualize design concepts for the elements of the system which must satisfy higher-tier functionality. Part of the customer interaction process is a question and answer activity that attempts to bring forth from customer representatives answers about their intended application. A key part of the mission analysis process is the development of alternative solutions and trade studies focused on identifying the preferred solution among those studied. Figure 4.8 illustrates this process.

The need is expressed by one or more life cycle master flow diagrams, such as Figure 4.6, and the difference between them is expanded as necessary to differentiate between them. We then allocate the top-level functionality to things that would accomplish that functionality, forming one or more architectures for each functional view. For each architecture, we should identify the corresponding top-level interfaces, environmental factors, a logistics concept (possible basing differences, for example), a brief scenario communicating pictorially or in the form of an event list what happens in employment of the system, and a set of descriptive data upon which the selection decision will be made. Through requirements analysis we uncover a set of requirements which all candidate solutions must satisfy and a set of selection criteria used as the basis for selection. A trade matrix is used to succinctly capture the intersections between alternative characteristics and selection criteria. A decision-maker selects the preferred system from the candidate systems studied based on their relative ranking in the trade matrix.

4.5.1.1.2 Manage requirements analysis at PIT level, F4113. Functions F4111 and F4112 are accomplished by the PIT for the system level and such other levels as necessary to arrive at a level that is appropriate for tasking the highest level IPT. These processes are repeated by the assigned teams for the items for which they are responsible until the corresponding problems are completely defined. All of the IPT requirements work is coordinated and integrated at higher system levels by the PIT. This may include team reviews chaired by the PIT or audits conducted on team results.

4.5.1.1.3 Integrate and optimize preliminary design at PIT level, F4123. The PIT does no design so is not involved in functions F4121, F4122, F4131, or F4132, but it must accomplish integration and optimization across the IPT boundaries and manage the program risk process. Integration work includes auditing cross-organizational interface terminals for compatibility, evaluating evolving designs for conflicts, and identifying and correcting system level suboptimization. Risks in cost, schedule, performance, and technology sets are identified and tracked toward reduced probability and reduced consequences.

4.5.1.1.4 Integrate and optimize detailed design at PIT level, F4133. The PIT accomplishes activities similar to those in the preliminary design phase with emphasis on detailed design compliance with requirements, evolving physical interface compatibility, manufacturing, quality, logistics, and verification (qualification and acceptance) detailed planning.

4.5.1.1.5 System design reviews. Generally, there are two levels of design reviews in a development effort: a system design review and item reviews. The former exposes the system design concept and determines if it will satisfy system requirements within a context of controllable risk. Item reviews are generally held in two sets: preliminary design review and critical design review. The former demonstrates that the performance requirements can be satisfied within cost and schedule constraints and commits to the development of the detail design. The latter approves the detailed design and commits to manufacturing the resources needed for test and evaluation. These two reviews are closeout actions for functions F4121 and F4131, respectively. In a simple system, one design stage and one review may be sufficient, but in a complex system it will be necessary to use a two-step process to encourage sound risk control. In either case, the PIT should hold item reviews presented by the IPT followed by a system review presented by the PIT for the program manager that closes out all of the item reviews.

4.5.1.1.6 System requirements validation.   All of the requirements in top-level specifications are passed through a risk filter to determine the degree of difficulty anticipated in complying with them. Where a concern is uncovered, tests, analyses, demonstrations, and examinations are accomplished to improve confidence that the requirements can be complied with. All of this work should be complete prior to holding the preliminary design review.

4.5.1.1.7 System verification planning.   Every requirement identified in every specification is screened for ways to prove the design is compliant. Each requirement which is extended into a verification requirement linked to verification tasks for which plans are developed, implemented, and documented. The documentation provides the evidence of compliance. This process begins with PIT development of system verification requirements and planning data, followed by IPT implementation of the same process for their areas of responsibility under the coordination, integration, and review by the PIT.

4.5.1.1.8 Maintain program plan.   The program plan, developed during preprogram work (proposal development or commercial program development), must be maintained accounting for changes driven by improved knowledge about program needs. PIT and the program office must cooperate to accomplish this work to ensure that technical work demand is coordinated with resources.

4.5.1.1.9 Manage configuration.   All program documentation other than memoranda should be formally released, controlled, and correlated to particular design baselines defined by specific document sets or computer database versions. The configuration of design representations (models, simulations, mockups, etc.) must also be controlled. During the work accomplished in function F41, there is not normally any physical product, so this work is focused on documentation.

4.5.1.1.10 Manage PIT.   Program performance of the PIT is tracked in accordance with the program plan, cost and schedule trajectories, and direction is provided to bring closure between the plan and actual performance.

4.5.1.2   Integrated product team (IPT) activity

The programs should be organized and oriented toward the product architecture. A given product may require several teams, even teams of teams as many as three tiers deep depending on the scope and complexity of the product system. The functional system engineering department should supply people to work the lower team levels, but the person who leads an IPT may come from any functional discipline. All of these teams apply the same process to identify appropriate requirements, develop compliant designs, and prepare for the verification of design compliance with requirements. We will describe the work of a typical team, Team L, responsible for product development process steps F4111, F4121, and F4131 in a program context as they relate to the assigned item and subordinate items.

In a large system development activity, the team expands item functionality in function F4111 and as a result may spawn one or more subteams, each of which repeats these activities at a lower level. As noted, this nesting process could continue for more than one level. When this happens, the parent team must take on some of the characteristics of a PIT for its subordinate teams. It must act as the system agent for requirements analysis and integration. Each team is fully responsible for its lower-tier development, whether subteams are applied or not, and this system responsibility flows down to each level for everything below it.

Figure 4.9 illustrates the expansion of function F4111, F4112, and F4113 to the next lower level. Some of these functions are explained in subordinate paragraphs in the context of an IPT.

4.5.1.2.1   Form IPT, F4115.   The PIT identifies an initial team leader who must form a team drawing personnel required from the functional organization or with the help of functional management from other programs. The PIT provides the team with a specification covering the top-level team item; a budget; and the nested portion of the work breakdown structure (WBS), statement of work (SOW), integrated management plan (IMP), and integrated management schedule (IMS) corresponding to the team responsibility. The PIT also provides the resources necessary to train the team to perform integrated product development, if necessary.

4.5.1.2.2   Define product system requirements and develop concepts, F4111.   The team continues to expand the functional model to gain insight into item performance requirements and the physical product entities at the lower tiers. Constraints analysis work completes the content of lower-tier specifications in terms of interface, environmental, and specialty engineering requirements. This process may have to be performed repeatedly for each of the several items for which the team is responsible. Each specification must be reviewed and approved prior to the beginning of any detailed design work.

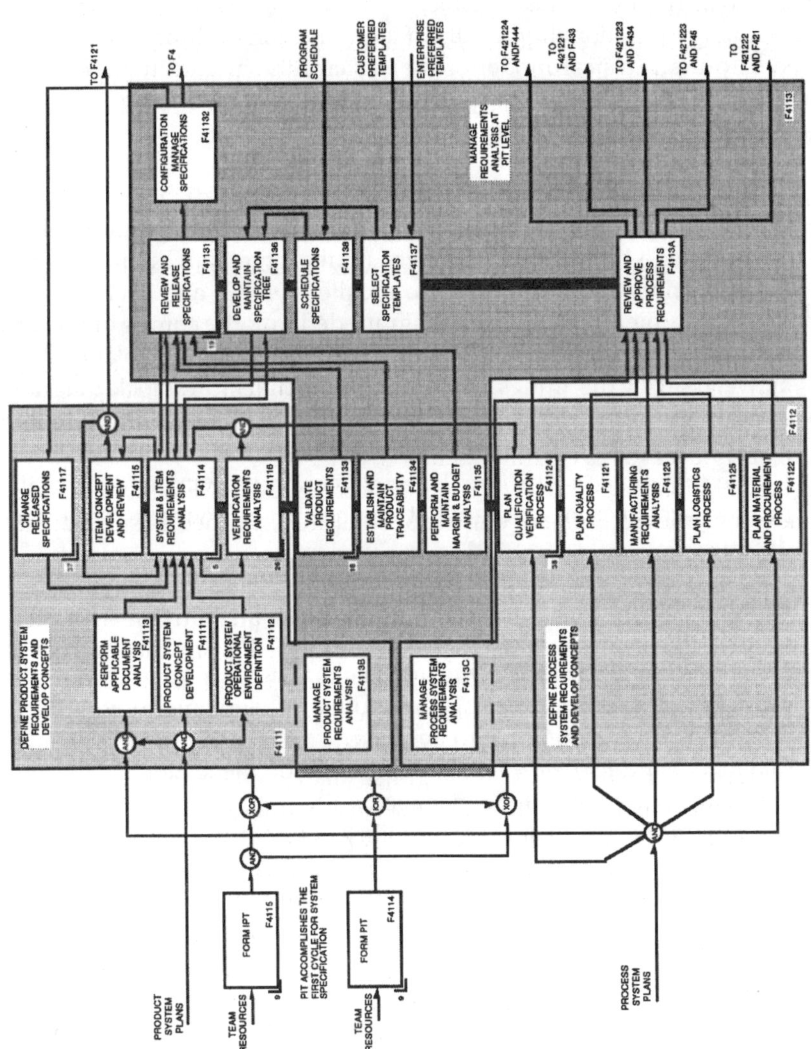

Figure 4.9 Grand system requirements analysis.

Specifications are developed for all things in the systems architecture, which is also determined in this process. This task also covers development and maintenance of requirements traceability, requirements management, and other related tasks. The understanding is that the PIT applies this process initially to develop the system specification and possibly the next tier specifications. Then the IPTs all apply this process to develop all of the specifications for which they are responsible.

Allocation of functionality results in identification of things that have to be fitted into the physical model of the system, the architecture. Requirements analysis sheets, RAS in paper or computer screen format, are the principal inputs. The outputs include:

a. *System Description Document.* Captures the system diagrams defining system composition. It includes: (1) a functional flow diagram, illustrating needed system functionality, (2) the aggregate RAS, capturing the allocations of functionality to architecture, (3) architecture block diagram, a hierarchical block diagram defining the things in the system, (4) schematic block diagram defining all of the relationships between all of the things shown on the architecture block diagram, and (5) any other documentation products of the system definition process.
b. *Drawing Breakdown Structure.* The engineering drawing overlay of the product architecture telling what engineering drawings will be produced.
c. *Specification Tree.* The specification overlay of the product architecture telling what specifications will be prepared and what format they will follow.
d. *Manufacturing Breakdown Structure.* The manufacturing overlay of the product architecture defining the groups of elements moving from one major production area to another.
e. *Configuration/End Item List.* Identification of the elements through which the program will be managed.
f. *IPT Responsibilities.* Team responsibility boundaries relative to the architecture.
g. *Work Breakdown Structure.* The product component of the WBS is a finance overlay of the product architecture upon which is based the whole program plan.

The top-level specifications are approved by team and program management as a prerequisite to development of lower-tier specifications and top-level designs.

4.5.1.2.3 Define process system requirements and develop concepts, F4112. Concurrently with program product development, the team must also develop the program process so both are mutually consistent. The teams work to understand the process problem in the form of process requirements expressed in process specifications and plans. The IPT must develop process requirements concurrently with its development of product, but there is also a process integration and optimization process that must go on through PIT.

4.5.1.2.4   Item team verification requirements analysis, F41116.   Item requirements are translated into verification requirements and identification of implementing verification tasks, resulting in development of task plans and procedures. During preliminary design work, this is focused on qualification planning and during detailed design the verification planning work is focused on acceptance planning work. All of the design and analysis documentation developed for an item is completed as a coordinated package and moved through a formal release process which ensures that all enterprise documentation requirements have been satisfied in the package and that the content is mutually consistent within the context of the item and the system at large.

4.5.1.2.5   Item team preliminary design, F412.   Design work is accomplished in preliminary and detailed sequences. The design work covers the product (F4121) and related processes (F4142). The design work of all of the teams is integrated and optimized by the PIT in task F4123. Item design and related process designs are reviewed and approved as a prerequisite to detailed design go-ahead. Figure 4.10 offers a detailed view of this process. A trade study process is encouraged where the preliminary design selection will not yield easily to sound engineering judgment. At the end of function F4121, the actual preliminary design work is accomplished in the task identified as Document Preliminary Design, F41217. All of the work in this function (F412) is contributed to by people from many specialized functional departments and must be managed as a cross-functional IPT.

4.5.1.2.6   Item team detailed design, F413.   Approved preliminary design work is expanded to describe the preferred solution and submitted to a detailed design review as a prerequisite to manufacture and related procurements. Figure 4.11 offers a view of this process. As in preliminary design, these tasks are accomplished by specialists from many functional departments coordinated by IPT management within the context of the team, and across the teams by PIT.

4.5.2   Acquire material, F42

All product material identified in program engineering documentation is acquired from approved sources through contractual and purchase relationships with other firms, associate contractors, and the customer. This material is made available in accordance with schedule needs for manufacturing purposes in task F3 and logistics support purposes in function F5.

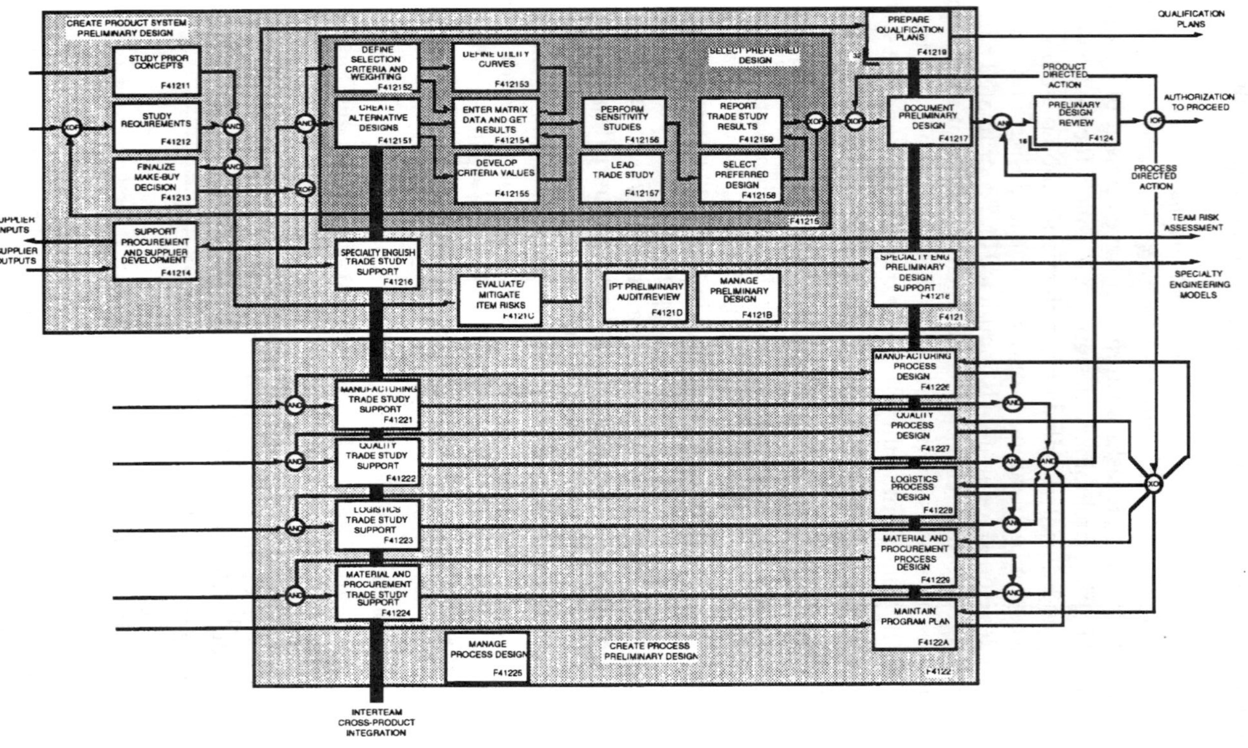

Figure 4.10 Preliminary design.

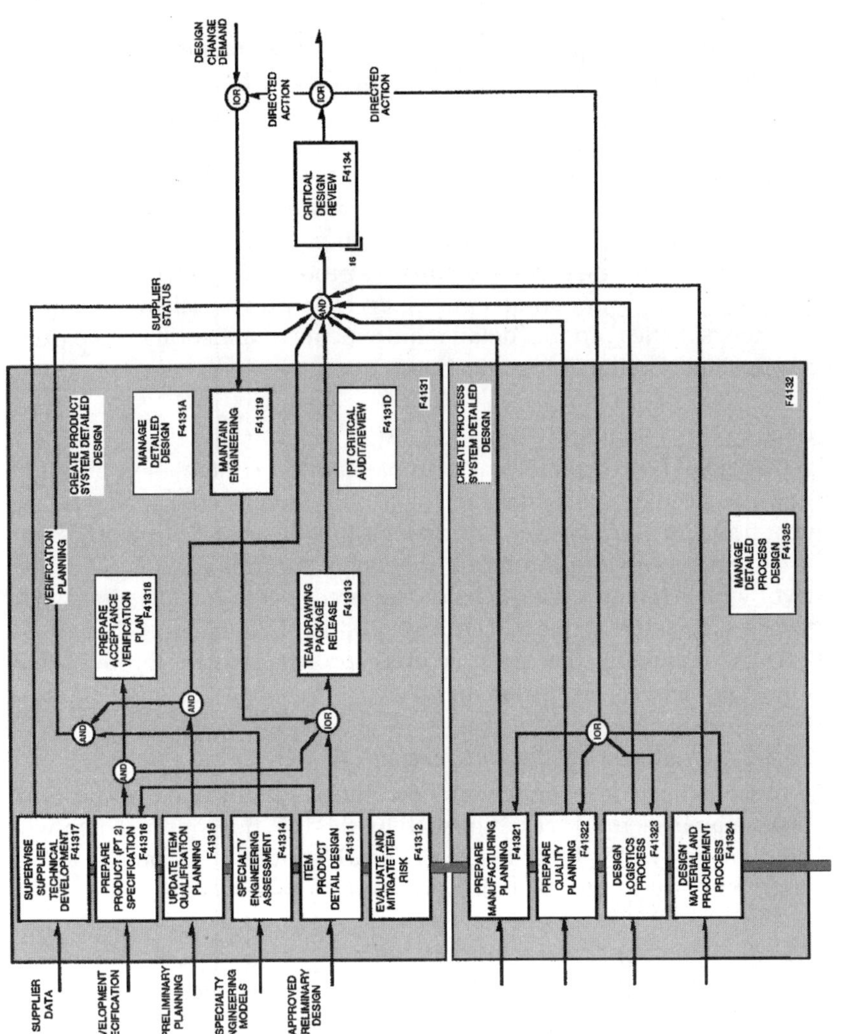

Figure 4.11 Detailed design.

4.5.3   Manufacture system, F43

The product is produced in a physical sense. Material is processed in accordance with manufacturing planning crafted in tasks F4123 and F4133. The manufacturing functional department is responsible for this activity.

4.5.4   Verify system, F44

The system engineering process involves three fundamental steps: define the problem (in specifications), solve the problem (through design), and prove that the design solved the problem (verification through test and analysis). Verification planning accomplished in tasks under function F41 are implemented to develop evidence of compliance which is accumulated, correlated with planning data, and status reports are developed. The verification process consists of two cycles: qualification to prove that the product design and manufacturing process produces a product that satisfies its development requirements, and an acceptance cycle to prove that the manufacturing process satisfies product detail requirements. Engineering is the principal responsible functional organization.

4.5.4.1   Verification management

Management skill is brought to bear to implement the content of verification plans and procedures, track status of the activity, report status, take action to prevent problems, and correct for problems that do arise. The qualification work is brought to a close through a functional configuration audit at the team level which reviews the evidence captured in the qualification work and reaches a decision about it. The first product acceptance cycle is used as a basis for reviewing the manufacturing process and the product that results in a physical configuration audit.

4.5.4.2   Qualification implementation

The PIT must manage the overall work flow during qualification verification. Some tasks may be actually accomplished by the PIT, but most of them will likely be done by an IPT. In all cases, this work should be accomplished in accordance with the careful planning accomplished as a prerequisite.

4.5.4.3   Functional configuration audit

A functional configuration audit is held to audit the evidence of compliance in the form of test and analysis reports created in function F4121 and F4131. The conclusions are captured in meeting minutes directing any additional work that must be accomplished and identifying any residual actions that must be completed prior to closure of the FCA. Generally, FCA closure is necessary to permit system-level testing involving the item in question. One or more FCA may be held for each IPT, concluding with a system-level FCA to close out any remaining open items. Alternatively, in a smaller system, a single FCA at the system level may be adequate.

4.5.5 *Logistically support product system, F45*

Throughout the development period, logistics support resources are defined as well as product resources. These resources from the manufacturing (F43) and material acquisition (F42) functions are applied to support field use of the product system through all or some portion of the product life cycle. At some point, the logistics support responsibility may shift from the producing contractor to the customer, which will require the resources to continue the support through product life. Logistics is the principal responsible functional organization.

4.5.6 *Deliver/deploy product system, F46*

The manufactured system resources must be moved from their point of manufacture to their point of operation and placed in initial operation. This will commonly involve a transportation function. Deployment is accomplished in accordance with plans made during function F41. Logistics is the principal responsible functional organization.

4.5.7 *Manage program, F47*

The overall program is managed toward program goals defined in program planning. The principal functional department is programs.

4.5.8 *Modify product system, F5*

The system is modified based on customer needs. This will commonly involve coordinated hardware, software, and procedural changes. This function includes the whole process of defining the change, implementing it in design and manufacturing, and verifying compliance. This is essentially a mini program focused on a specific change that may involve a minor change or a substantial update to the system. The principal functional organization is programs.

4.5.9 *Use product system, F7*

The system is placed in operation as planned in function F1 and verified in function F4. Over the life of the system, a need will materialize to modify the system for improved performance or to correct problems observed in operation and maintenance. Logistics is the principal responsible functional organization.

4.5.10 *Dispose of product system, F8*

The system is analyzed for hazardous materials, and materials identified are disposed of in a safe manner. Ideally, these materials should have been identified and minimized in functions F2 and F8. Some system resources may have residual capabilities that can be applied in new systems, either as-is or after modification or update. Other system resources may be scrapped. Logistics is the principal responsible functional organization.

4.6 *Formal allocation of functionality to the functional department structure*

The functional decomposition process can continue to expand the functionality identified in this chapter and illustrated in Appendix A. Table A.1 identifies the detailed allocation of functionality to functional organizations listed in Table B.1 and illustrated in Figure B.1. Table A.1 is essentially the requirements analysis sheet (RAS) for the enterprise system.

chapter five

Functional organization charters and relations

5.1 Functional department formation

The functional departments of the enterprise should be formed based on the aggregate collection of generic tasks mapped from the generic process steps covered in Chapter 4 to the functional organization structure. The functional department managers are made responsible for documenting their department charters based on the management's directive to provide programs with well-qualified persons, good tools, and proven best practices. They must determine the knowledge base needed to satisfy this requirement and thus the qualifications of the people appropriate for membership in their department. Where multiple departments share responsibility for a given task, a lead department is identified that should provide charter leadership for that task in terms of tool selection, training, and practices.

As discussed earlier, in a more realistic situation where the enterprise is taking the first step toward implementing an improvement process moving from a prior *ad hoc* department charter structure, there must be an orderly movement from the current methods to a continuous ribbon of improvement toward perfection. There will be some initial resistance observed from functional managers because the change will represent instability, a fear-inducing force in any organization. The enterprise executive must prepare current functional management for the transformation by causing them to recognize that there will be some instability but that this offers new opportunities rather than the seeds of career disaster.

The current functional managers must come to understand individually what drives them professionally and personally. If the organization has had a strong functional axis in the matrix where the functional managers actually managed their department's portion of program work either directly or through department surrogates, they have to realize that the new role of functional management is going to carry with it much less organizational power and prestige. If these are important to the individual, one should probably consider a change of assignment, and top management must be tuned in to the need for some assignment changes. Power and prestige will probably, in this situation, migrate to some extent from the functional organization to the program organizations in the form of program managers and team leaders.

5.2   First-tier functional departments

In Chapter 3 we identified the first-tier functional departments in Table 3.1, and they are repeated in Table 5.1 for convenience. These represent our top-level view of the partitioning of man's knowledge into common sets for mastery by people. Each of these departments will presumably include sub-departments, reflecting a finer degree of specialization necessary to master the whole field of system development skills and knowledge.

5.3   Expansion of the functional organizational hierarchy and charter development

You will need to perform an expansion of your enterprise functionality based on your product line, customer base, and preferences, all summarized in the enterprise vision statement shown in Figure 5.1. The other beginning point for this structured process for defining an enterprise infrastructure begins with the knowledge base necessary to solve the kinds of problems anticipated in light of the vision statement. One way to organize this knowledge in a functional organization structure for an enterprise involved in solving complex problems requiring the application of engineering skill is shown in Table 5.1. We now need to fill in the details covering what these functional organizations should bring to program work.

The author has created an example of these details based on having worked in several large engineering organizations, and observing many more. The results are captured in Figures A.1 through A.46, a functional process flow diagram; Table A.2, which allocates enterprise functionality to functional organizational entities; Figure B.1, which shows the resultant functional organizational structure; and Table A.1, which defines the charter for each functional department.

The department charters were derived by aggregating all of the functionality allocated to them in Table A.2 and working back and forth between the needed functionality and physical organizational forms a few times. This is an iterative process that you will find very difficult to complete in one analytical cycle. While doing this work you must keep several documents in a state of agreement: the enterprise functional flow diagram (Figure A.1) identifying needed functionality, the function allocation matrix (Table A.2) marrying the needed enterprise functionality to the organizational structure, the functional department diagram (Figures B.1 through B.3), and the related department charter matrix (Table B.1). You will find that the discipline of the

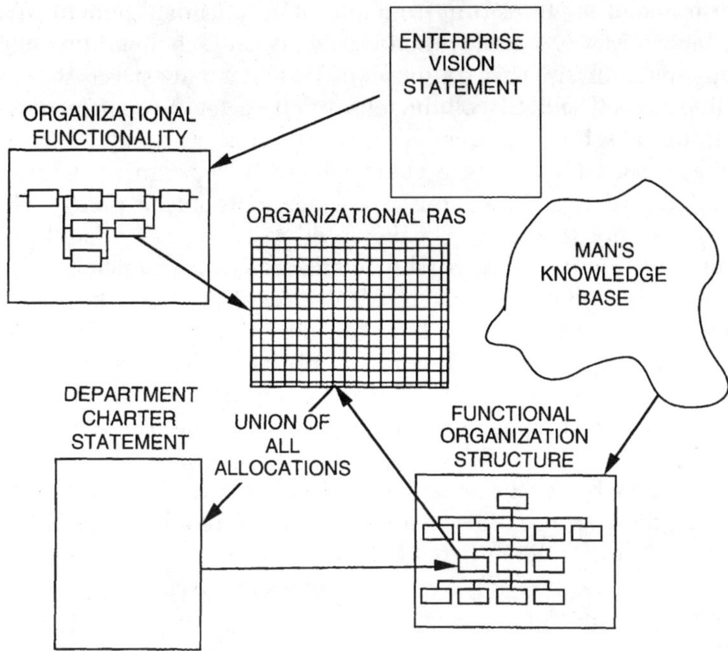

Figure 5.1   Organizational structured analysis flow.

Table 5.1   Top-Tier Functional Structure

Org ID	Department Name
000	Enterprise Executive
100	Business Functional
200	Technical Functional
210	Engineering
220	Manufacturing
230	Logistics
240	System Quality
300	Enterprise Integration Team
400	Programs Management
900	New Business

process suggested will evolve a consistent set of data that comprehensively addresses all of the work that will have to be accomplished on any program in your company's normal product and customer mix.

The charter statement is a high-level summary description of what the department is responsible for, the department version of the enterprise vision statement. The details are provided in Table A.1, and this should be the basis for development of department practices supportive of the department charter.

As an example of the transform from allocated functionality to department charter, Table 5.2 lists all of the functionality allocated to department D216-2, Requirements Analysis. This is a subset of Table A.2 that extracts department D216-2 line items. The RESP column tells the character of department responsibility from the set: L = Leadership, C = Contribution.

One approach to crafting a charter for this department would be to simply aggregate all of these function statements into a paragraph. It is recommended though that you make an effort to digest or distill this raw input into a very few sentences. The details are always available to anyone interested in your allocation matrix. Ideally, these data have been captured in a relational database so that they can be brought up on the screen or printed out in any way desired. For example, we should be able to call up the Table 5.2 view of the data in Table A.1 from the same database system, just different views of the same table filtered in accordance with some criteria.

Refer to Table B.1 for the department 216-2 charter driven by the Table 5.2 content. Yes, there appears to be a lot of functionality missing in the charter. You have a choice of setting up the charter statement on any level of allocated functionality you choose. You can try to combine dozens of lower-tier allocations into a synthesis or evolve a charter statement from a relatively few high-level functions.

5.4   *The popular knowledge map*

It is important to keep in mind while building the picture just sketched how the rest of the world has partitioned the knowledge base to which your enterprise will need access to staff the organization. True, the enterprise can organize itself however it wants, but there is some advantage in reflecting the norm in terms of being able to hire people who do not have to be extensively retrained to fit your mold. The point is not that you should *always* respect the job marketplace, but it should be one of the factors considered in building the enterprise charter statements.

5.5   *Program knowledge interfaces*

It is true in developing product using the structured analysis process that you predetermine the interfaces that will have to exist in the physical product by the way you allocate functionality to the things in the product architecture. It is also true that you will predetermine the communication interfaces that must exist on programs between the people from the different departments by the way we allocate enterprise functionality. The organizational RAS in Table A.2 identifies all of the allocations of functionality to functional departments. This does not entirely tell how people from these departments will have to cooperate in performing these tasks on programs, but these interactions are predetermined by Table A.2 content.

Table 5.2   Department 216-2 Functionality

Function ID	Function Name	Resp
F4111	Define Product System Requirements and Develop Concepts	L
F41111	Product System Concept Development	C
F41112	Product System Operational Environment Definition	L
F41113	Perform Applicable Documents Analysis	L
F411141	Study System/Item Concept	C
F411142	Select Specification Template	L
F411143	Item Applicable Documents Analysis	L
F411144	Traditional Structured Analysis	L
F4111441	Functional Analysis	L
F41114411	Function Diagramming Steps	L
F411144111	Identify Functions	L
F411144112	Functional Flow Analysis	L
F411144113	Process Analysis	L
F411144114	IDEF 0 Functional Analysis	L
F411144115	Hierarchical Functional Analysis	L
F41114412	Functional Requirements Analysis	L
F411144121	Define/Refine Measures of Effectiveness	L
F411144122	Unprecedented Performance Requirements Analysis	L
F41114413	Function Allocation Steps	L
F411144131	Form and Brief Team	L
F411144132	Review Prior Allocations	L
F411144133	Apply Brainstorming Technique	L
F411144134	Consolidate Allocation Statements	L
F411144135	Make Changes	L
F411144136	Engineering Review	C
F411144137	Perform Trade Study	L
F411144138	Consider Direction Impact	L
F411144139	Capture/Document Allocations	L
F41114414	Prepare/Maintain SDD Inputs	L
F411144141	Capture/Document Diagrams	L
F411144142	Prepare Descriptive Function Text	L
F411144143	Format SDD Content	L
F4111442	Performance Requirements Analysis	L
F4111443	Timeline Analysis	L
F4111444	Architecture Synthesis	L
F4111445	Interface Analysis	L
F41114451	Evaluate Allocated Functionality	L
F41114452	Perform Interface Analysis	L
F41114453	Evaluate Cross Organizational Interface Impact	L
F41114454	Document Interfaces	L
F4111446	Process Analysis	L
F4111447	Specialty Engineering Requirements Analysis	L
F4111448	Service Use Profile	L
F4111449	End Item Zoning	Rel
F411144A	Interface Requirements Analysis	L
F411144B	Write Requirements Paragraphs	L

Table 5.2 (continued)   Department 216-2 Functionality

Function ID	Function Name	Resp
F411144C	Lead System/Item Requirements Analysis	L
F411144D	Edit/Update Current Content	L
F411144E	Integrate Item Requirements	L
F411145	RDD	L
F411146	Modern Structured Analysis	L
F411147	Database Modeling	L
F411148	Object Oriented Analysis	L
F411149	Strings and States Analysis	L
F41114A	Publish Specification	L
F41114B	Clone Specification	L
F41114B1	Select Cloning Mode	L
F41114B2	Select Like Item	L
F41114B3	Select Standard	L
F41114B4	Identify End Item	L
F41114B5	Identify Zone	L
F41114B6	Perform Item-Unique Environmental Requirements Analysis	L
F41114B7	Perform Item-Unique Performance Requirements Analysis	L
F41114B8	Perform Item-Unique Interface Requirements Analysis	L
F41114B9	Perform Item-Unique Specialty Engineering Rqmts Analysis	L
F41114BA	Copy Standard Specification	L
F41114BB	Analyze Content Relationships	L
F41114BC	Copy Parent Specification	L
F41114BD	Analyze Content Relationships	L
F41114BE	Copy Like Item Specification	L
F41114BF	Analyze Content Relationships	L
F41114BG	Modify Content	L
F41114BH	Manage Cloning Task	L
F41114BI	Integrate Content	L
F41115	Item Concept Development and Review	C
F41116	Verification Requirements Analysis	L
F41117	Change Released Specifications	L
F411171	Perform Change Requirements Analysis	L
F411172	Revise Specification	L
F411173	Prepare Page Change Notice	L
F411174	Prepare List Notice	L
F411175	Review Specification Change	L
F4112	Define Process System Requirements and Develop Concepts	C
F41131	Review and Release Specifications	L
F411311	Prepare For Requirements Review	L
F411312	Review Specification	L
F411313	Peer and Informal Reviews	L
F411314	Perform Post Review Actions	L
F411315	Acquire Approval Signatures	L
F41133	Validate Product Requirements	L
F411331	Evaluate Requirements	L
F4113311	Assess Specification For Completeness	L

Table 5.2 (continued)   Department 216-2 Functionality

Function ID	Function Name	Resp
F4113312	Screen Requirements For Necessity and Correctness	L
F4113313	Rewrite Requirement	L
F4113314	Discard Requirement	L
F4113315	Screen Requirements For Structural Correctness	L
F411332	Change Requirement	L
F411336	Lead Requirements Validation	L
F411337	Technical Performance Measurement	L
F4113371	Track and Manage TPM	L
F4113372	Terminate TPM Track	L
F41134	Establish and Maintain Product Traceability	L
F41135	Perform and Maintain Margin and Budget Analysis	L
F41136	Develop and Maintain Specification Tree	L
F41137	Select Specification Templates	L
F41138	Schedule Specifications	L
F4113A	Review and Approve Process Requirements	L
F4113B	Manage Product System Requirements Analysis	L
F4113C	Manage Process System Requirements Analysis	L
F41212	Study Requirements	C
F412351	Technical Performance Measurement	L

5.5.1   Intratask relationships

So far, we have used a process flow diagram to define what the development organization must do, and we have allocated this functionality to a development organization architecture called a functional department organizational structure. It remains to give some idea about how the elements of the resulting organization will work together to accomplish the process goals, and this requires some understanding of the interfaces between the people from the functional departments when they ply their trade on programs. Part of this story is told by the allocation pattern in Table A.2, which identifies organizational sets jointly responsible for process steps. Where this matrix shows a single functional organization as the only entry for a particular process, there is no difficulty in understanding organizational behavior. They simply have to do it all and the process dictionary in Table A.1 tells this story. The problem appears when multiple functional departments are called upon to contribute to a process step. Where this is the case, one of these should be identified as the lead functional department, and this is the beginning of the behavior definition activity.

Before continuing with this work relationship discussion, the idea should be reinforced that the program work should be accomplished by cross-functional teams, referred to in this book as integrated product teams (IPT). These teams are staffed by people assigned from functional organizations, and they contribute the skills and knowledge related to the functionality component allocated to their functional department. The team is physically

FUNCTION	FUNCTIONAL DEPARTMENTS	ROLE
F41114	D212-3	C
F41114	D212-4	C
F41114	D213-4	C
F41114	D213-5	C
F41114	D214-3	C
F41114	D213-4	C
F41114	D215-1	C
F41114	D216-1	C
F41114	D216-2	L
F41114	D216-4	C
F41114	D216-5	C
F41114	D216-6	C
F41114	D216-7	C
F41114	D216-8	C
F41114	D221-1	C
F41114	D231-1	C
F41114	D241-1	C
F41114	D242-1	C

a. Organizational RAS

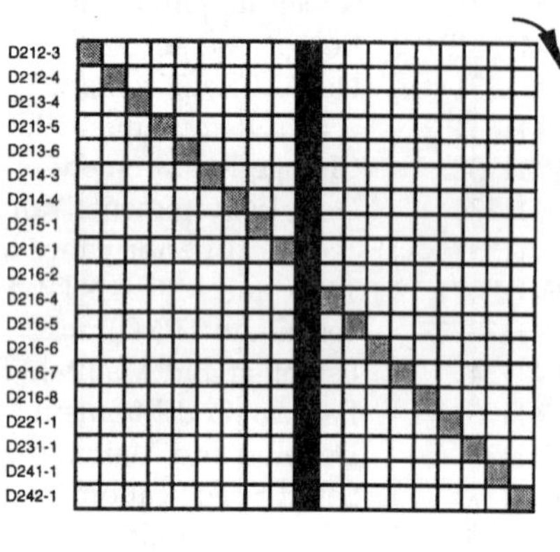

b. Organizational N-Square Diagram

Figure 5.2 Enterprise organizational interrelationship definition models.

collocated and led toward their program goals by a program team leader. The interfunctional department relationship that must be identified is intended to be accomplished in the context of this team structure.

The intended behavior could be communicated through the organizational RAS only, functional organization n-square diagrams or functional organization schematic block diagrams, as illustrated in Figure 5.2. In Figure 5.2a, the fragment of the RAS provides an indicator of needed cooperation but does not communicate what must be communicated. It does define who is the responsible agent for process development and implementation on programs in that D216-2 is identified as responsible for leadership (L) and others are to cooperate (C). For a given process involving several functional departments, we could craft a schematic block diagram or n-square diagram for department contributions for task F41114, as illustrated in Figure 2.5b. The contributing departments that must cooperate are listed down the left-hand axis, and this same listing is assumed on the horizontal axis. Normally, these elements would be identified down the diagonal but the diagram used here is too small for that. The diagonal intersections are marked to indicate the internal actions necessary for them to do their work. In this example, people from all of the other departments listed must accomplish work to achieve their goals for process F41114, and all of these results are used by department 216-2 in producing the final product from this task, in this case a specification.

If we look more deeply into the subtasks for F41114, we find progressively fewer departments involved and a need for less interaction between them. We would also find one task, F4111447122, Reliability Allocation, that

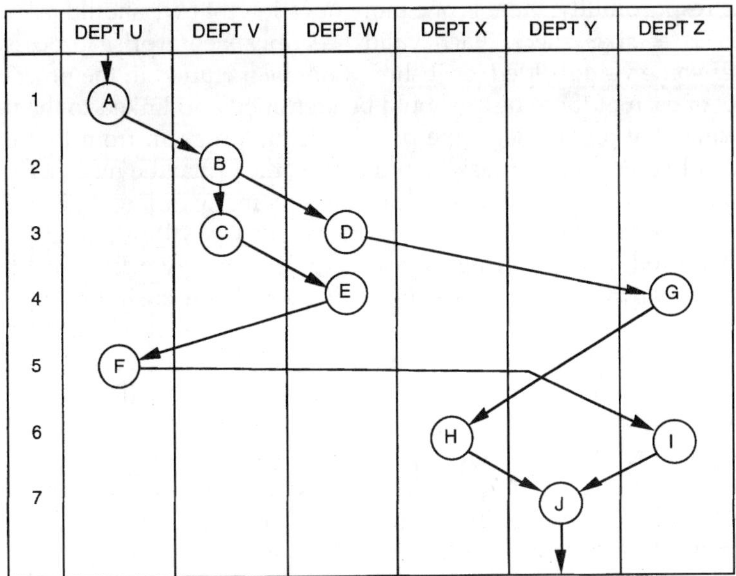

Figure 5.3   Operational sequence diagram.

will be accomplished entirely by the RAM Engineering Department, D216-4. Its output is a reliability figure for each item in the product system, and these figures flow into the specifications.

Figure 5.3 offers a fragment of another intratask exposition model. It is an operational sequence diagram illustrating that department U begins a process, passing on some results to department V. The reader can follow the sequence from that point. Each entity (department in this case) has a separate column. The analyst places activity symbols in the columns corresponding to work responsibility and connects the activities in sequence. This kind of diagram identifies two or more channels in one axis, and time flows in the other axis (down in this case, though the diagram could be laid out horizontally).

All work may be in a simple serial pattern or require concurrent activity by people from two or more departments. The analyst may use different symbols to indicate different kinds of actions such as do physical work, make a decision, or do analysis. This diagram could be supplemented with a task dictionary to explain what it means for department W to do task D. The numbers down the left column suggest a clock running in some units of measure like hours, days, or weeks. Vertical bars could be used to indicate period of performance of the tasks where they last longer or shorter than one time increment.

5.5.2   *Intertask relationships*

While the process definition and allocation activity described in this book should produce a thoughtfully complete accounting of needed work linked to clear responsibility, there is one more question that we should ask of each process as a cross check. Each valid task in an enterprise process must produce some value added, or it should not be included in the process. The outputs of each of these tasks should be identified and linked to the responsible functional department (the people on the program from that department). At higher levels of task identification, there may be many task products that must be contributed to by people from many different departments, but as we flow further down into the hierarchy, we will find a point where most task products appearing as task outputs are the work of people from a single department, or at least the clear overall responsibility of a single specialty.

In that the process diagram recommended in this book is constructed as an expansion, it stands to reason that higher-tier tasks will require a great deal more intratask, interdepartment coordination and activity than lower-tier tasks that tend to be single department oriented. There are several ways one can study the needed intertask relationships. The process diagram itself gives us some information by virtue of the flow indicated, but this format is not helpful in showing all of the data flow because it adds so many lines that the sequence intended is hard to spot visually.

A process N-square diagram could be used, but where the tasks in question are contributed to by people from different departments, it may not be clear who is responsible for the interfaces indicated. This could be listed in a separate dictionary.

Table A.2, the organizational RAS, identifies for each line item, which is a program process step hooked to a functional organizational responsibility, one or more task products that must result from the related work. These are the significant accomplishments of the tasks. These are the things a manager should expect for the cost of supporting that task with money and schedule time. These products flow to other tasks, and a thorough task plan would also identify the terminal tasks for these task products. That step has not been included in the tabular data supplied with this book but could be linked to the tasks as input requirements, illustrated as data flow in an RDD diagram, or illustrated in an IDEF 0 portrayal of process either as resource inputs on the lower edge or controlling influences on the upper edge of terminal task blocks.

The two other diagrammatic treatments referred to in the previous paragraph lend themselves to illustrating not only the process flow but the interfaces between those responsible for accomplishing the indicated work. They are Input Process Output (IPO), or Requirements Driven Design (RDD), and SADT, or IDEF-0. IPO and SADT were originally created as software analysis models but were respectively adapted by Mack Alford as the basis for the Ascent Logic system engineering tool RDD-100 and the U.S. Air Force as the basis for integrated computer aided manufacturing (ICAM) definition (IDEF) language.

In the IPO/RDD application, the process diagram runs down the page like an old computer programming flow chart, as shown in Figure 5.4. In

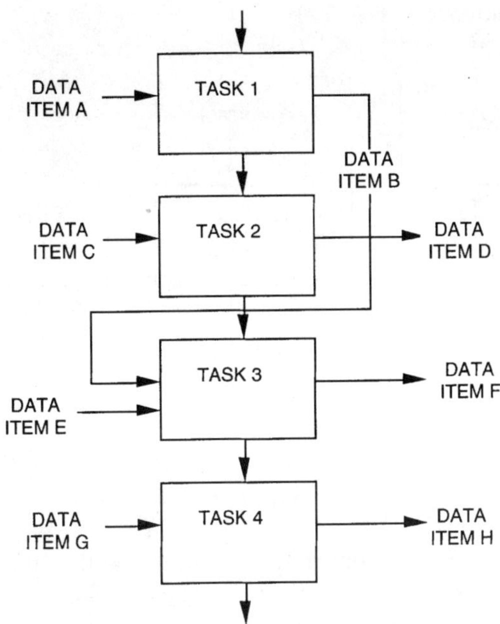

Figure 5.4   IPO/RDD application.

addition, we include a lateral connection for data flow into and out of these processes, which have been derived from outside the process illustrated or from a prior process step within it. Note that the flow diagram is run vertically on the page rather than horizontally as in Appendix A. When the work is performed, there is a work product that flows out the right side of the block. The work products, which often will be documents in whole or in part, flow from source tasks to tasks where that work product is consumed or applied.

It should be noted that RDD-100, which uses this behavioral model, can be used to capture the generic and program process flow definition with associations to the contributing functional organizations just as it can be used to capture product functionality and allocation to product architecture. When the program process is thus defined, RDD-100 will permit simulation of the program before you attempt to implement the program (as in the proposal period), yielding information useful in verifying the plan.

The SADT or IDEF 0 approach is illustrated in Figure 5.5. It consists of a flow diagram that steps across and down the page. Directed line segments can also enter the process blocks at the top to control the activity or at the bottom to provide resources needed in that process step. IPO/RDD and SADT/IDEF 0 offer ways to link functionality with data or work products, but one might still ask, "How do the people from those five different departments required for this task accomplish the work within the box?" It may be that an expansion of that block to a lower level, using one of these two techniques, will make that answer clear. Otherwise, an application of

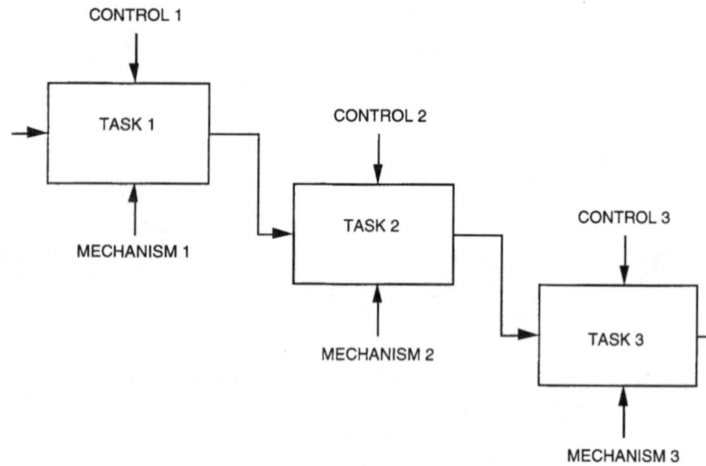

Figure 5.5   SADT/IDEF 0 application.

n-square or operational sequence diagramming within the lowest-tier task may achieve the desired degree of process specificity.

There may be a balance point between detailed process specificity and intuitive common vision that would make it unnecessary to expand the process responsibility definition beyond some point determined by good judgment or common sense, and the practitioner should be alert to that acceptable level of detail.

Readers may say that they have experienced many of the tasks included in the process diagram included in this book and can think of many cases where there simply is not any tangible task product that can be identified. The author rejects this conclusion. A tangible product can and should be identified. It may be as simple as a memo of conclusions that should be captured in the program library and that documents what the person who did the work communicated verbally to the person who needed that infor-mation. If there is no tangible product, then there is no task and no budget. This message will cause even the most reluctant engineer or manager to invent a value-added reason for their existence in the process.

5.6   *Process integration*

As the generic process is crafted, the work recommended for the many processes by the several functional departments may be right. Then again, some of this input could be unnecessary. Other necessary input may be missing. The integration agent encouraged is the Enterprise Integration Team (EIT), the enterprise process owner. When first creating this kind of infor-mation, the PIT should evaluate all of the task outputs to ensure they are necessary, that someone is using the work products created, and that there are no omissions.

chapter six

Generic practices documentation

6.1 Enterprise generic documentation tree

The enterprise process could be documented in many different ways. The author suggests the development of an enterprise process specification containing the flow diagram (Figure A.1), functional organization diagram (Figure B.1), function allocation matrix showing the correlation between process steps and the functional organization departments (Table A.1), and a sketch showing how programs shall be organized using integrated product teams and program integration teams (Figure B.2). This document becomes the driver for functional department action to document their components of the enterprise identity.

The union of all functionality mapped to a particular department must be organized into a department charter defining the responsibilities of the department. At some level all of these charter statements should be collected into one or more documents telling what work must be accomplished by department personnel on programs that require that activity. All of these charters could also be included in the process specification. Figure 6.1 shows a top-level functional department documentation tree, beginning with the enterprise vision statement.

The numbers in the lower right-hand corner of the document blocks (Figure 6.1) indicate the responsible functional organization drawn from Figure B.1. All of these responsibilities are obvious from the organization diagram except perhaps for the Enterprise Process Specification. As mentioned earlier, the enterprise process owner is the Enterprise Integration Team (EIT), also responsible for the Enterprise Development Manual. Other lower-tier manuals not shown, in the interest of illustrating all of the engineering-related manuals in one simple figure, may also be necessary.

Some critics will complain that ISO 9001 encourages that all enterprise documentation be subordinate to a quality manual, but the author believes this to be an unrealistic and flawed interpretation. However important the quality function is, it is not everything, and all of the enterprise responsibility should be recognized at the higher levels of documentation. Another person

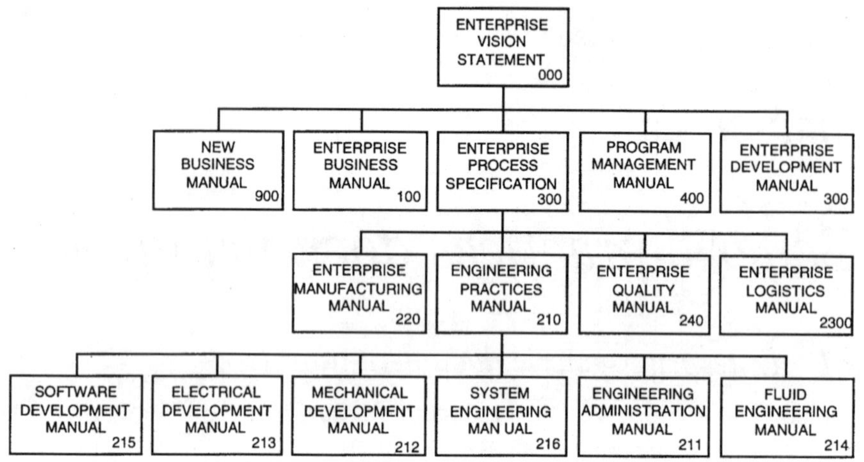

Figure 6.1   Enterprise process documentation tree.

may well elevate the quality manual one level to the first tier, and that is simply an individual choice, as is the configuration illustrated in Figure 6.1. The precise configuration of this diagram is not important in isolation. It is important how this documentation is structured relative to the process responsibilities and functional organizational structure.

6.2   *Practices documentation content and responsibility*

The manuals shown in Figure 6.1 must contain information telling what must be done on programs. They are internal standards for the performance of program work. They are an extension of the department charters contained in the Enterprise Process Specification. The arrangement of content in the overall manual structure is determined by the way enterprise functionality has been allocated to the functional organization. Every function should be allocated to one or more functional departments and, where it is allocated to more than one department, one of the departments must be selected as the principally responsible department, as illustrated in Figure 6.2. Function F has been allocated to departments W, X, Y, and Z. Department X has been identified as the principal department.

Department X has the responsibility to lead the effort to document Function F work requirements, with the support of departments W, Y, and Z, and the extension of that work definition to their individual departments. Department X is also responsible for selecting or developing the tools to be used by all four departments in the accomplishment of this work on programs and referenced in the practices manuals. Finally, department X must also develop or select the means by which persons in all four departments will acquire the skills necessary to perform the work corresponding to function F. This may be the selection of a textbook covering the activity and

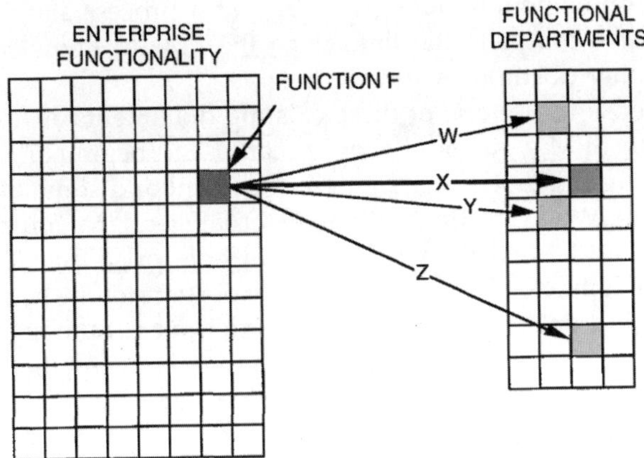

Figure 6.2 Principal functional department.

purchased for enterprise of department libraries or development of a training course. Department X becomes the project engineer for function F, leading the efforts of several departments for that task.

6.3   *System engineering manual*

This book is focused on system engineering and the deployment of it into programs so we will discuss only the development of the system engineering manual. The development of the other documentation can follow a similar pattern. There are several sources of manual content. One could begin with one of the author's earlier books titled *System Engineering Planning and Enterprise Identity*, which includes a complete SEM in paper and computer disk formats. The disk copy can be edited as seen fit by the company or it can simply be used as is. Alternatively, one could turn to an available national system engineering standard and use it as a model. IEEE P1220 was intended as a commercial system engineering standard, and EIA 632 was initially intended as a replacement for MIL-STD-499B, which was prepared but not approved by DoD. MIL-STD-499A or 499B could be used as well since the basic process discussed in all of these standards is essentially the same. The Defense System Management College (DSMC) has crafted several versions of its system management guide that could also act as a guide for one's SEM development.

Appendix C of this book includes a complete System Engineering Practices Manual (SEPM) in paper media and an enclosed disk provides this document in Microsoft Word for Macintosh and Windows machines on an IBM-formatted disk that can be used in either kind of machine. This document is different from the SEM included in the author's earlier book in that it is actually a process specification rather than a textbook providing how-to information. The author reluctantly concluded that an enterprise would be hard pressed to create what is essentially a textbook in this field and should focus on something along the lines of a process standard, such as IEEE P1220, and complete the linkage to the how-to knowledge through reference to other documents and sources.

If you choose to use one of these existing documents as a standard for preparing your SEM, you could simply scan it and begin editing it to your needs, assuming it is not a copyrighted document and some of those identified above are. When starting with an external standard you should make sure you have addressed the results of your enterprise functional analysis and allocation. It may be necessary to rewrite portions of the source material, add material, and delete other material in order to craft the final document. An organized way to do this is to develop a map between the functionality uncovered in your analysis and manual content, thus establishing traceability between the two.

Rather than begin with an external model, you could begin with the list of allocations made to the system engineering organization. A brainstorming session, attended by several respected system engineers, can be very helpful in focusing these allocations into an outline for the SEM. This exercise can be implemented by inscribing each function on a yellow stickyback note and initially placing them in some kind of outline arrangement rather than beginning from scratch. The team should try to move to consensus on the way the functions have been hierarchically organized, which ones are major points and which are subordinate.

Another way to do this is to apply computer projection where the functions are objects on a desktop that can be moved around relative to each other. Periodically, you should save numbered baselines so that you can go back to earlier versions if necessary.

Once the outline is agreed upon and it has been verified that it covers the allocated functionality, you can start writing the document. Make assignments against the outline and schedule the document development.

A third approach for developing the SEM is to draw upon some of the past program SEPMs that you have prepared. Even if you never followed them, they may have good content. You may be able to cut and paste a quality SEM from all of this material with a few remaining original writing or editing jobs remaining.

6.4   External standards mapping

As you build your SEM, you should find out what standards of system engineering your customers prefer and map the content of those standards to your evolving SEM. You should end up with one or more matrices that show for every element of your customer's preferred standards where the customer can find the corresponding coverage in your SEM. It is possible that you will choose not to do some work covered in the customer preferred standard, so when you quote a job to the customer you should make it clear what parts of their preferred standard you will apply and what parts you will not. You should have a reasonable explanation why you have chosen not to perform deleted parts of their preferred standard.

Complete this job by tailoring the customer-preferred standards to agree with your manual. This can be done using the legalistic or in-context tailoring methods. In the legalistic approach, simply make a list of the changes you would make in the document, 1 through N. Each of these statements covers a change such as, "13. Delete paragraph 3.4.5.3." or "Change 13 pounds to 15 pounds in paragraph 4.5.3.2.1." Your traceability data should be to the tailored standard. It should be noted that tailoring should not be applied to any laws or regulations that must be observed in a contract. This is called *breaking the law.*

6.5   *The discipline for process compliance*

If your organization has no history of a written process definition or of following its own process documentation, something has to change, preferably before you commit resources to preparing a new generic SEM. If you prepare an SEM, what will ensure that its content will be followed? Top management must be prepared to expend some energy to cause the content to guide behavior. If management is not prepared to expend the necessary energy, then the attempt to improve process control will fail. One of the secrets of effective management of process improvement actions is to limit the scope of the change at any one time to that which can be easily communicated and accomplished. Management must communicate a simple message repeated over and over with conviction, transmitting some evidence of displeasure where the direction does not have the correct effect and offering some form of positive encouragement where people perform as desired.

People at the working level in engineering organizations often endure periodic intense management devotion to new initiatives sweeping through industry, and contribute their interest and energy to comply, only to find after a brief time that management is no longer interested in that initiative. If this process is repeated often enough, people become reluctant to participate energetically. As you prepare the SEM, also prepare management to support the process defined therein.

6.6   *Practices deployment to programs*

Later in the book, we will introduce a document called an integrated master plan (IMP) that identifies all program work. Some customers encouraging the use of this kind of planning document suggest that it contain two sets of information called the *work definition* and *narratives*. The latter includes text and graphics that explain how the program work is going to be accomplished. The author maintains that the enterprise should build a generic set of practices and supporting how-to information that is essentially the same on every program. The program planning will tell what has to be done, by whom it shall be done, when it must be completed, and perhaps why. Each program task is an implementation of a generic task identified in the enterprise process specification and about which generic practices exist, extending to tools to be employed and how-to information references. Where a customer requires development of an IMP or other form of program planning data, the contractor should reference generic data for access to these details rather than creating a new process description for every program. If you must craft narrative information as an integral part of the IMP, these data should be created by simply cutting and pasting from the source documents.

For each program task, we should identify a generic process description. One can picture a matrix making this transformation of generic planning data into program planning data. In that the program data are going to employ tasks from your generic process diagram applied to items that must be developed in this specific systems, much of this matrix can be canned and simply edited for the program. You should not have to create any new how-to data for a new program. Just apply the generic data. If a contract requires work never before accomplished on prior contracts and not captured in your generic data, you will have to build new descriptive coverage, and you may wish to embrace this as part of your generic process description.

6.7   *Process quality assurance on programs*

All of the maturity models encourage an enterprise to have a written process description and to follow it. If the reader follows the prescription in this book, the written process will be made available. It is even harder to follow the second step of actually doing the work in accordance with the written instructions. An enterprise requires a process quality agent to ensure this happens not just for the manufacturing process but for the whole development process as well. The author has mapped that responsibility to the Quality functional organization. The Quality member of the PIT should periodically assess program compliance with generic and program-specific planning data and report the results to program and functional management for possible corrective action.

6.8   *Functional metrics*

Concurrent with the development of the practices, we should consider what measurements can be made when implementing the practices on programs to indicate the quality of performance of the process. When writing requirements for inclusion in specifications, it is a good idea to also craft the verification requirements at the same time. The reason for this is that it forces you to tell how you will determine whether or not the design that evolves satisfies the requirements. In the process of trying to phrase the verification requirement you find out a lot about the quality of the product requirement. If the requirement is not quantified, for example, you will find it very difficult to write requirements for proving that the product satisfied the requirement,

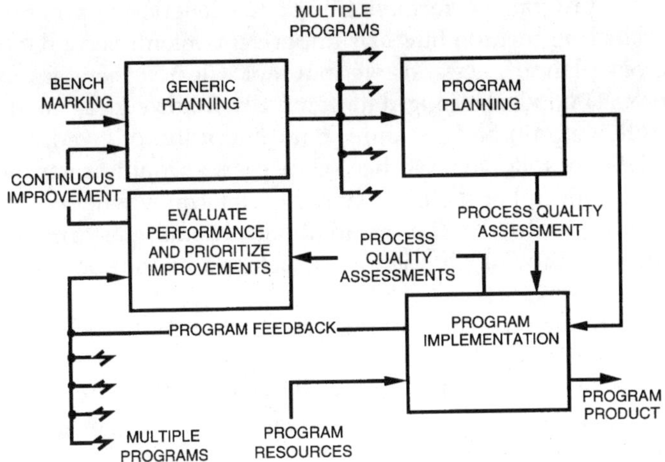

Figure 6.3   Continuous process improvement.

to say nothing of the difficulty the designer will have trying to craft a design responsive to it. The ultimate bad requirement might be, "The product must work well and last a long time." Compliance with this requirement would be very difficult to verify.

The author recommends you apply this same approach in creating your practices documentation. As you create a paragraph for the SEM, write another describing how to verify that the program complies with that content. If possible, craft a metric defining with mathematical precision how to make the measurement and what it means. The author has never seen a practices manual that included a verification section along the lines of a specification, but it might not be too bad an idea.

6.9   Process improvement

Programs present the enterprise with tremendous opportunities to observe the quality of their performance, and all too many enterprises disregard the information available to them about their quality of process performance. Sure, you need metrics that act as process test points feeding you information about process performance details that can be tracked and compared over time to detect changes in performance, but you also need to apply human assessment, using quality engineers who can compare generic and program-peculiar planning with actual performance and identify any differences between them. This is the error signal of continuous process improvement. The program feedback lines in Figure 6.3 provide this error signal characterized by the metrics supplying data for review.

Ideally, we would have perfectly crafted practices and plans, and we would implement them absolutely faithfully. The real world, however, is not built to match this perfect model. There will be two sources of errors in the observation of program performance. First, the generic and program-specific planning data will include flaws or imperfect content. Second, we may be following our planning data, or we may not. Clearly, there are four possibilities here: (1) good planning data faithfully followed, (2) good planning data not followed, (3) bad planning data faithfully followed, and (4) bad planning data not followed. Of these choices, we would prefer the first.

We have to decipher the error signal to determine which of these cases applies in order to decide what to do about program performance that we believe can be improved. Our action in each case should be as follows:

1. Kudos to the program and the functional departments providing the generic planning data covered by the audit.
2. Critical feedback to the program to follow your practices, with possible replacement of one or more team leaders and even the program manager in flagrant cases.
3. Improve the planning data as quickly as possible and caution the program to be more watchful for practices in need of improvement.
4. Improve the planning data, either generic or programmatic, as quickly as possible and caution the program that their performance would be incorrect if their planning data had been satisfactory. Encourage them to come forward in similar situations to report imperfect data rather than simply not following it.

By following this prescription, the enterprise will progressively improve its performance. The scenario above applies not just to written practices. It also applies to improving staff competence, encouraged through process repetition, but this also requires seeking out talent in the marketplace, weeding out and separating or reassigning those who cannot perform in their current assignment, and improving the tools applied on programs.

chapter seven

Organizational structure for work

7.1 Overview

The cross-functional integrated product team is advanced as the optimum engine for program work. The teams are identified based on the evolving product structure functionally derived through structured analysis following the "form follows function" paradigm. WBS, MBS, DBS, specification tree, and other views are integrated into the system architecture and associated with team responsibilities. How to structure the relationship between the teams and troublesome disciplines is covered. Cross-functional interfaces, one of the principal problems in system development, are highlighted and ways to identify and adjust are covered.

7.2 The two axes

The matrix has been offered for our enterprise structure despite its well-known flaws because of its recognition of the need to organize in two fundamental ways that are in conflict. The functional structure corresponds to the way we humans and our organization have chosen to partition knowledge of use to the enterprise. These are the fields within which specialists develop domain knowledge. By grouping like specialists together in the functional departments, we encourage enterprise mastery of each of the specialized fields needed to develop product in our chosen product line and customer base.

This functional structure is not well suited for running programs, however. Many companies have made the mistake of permitting functional departments to manage big pieces of programs and found that this approach is badly flawed. The problem is that the system does not commonly follow the organizational structure of the knowledge base. Also, every item development effort is going to require the services of many specialists working together because all of us are narrowly specialized. If we organize by functional disciplines, we make it very hard to communicate across the disciplines.

Thus, we require, in addition to the structure for knowledge mastery, a means of organizing for program excellence. The integrated product team (IPT) notion is perfect for this purpose. This structure is overlaid upon the product architecture, and teams thus oriented are staffed by people from the functional departments based on the technology they will appeal to in solving the design problem they are presented with. Thus, the matrix has two axes: the functional department axis providing programs with good people, good tools, and good practices all coordinated around a commonly held view of the development process, and a product-oriented team structure.

7.3   Team structures

Figure 7.1 illustrates the recommended enterprise structure showing the two axes reporting to the enterprise executive. In this structure, the functional departments exist to supply programs with the resources they need. The program managers are assigned from the Program Management functional department, and team leaders are selected by the program manager from among the people assigned to the program by the other functional departments, or the program manager can request assignment of particular people to serve as team leaders.

As a program begins in a proposal or study effort it very likely requires only a business team and one technical team responsible for the whole system. The author refers to this latter team as the program integration team (PIT). It is staffed from the functional departments, with heavy participation by the Advanced Development department.

7.4   Team formation and staffing on programs

As the PIT identifies the product architecture as a result of a functional analysis that allocates functionality to things synthesized into a physically oriented product architecture hierarchy, it determines how that architecture will be overlaid with team responsibilities. This may require a few teams all at one level or involve two or more layers of teams, depending on the complexity of the system. The right mixes of personnel are brought on board from functional departments based on the requirements and design concepts associated with the items selected for team coverage. As shown in Figure 7.2, program work is assigned to the teams in especially crafted patterns which, in combination with the nature of the functionality identified for the item, determines the staffing required from the functional organizations.

The first step in forming a team is to identify its leader, presumably the program manager, though in some companies the team is picked first and allowed to choose its leader. The person should have considerable experience with the product line and customer involved. It is not possible to select someone who has mastered every facet of the design problem because of human knowledge span limitations, so we can pick the leader from any

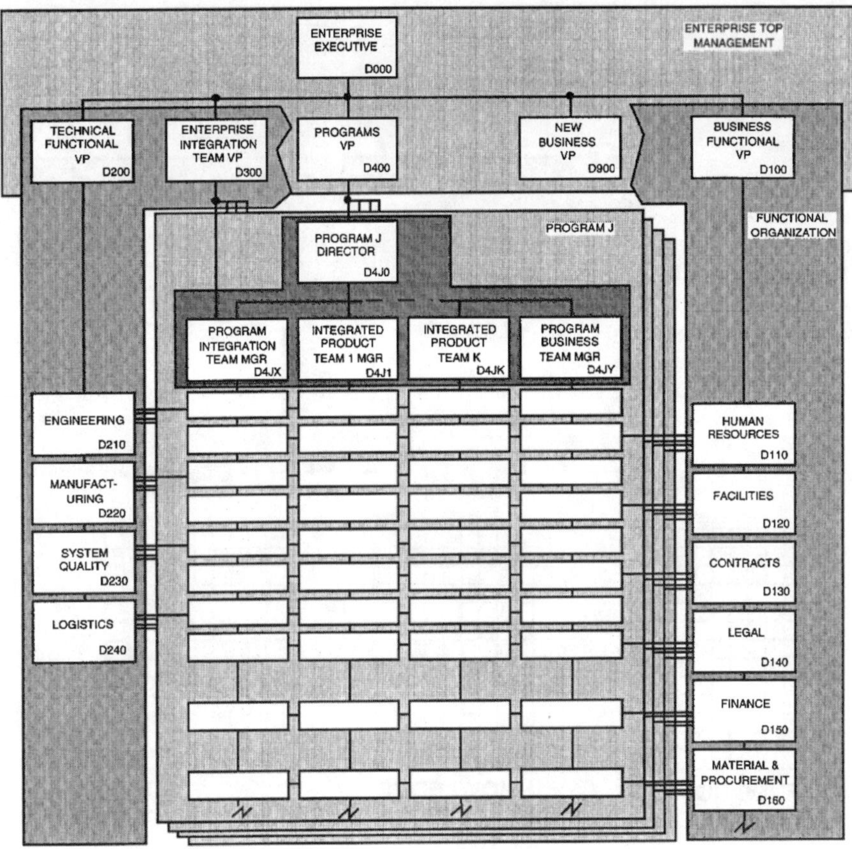

Figure 7.1   Matrix structure.

discipline, but it would be best to select an engineer to lead the early development work, someone from system engineering or the lead design function.
The leader should have good interpersonal skills, because this is a leadership
position, and have a broad outlook rather than a narrow single-discipline
view.

The team leader must then take on the responsibility for creating the
team. Figure 7.3 offers a sequence of steps leading to a state of readiness to
perform the first responsibility of the team — requirements analysis. The
leader can partition all needed personnel into core and extension components. The core people will work full time on the team, whereas the extension
component is part time. As the core people are identified and acquired, the
team leader and team members study the team assignment and form a
common vision of the task before them.

In support of this first team effort, the PIT should provide the team with
a particular collection of information (available team item data in Figure 7.3)
as they begin to form:

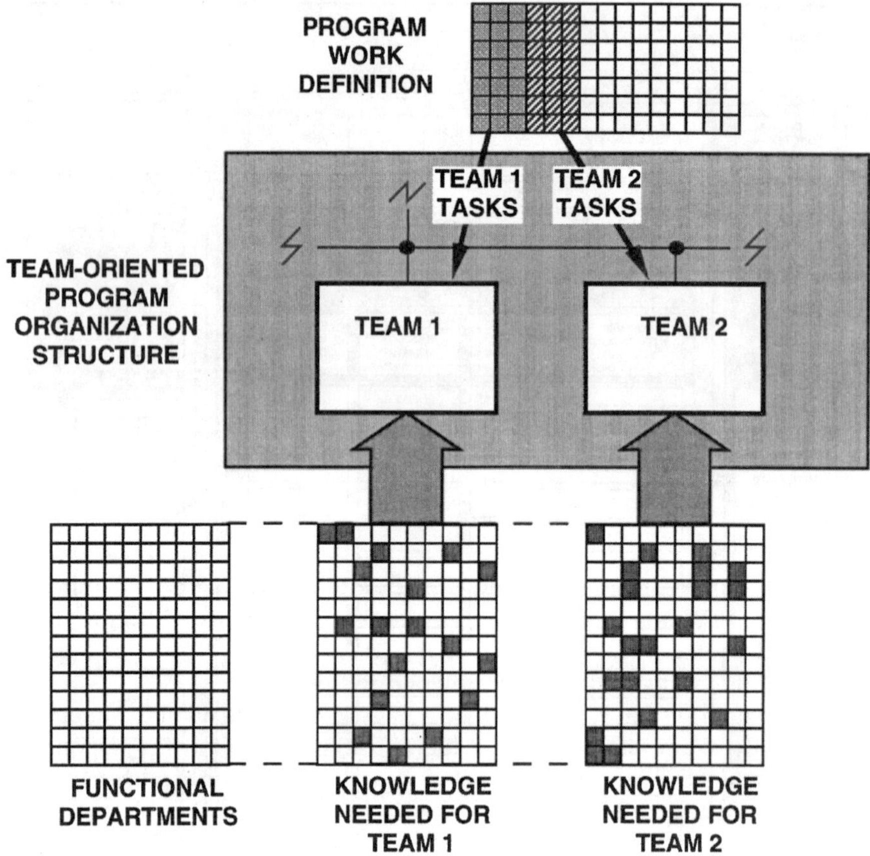

Figure 7.2   Program team formative factors.

a. The system specification, a mission analysis report telling what the system must do in a mission context, and a system definition report, providing the system requirements analysis products to date (functional analysis diagrams and data, architecture block diagram, and schematic block or n-square diagram).

b. A WBS assignment linked to a particular piece of system architecture. In the ideal case, this should be a single item (it might be a whole aircraft with subteams or the avionics system of the aircraft), but it could involve a contiguous or closely related collection of things (such as all of the product software, in which there are eight software entities in eight different pieces of equipment including two on-board computers, three pieces of support equipment, and three avionics packages).

c. A contiguous fragment of the program statement of work (SOW), plans, and schedules covering, ideally, the one product architecture item involved in their responsibility. It should be possible to simply

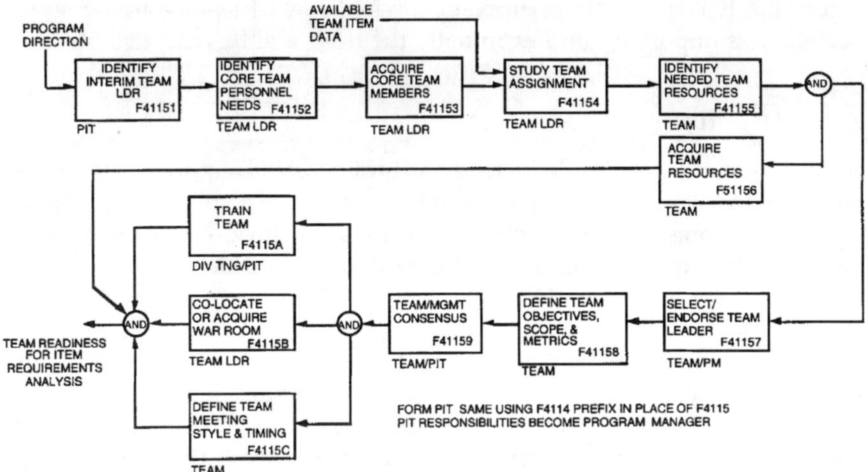

Figure 7.3   Team formation process.

lift out the team's planning data from the whole of the program
planning data without any complex entanglements with other planning data.

d. A budget tied to the WBS assigned to the team.
e. An item system schematic block diagram showing all of the interfaces
of the team item with clear identification of the parties responsible
for the other terminal of each of the interfaces that touches the team
item.
f. A released or nearly complete specification covering the top-level item
for which the team is responsible.

The team must then identify the resources they will require to carry out
their responsibilities and undertake action to acquire these resources (see
Chapter 8 for those dealing with communications). Some organizations
believe that the team should select the leader, so Figure 7.3 includes an
opportunity for the team to endorse the leader selected by program management to get the team started. We are now ready to identify team objectives, the scope of the work to be performed, and any metrics to be used by
the team to monitor performance. The program manager should be briefed
on the team conclusions (F41157) to make sure the team's shared vision is
in agreement with the program manager's. Finally, the team should decide
on a meeting style and the timing of meetings and arrange for a war wall
or war room upon or within which large sketches can be hung and computer
projection used for group access to all information on the network. If the
team members have never worked on an IPT previously, it may be necessary
or useful to pass the team through team training covering consensus building, interpersonal relationships, and group cohesion, as well as how to avoid
group problems such as group think. As the enterprise culture comes to

accept the IPT approach to problem solving, special team training should become less important, and eventually the need will be nonexistent.

7.5   IPT work

At this point, the IPT should be ready to apply itself to requirements analysis work for the item. This should include a rapid but complete study of the specification presented to them followed by a conclusion that it is complete and valid or requires one or more corrections or additions. These changes should be approved by the PIT. The functionality allocated to the top team item may require further decomposition and allocation to identify the lower-tier items of which the item consists. In the process, the team should identify those items, see that they are added to the architecture diagram maintained by PIT, and expand the interface definition relative to the item and its parts also maintained by PIT. Depending on the complexity of the item for which they are responsible, the team may have to recommend to the PIT that subteams be formed subordinate to their team.

Should this occur, the IPT becomes the PIT for those subteams, with all of the related responsibilities within the scope of their top-level authority. Alternatively, the team will identify several items subordinate to the top-level item and assign principal design engineers to them from available team members. These principal engineers then continue work as mini team leaders applying the available team personnel resources jointly to their problems. Obviously, conflicts will develop in the joint use of the team members to serve several masters and this will have to be adjudicated by the team leader from time to time and possibly lead to a request for additional specialists during brief time frames.

Another possibility in addition to subteams and principal engineers is that the item in question will be procured. In this case, the team must clearly identify the product and process requirements for the item in a procurement specification and a statement of work if it is to be procured via a specification. The team may choose to implement the design, as well as prepare a specification, followed by procurement of the item to a set of drawings.

As the requirements for lower-tier team items are documented and approved by the team leader, the principal engineers or subteam leaders should be authorized to proceed with preliminary design work, related analyses, and requirements validation activity possibly involving some testing or demonstration. The team should stage a preliminary design review chaired by the PIT followed by closure of any action items assigned and movement into detailed design, probably with an increase in budget and team size. The detailed design should be concluded by a critical design review presented by the IPT and chaired by the PIT. Upon release of all of the team product design packages, the team task is essentially finished except for manufacturing and verification support activity.

At some point, the team can be disbanded and residual work handled by the PIT. Once the whole product is in a manufacturing or procurement status, it may be possible to collapse all of the IPT back into the PIT with manufacturing-facility-oriented teams formed according to the way the enterprise has facilitized their manufacturing process, possibly called *integrated product manufacturing* (IPM) teams. In these teams, manufacturing people should provide leadership with engineering support, just the opposite of the IPT. So rather than some static team arrangement on programs, the program team structure is a living thing that is born, grows, adjusts to changing conditions, and retracts, passing from the scene entirely as part of the produced system life cycle and the life cycle of programs within the enterprise.

7.6   PIT activity

Two teams that should exist as long as the program does are the business team and the PIT. The PIT is staffed exactly the same way the IPT is, from the functional departments based on the technologies involved in the system being developed. If a system will appeal primarily to mechanical engineering, we should expect it to be dominated by mechanical engineering functional department people supported by specialists from other departments. If it is mainly electrical in nature, we would expect the electrical engineering department to be heavily involved. Today and in the future, every program with any complexity will involve software people. The PIT will require not just a crowd of system engineers but specialists from many of the other disciplines. The PIT can be effective if it is staffed to understand the ongoing work in all of the IPT so that the results can be assessed from an integration and optimization perspective, and the cross-organizational interfaces between the elements the teams are responsible for can be audited for compatibility.

The PIT should have a reliability–availability–maintainability (RAM) engineer to first develop the RAM models and flow down the RAM values to the team level·and thereafter audit the work of RAM engineers on IPT, ensuring that everyone is working to the same RAM models. A mass properties engineer on the PIT should do a similar thing for system end item weight and balance data allocating allowable weight values to each item at the top end of IPT structures. Life cycle cost is another specialty needed on the PIT, and we may need many others to identify system-level values and allocate values to team product entities.

In some cases, we may only need a particular specialist at the PIT level who allocates requirements throughout the IPT architecture in each case and monitors team design efforts relative to compliance with these requirements. In other cases, it may make better economic sense to place specialists on the IPT as well to distribute the work between the teams in a kind of federated structure. In all of these cases, however, the PIT should own the system models jointly used by PIT and IPT.

Wherever possible, work should be delegated to the IPT, with PIT auditing and technical oversight. Members of the PIT should participate in all IPT meetings to gain information on the ongoing design effort, being alert to conflicts that may emerge between the work on two or more teams without the knowledge of IPT members, who tend to become isolated and focused totally on their area of responsibility. For this reason, the PIT should require a staggering of the IPT scheduled meeting times throughout the week. This will permit cross-team meeting attendance and avoid simultaneous team meetings.

7.7   The PIT–IPT team matrix

All of the teams are staffed from the functional departments, forming a program/functional matrix at the enterprise level, illustrated in Figure 7.1, but the functional departments should not be permitted to supervise these people within the context of the program. Functional managers should be encouraged to support their personnel assigned to programs from a discipline perspective but not for the purpose of controlling their work assignments. There is, in a program organized in accordance with the principles covered in this book, another matrix involving work direction from outside the IPT, however. The PIT is composed of most or all of the same disciplines in the aggregate as all of the IPT. These PIT member specialists are responsible for these specialist areas for the complete system, whereas the IPT members in those disciplines are responsible for the system subset corresponding to the team responsibility. These PIT specialists must interact with their counterparts on the IPT, forming another matrix totally within the program organization.

PIT specialists are networked into all of the IPT for information exchange purposes, but a question remains as to what authority they should have in directing IPT-member work. The understanding encouraged is that the PIT specialists are leaders of their specialty on the program and should be the most knowledgeable in their field within the company relative to the people assigned to IPT. The PIT specialists should cultivate a good working relationship with their counterparts on the teams and organize brief meetings periodically to discuss the evolving work. Generally, they can probably influence the program work in the right direction through peer pressure when things appear to be off the mark on a particular IPT.

The formal rules of IPT and PIT team behavior should, however, prevent IPT members from feeling obligated to take technical direction from their counterparts on the PIT. The IPT must be allowed to build and sustain a team integrity between its leader and staff. Formal direction of the work should flow from the team leader, who is responsible for team cost, schedule, and performance metrics. Therefore, in this environment, a PIT specialist may find that it is not possible to move an IPT member in that specialty field into agreement on actions to be taken in that discipline in the interest of the system. In this case, the situation should be elevated to the PIT level and worked out between the teams at that level.

It is true that ego could be the cause of a lot of this, but these situations can also represent potentially valid, strongly-held views by well-qualified engineers who happen to disagree. In this case, the different views should be carefully considered as alternative solutions to a problem, and a preferred solution should be selected, with critical comment allowed by all parties. It is possible, in these situations, that a particular IPT happens to have a good idea that should be adopted as the norm even though it is in conflict with the current group think position.

The PIT should manage a formal action item system to track actions that are identified outside of the coverage of the program planning documents. These may cover actions that an IPT needs the PIT to accomplish, as well as the other way around, or actions across IPT boundaries. In addition, the PIT should manage a formal issues list. This should be a list of problems not yet well understood in terms of causes, effects, or responsibilities. The issue category focuses the attention of all parties on the problem during its maturation process, whereupon it becomes a PIT direction for a particular IPT to accomplish, or to share in the accomplishment of, certain work to resolve the issue. The PIT should be working to zero out the action items and the issues list at all times, while also encouraging resolution of both kinds of program discontinuities.

Program communications

8.1   Communication dichotomy

Programs receive many resources from the functional departments, including human specialists, sound practices, and tools, and they must mold them into a powerful engine of development. One of the most important parts of this process is to provide effective ways for specialists to communicate so the program can benefit from the effect of one great mind composed of all of the specialized knowledge of all of the people working on the program. In Chapter 2 we recognized that this communication network consists of human, computer data, and physical data (paper, tape, and disk) nodes as well as the synergism of the group. Now we need to put these ideas together with effective methods for organizing people.

Those who take on the responsibility for setting up programs are obliged to find ways to maximize the opportunities for communication on those programs. We will cover several of these techniques in this chapter, but we also have to recognize that the specialists we employ to develop complex systems need time to think within the bounds of their discipline. When you are trying to concentrate on work and three people are jabbering about their weekend of fun, the conversation is a distraction and prevents you from giving your very best to the program. This would also be true if the three people were talking about nuclear physics on a program needing that kind of thinking and you were an electrical engineer working on some other facet of the program. It is still noise from your perspective. So there can be such a thing as too much communication.

Program managers and team leaders should recognize a need to maximize the *opportunities* for communication, while minimizing the *need* for communication. The result will be optimized synergism, leaving the specialists time for independent specialty thinking — the results of which will flow into the program conversation as needed.

8.2   The critical nature of communications

Successful system development can only occur when the humans performing the work communicate effectively. This simple idea is the single most important notion in this book. It is so because the work must be accomplished by many people, none of whom knows everything. The knowledge and conclusions held by each member of a development team must be shared with other team members. This process of sharing information is called communication. Communications connects the many specialists into a network of ideas that all can tap and transforms the individuals into a team and one very powerful development force.

We seek to build, in our IPT, a series of subsystems, not unlike those of the product system we intend to develop. Our product system will include subsystems that must interface (communicate) with other subsystems to exchange information, and the components of each subsystem must interact in the same fashion. We seek to weld the humans working on the IPT into a temporary system as well. We have already explained the second most important point in the book, that there should be a strong correlation between the product system composed of things and the program system composed of IPT.

The principal mode of communications between humans is language composed of words and symbols, about which we have reached agreement, arranged to express ideas. As we know, these ideas can be expressed as marks on paper or in their spoken equivalent. So communication takes place through our power of speech and sign and our senses of hearing and sight. It is hard to imagine effective communication appealing to other senses except in the isolated but important case of the blind using touch. The media that we have found most effective in communicating include: language on paper, speech (direct, recorded, transmitted), and, more recently, computers using keystroked and video-displayed language and even the synthesized spoken word.

The single most important communication channel on a program is probably direct, face-to-face human interchange. In order to get the full benefit of this mode of communications, with the technology available at the time this book was written, the team must be located so that members are close together. It is possible to get some small benefit from excellent telephonic communication between physically dispersed team members, but there is an added dimension to actual direct conversation in which eyesight as well as hearing can come into play.

Unfortunately, it is not possible to develop complex systems using only verbal communications. We need records and we need instruments through which to communicate complex ideas. It would not be possible to communicate the design of a rocket engine propellant turbo pump from the design community to the manufacturing community using only verbal conversation. The richness of the idea defies simple treatment. It is true that computer networks in the workplace are becoming capable of handling video images as well as the spoken word, but there is still no match for face-to-face communication in the same physical place.

In addition to direct voice communication, engineering drawings and reports are needed to communicate the designer's conclusions for translation into manufacturing instructions appropriate to the operator of a machine tool, whether it be a human being or a computer program. Even the requirements for a valve are too many and too complex to trust to conversation between a buyer and a supplier. Without belaboring the point with a blizzard of examples, there is a valid need to capture a tremendous amount of information in the development of a system. This information must be organized and stored in a fashion that permits quick and easy retrieval. It must also be protected against unauthorized change, loss, or damage.

Those familiar with the great fictional detective Sherlock Holmes will recall that he felt a detective should not burden his mind with useless information. Nor should system engineers or the information systems that serve them. We must be careful to select essential information for retention and allow the useless to be lost to ensure that our capacity to retain information is not overcome and that the process of retrieving needed information is not made unduly complex.

The big question is, of course, "What is important and who decides?" Clearly, the customer has a big stake in what is retained. A DoD customer will define their formal needs in a contract data requirements list (CDRL). For other data that exist, the customer may require the contractor to list it in a data accession list (DAL) and share the list with them. For an agreed upon fee, the customer is allowed to obtain any item on the list. The things on the DAL should be every significant data item that is not a CDRL item. The criteria for significance could be a matter of judgment on the part of the program data manager or could be defined by a written DAL criteria. Here is a partial list for such a criteria (all of these items are listed on the condition of not being CDRL items). The reader will be able to add many other items to the list and should do so as an exercise.

   a. Procurement specifications and in-house requirements documents, none of which are configuration item specifications so are not CDRL items
   b. Minutes of internal design review meeting
   c. Supplier correspondence and supplier data requirements list (SDRL) items
   d. Engineering drawings
   e. Analysis reports
   f. Test requirements documents, test plans, test procedures, and test reports
   g. Program memos
   h. Manufacturing planning and stamped planning cards
   i. Quality inspection data, including deficiency reports and responses
   j. Lower-tier program plans (assuming the top tier are CDRL items) and procedures
   k. Generic company procedures that apply to the program
   l. Program plans, controls, and schedules
   m. Design decision traceability database
   n. Requirements traceability data

Many program managers have agonized over whether particular items should be placed on the DAL due to their sensitive nature. An internal memo might indicate a serious flaw in the design that would be damaging to the company's position with the customer and even end up costing money when there is already a thin profit margin. It is very hard, under these circumstances, to do the right thing. It is very hard to accept that it is in your long-term best interest as a company and a person of integrity to provide this kind of information to the customer. A habit of openness with a customer will generally be repaid, if not in this life (program) then in the next. Any deviation from this position requires you to keep two sets of books with the ever-present possibility that the second set will come out in a very damaging fashion during a subsequent investigation or litigation over the consequences of the design flaw. There are, of course, sensible as well as self-destructive ways to initially inform a customer about the content of a damaging memo or report.

This kind of memo contains valuable design rationale information that completes the design decision trail for future research on the same product or another. One of the side benefits of being open with the customer in defining DAL content in the broadest possible way is that the combination of CDRL data, DAL data (including SDRL), and contract correspondence defines the materials that should be available in the program library for internal access by everyone within the bounds of security prescribed by the customer for the program. It is possible that another program library category could be identified for private company information dealing with competition-sensitive matters of value in the next phase of competition, such as marketing reports. However, this category should not be extended to accommodate a place to hide cover-ups of our mistakes.

Okay, so we have decided that we will retain everything of lasting value, and we have a criteria for deciding what is important. Now we need to decide in what media we shall retain the data. Those familiar with data management understand that the volume of information just from suppliers can become overwhelming in a paper media. The Data Manager on the General Dynamics Space Systems Division Titan–Centaur Program had to move into the electronic age when the paper-filled file cabinets were about to literally push the desks of the data management people out of the available space. The obvious alternative today is to place all data in electronic media and make it available to all via networked computer workstations. It is possible to purchase workstations for scanning supplier data from paper media to CD ROM. Your first reaction might be to reject this solution since the scans are bit mapped and not editable, but these data are contractual in nature and you should not be changing them.

Once you are in electronic media, whether you arrived there driven by physical space limitations or by a more conscious plan, you will find that many new possibilities present themselves that might not have occurred to you in the good old paper days. In the data management example, instead of physically routing supplier data in serial fashion through your departments for review, you will find that it is possible to make it available in parallel to all reviewers via the network with no reproduction or distribution delays.

Communication is the exchange of information and, while the most important communication that takes place on a program is direct verbal conversation, the prepared material that will become part of the program library is an essential part of the complete program communication picture. Information can become lost in our paper filing system, and the same can happen in our electronic equivalent, so not only do we need to capture information for future use, but we also need to be able to access it on demand when that time arrives. This often drives interest in file and directory naming conventions.

8.3   A common database

Some companies have made serious and costly attempts to establish a relational database that contains every piece of information related to their business and product line. One term to describe this entity is a *common database*. Few, if any, of these efforts have been successful. It is very hard to know at the macro level what information should be retained and how all of that information should be interrelated. Most companies have settled on a less ambitious approach involving growth to a common database condition through evolution of databases useful for subsets of the total set of information about their process or functional departments. As time progresses, interfaces between these subset databases, initially accomplished by human communication and sneaker networking, become automated, leading to an increasingly grand and complete database.

Clearly, the medium to use for information systems involves computers and computer networks. The original computer systems featured very large computers operated by specialists in a central facility in batch mode. People who wished to use these machines had to bring their work to the central point, in some cases in the form of a box of punched cards. The development of microcomputers, powerful engineering workstations, and effective networking hardware and software has radically changed the way information systems are organized. Today it is not uncommon to find Macintosh and IBM-compatible machines distributed throughout an organization. Many companies have interconnected these machines with file servers, powerful mainframes, and with each other via computer networks, enabling a degree of information sharing. Few have taken full advantage of these networks to encourage the degree of imnformation sharing that is needed to drive integrated product development into the success region.

Is there some useful prescription for employing these resources which also encourages growth toward a final solution? One approach the author has found effective is called an *interim common database,* suggesting that it is on the technology growth path toward a true common database. This approach is also tuned to the dynamic period encountered during the early program phases (where rigid configuration management of data normally

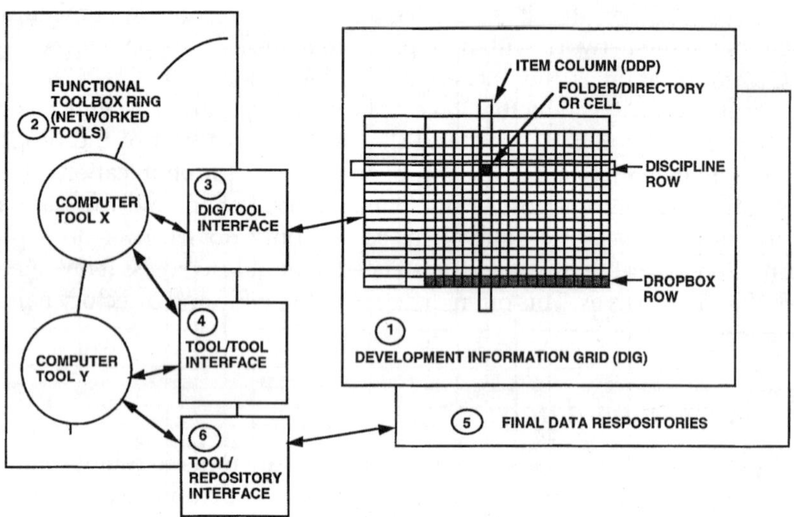

Figure 8.1   Interim common database components.

is not applied), with a built-in transition to production information retention. We will also look at an application of the world wide web for the same purposes.

8.3.1   Program interim common database (ICDB)

The ICDB has six fundamental components, illustrated in Figure 8.1. A grid of organized storage capability, referred to as a *development information grid* (DIG), provides a baseline set of information derived from all of the program specialists organized about the product structure. It is a matrix of information with a human interface. This information is accessible by everyone on the program in read-only mode from their networked microcomputer/terminal. It is protected on a network server storage device from being changed except by the assigned expert. The information is entered into and maintained from workstations forming a toolbox ring surrounding the DIG and connected to it via the network. The six components of the ICDB are: (1) the DIG, (2) the toolbox ring workstations, (3) the network resources that interconnect or interface the DIG and toolbox ring, (4) any connections between the tools, (5) the final data repository, and (6) a dataview connection between the toolbox ring and the final repositories.

8.3.1.1   The DIG

The ICDB centers on a core development information grid (DIG) organized into columns by product architecture (and, therefore, IPT) and into rows by specialty disciplines, as illustrated in Figure 8.2. The vertical columns are called *development data packages* (DDP) under the responsibility of the IPT leaders. The horizontal rows correspond to the several specialty disciplines

SUB FOLDERS		ARCHITECTURE/IPD TEAM ITEM FOLDERS (DDPs)
LONG NAME	SHORT	
TEAM PERSONNEL	TEAM	
GUIDELINES	GUIDE	
SCHEDULE	SCHEDULE	
REQUIREMENTS	RQMTS	
APPLICABLE DOCUMENTS	ADOC	
ARCHITECTURE	ARCH	
INTERFACE	IF	
TRADE STUDIES	TRADES	
DESIGN CONCEPT	DESIGN	DISCIPLINE ROW
DECISION RATIONALE	RATIONALE	
RELIABILITY	REL	
MAINTAINABILITY	MAINT	
SAFETY	SAFETY	
EMI/EMC	EMC	
DTC/LCC	COST	
DET	DET	
QUAL TEST	QUAL	
ACCEPTANCE	ACCEPT	
STRESS	STRESS	
MASS PROPERTIES	MASS	
OPERATIONS	OPS	
LOGISTICS	LOG	
QUALITY ASSURANCE	QA	
MANUFACTURING	MFG	
TOOLING	TOOLS	
MATERIAL	MATERIAL	
RISK	RISK	
DROP BOX	DROPBOX	DROPBOX ROW

Figure 8.2   Development information grid (DIG) organization.

assigned to the teams such as: reliability, mass properties, manufacturing, quality, and so forth. Compare the structure of Figure 8.2 with the structure in Figure 8.1 and you will see a strong correlation between the cells of the DIG and the cells of the organization performing integrated product development on the program. The DIG may have voids at its intersections, as the matrix organization diagram does, corresponding to IPT that do not require particular specialties. The program panes of Figure 8.1 suggest that a company actually needs not one DIG but DIG panes to serve its several programs.

The DDPs are computer folders/directories installed on a computer server not necessarily located within the physical spaces used by the team/project. These share areas are specifically set up for whatever kinds of machines the team has available: MAC only, IBM only, integrated MAC and IBM access, or combinations also involving engineering workstations. Each DDP folder/directory is composed of a column of subfolders/subdirectories so that everyone on the team has a place to put their team information product. The DDP, therefore, becomes a kind of joint, public engineering notebook for the product item for which the team is responsible.

This engineer's notebook analogy forces recognition of a real problem with the ICDB that must be disposed of as a prerequisite to success. The information in the DIG is accessible to everyone all the time, and that is its strength from a communications perspective. However, many of us find it difficult to share our work product until it is complete and perfect. This attitude is encouraged by management persons who are quicker to criticize than to encourage. Before we inaugurate our ICDB, we must deal with the human issue of access to imperfect and incomplete work. The DIG content should reflect, at any moment, the current information baseline, the best that we know at the time. People will be reluctant to share their evolving work in this open fashion unless they can be assured that it will be a positive experience.

The information contained in the DIG cells may have been placed there by any number of different applications: spreadsheets, word processors, graphics packages, or analytical tools. Naturally, the program, and perhaps the company/division, will be well served to encourage the smallest number of different software products for the same function, while still permitting freedom of choice in applications with which people can work efficiently. There is a broad continuum here with a large overlapping mutual conflict space. Frankly, some companies have found that it has been advantageous in the long run to allow employees and departments a degree of freedom in selecting applications with which they could be efficient. It has forced them to become more expert in data connectivity skills than they might otherwise have become; but ideally it is probably advantageous to work toward a condition of minimized applications with maximum interoperability. All windows and MAC applications can share information via a clipboard, for example. There is an increasing number of applications that can also share data across the IBM/MAC valley.

The immediate past suggests that the future of computer software will continue to be characterized by dynamic change. Some companies have attempted to adopt rigid software application standards for word processing, spreadsheet, and graphics applications only to find that other applications arriving on the market were more effective. The U.S. Air Force and NASA managers of the National Launch System program in the late 1980s and early 1990s encouraged an excellent standard for software from which we could all benefit. They encouraged the several contractors, competing in early stages of the program for later hardware contracts, to evolve program and product information system components that consisted, to the greatest degree possible, of off-the-shelf or over-the-counter software modules that could be replaced easily when newer, better software became available.

8.3.1.2 *The toolbox ring*

The computer applications that your enterprise very likely already owns, like word processors, spreadsheets, graphics packages, and specialized analysis tools, reside on the toolbox ring machines. The toolbox ring is formed simply by having them on a network and saying, "Hey machines, you are a toolbox ring." These machines are used to create and maintain program information initially populating the DIG. The engineer uses a particular tool to create a data file and copies the data file into an assigned cell of the DIG via the network interface. This information is immediately available to everyone in read-only mode from all other workstations on the toolbox ring.

The machines need not all be the same type, nor must the software be standardized initially. You can simply start with what you have. The machines must be networked together, however. You should work toward an early point where the software does follow some minimal standards to assure maximum access and use of the information. It is possible to configure servers so that both IBM and MAC machines can access the same data. Someone can write a report on a MAC in Microsoft Word and someone else on an IBM can call it up for review in Microsoft Word, even cut a portion of it and paste it into a different document.

There is no science to creating the toolbox ring. You can start, if your machines are already networked, by simply inventorying the machines you have and the software they are licensed to operate. Otherwise, you need to first introduce networking. Then you must decide what software you feel you should be using for each task that commonly has to be accomplished on programs and comparing that with what you have in each application type (such as word processing or spreadsheet). The result is an error signal that can drive your efforts in continuous process improvement.

The functional organization should be held accountable for developing or selecting appropriate computer tools for use by their department members and ensuring that their department members are skilled in the use of these tools in accordance with a standard procedure. Over time, the functional organizations should be improving their tool set in concert with other departments with which their tools have information interfaces. Each program will provide opportunities to experiment with slight changes in the department toolbox, and these opportunities should be seized as part of the department's continuous process improvement. As new tools are proven on programs, they should become new department toolbox standards for future programs.

8.3.1.3   Toolbox ring to DIG interface

Early in a particular program, the DIG is used to capture requirements and design (product, manufacturing, logistics, and operations) concepts for joint use throughout the project or during some fixed time span. As the program progresses to a more stable condition of the design and plans, information gradually drains from the DIG via the toolbox ring into the final repository databases where rigid configuration management policies restrict changes. Team members reach agreements on who is responsible for creating and maintaining each piece of information needed by the team. Further, they agree on which team members must be current on the content of the different team DDP folders. The DDP, therefore, becomes a means to communicate information about the product baseline and encourages consensus. Team members approach their concurrent engineering activities from a position of common knowledge of where they have been, where they are at any moment, and where they are going.

For example, early in a program, the designer for the vehicle hydraulic system uses the folder in the designer row and the hydraulic system column of the DIG to retain his/her concept sketches (using a simple tool like Mac

Draw, Designer, or AutoCAD). This information is available to all of the specialty engineering disciplines concurrently as a basis for their analytical work. The specialty engineers prepare their data and load it into their corresponding folders/directories within the hydraulic system column. The manufacturing people working concurrently with the designer, define a manufacturing process for the hydraulic components that will be manufactured and assembled by the company, and this is placed in the manufacturing row folder of the hydraulic system column. Procurement does the same for things that will be purchased.

IPT leaders reach agreements with team members on who is responsible for each folder/directory for that team. At any moment in time, the DIG contains within its directories/folders the joint conclusions of the IPT responsible for the vehicle hydraulic system. Everyone on the team and on other teams can gain access to any of this information from their workstation. Two human interface modes are suggested for the DIG: (1) an open house mode and (2) a control mode.

The open house mode is useful on early program phases when there is a great deal of volatility and a real need to initially load and change the data often to reflect changing concepts. The IPT leader, in the control mode, will allow people free access to the DIG folders corresponding to their area of responsibility in order to quickly load team data into the folders/directories. Once a baseline is established, reviewed, and approved by the team leader, chief engineer, or PIT, DIG access should be controlled. You reach this point when you become more concerned about unauthorized changes to the data than you are about stifling creativity. This change is accomplished by retaining read only rights for everyone and restricting writing rights only to the IPT leader or a person of his/her choosing for their corresponding column of the DIG, that is, their DDP.

Any team member who wishes to change the content of his/her folder, based on the results of new work, after a baseline has been established (control mode), must copy the contents of his/her folder/directory to their local workstation over the network, make needed changes using their toolbox, and then copy the folder or file into the appropriate DDP Dropbox, as illustrated in Figure 8.3. The Dropbox is a folder/directory that can only be viewed by the team leader or his/her designee but one into which anyone may deposit (write) a file. It is set up so that anyone may write but only the authorized person can see into its contents. This is essentially the reverse of the controls on the discipline cells in the DIG.

The authorized person opens the Dropbox periodically and opens new files deposited in the Dropbox. He/she reviews them for acceptance. Notes may be written upon this material as to its quality, appropriateness, and timeliness and then the file is copied back into the appropriate DDP subfolder from whence it came before modification.

Clearly the DIG depends on a prescribed file-naming convention that is strictly enforced. Some operating systems that permit only very short file names, such as DOS up through at least version 5.0, provide a real challenge

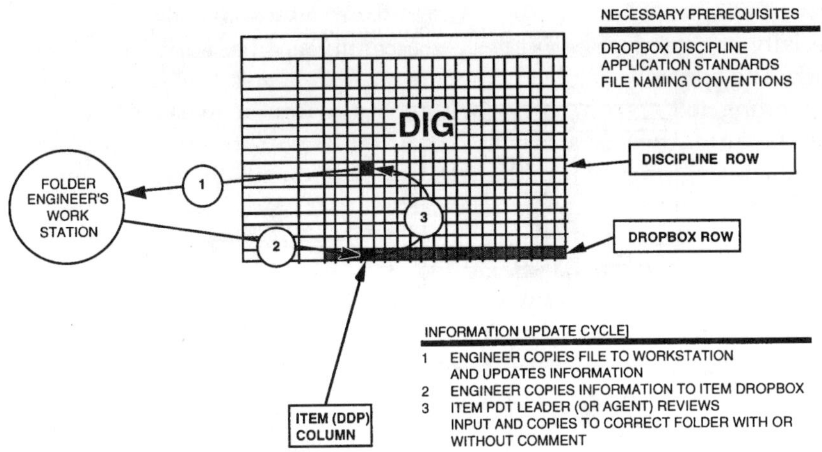

Figure 8.3   Data update in control mode.

to make file content and proper DIG cell location obvious from the file names. The Macintosh is much less restricting in this regard and later DOS, Windows, and/or OS2 systems are as well.

Now, how might the IPT members take advantage of the DIG to support concurrent engineering? The team members can periodically gather around one workstation and discuss DIG content, making changes in real time based on their conversation. If the team is too large for this but they are physically collocated, they can get together around several workstations in their common work area to view the data while carrying on a somewhat louder but no less effective conversation. If the group is too large for this, they can assemble in a meeting room and project the DIG images on a screen, making real time changes to content. The projection scheme could also be used in their work area using a clear wall as a projection screen. If some or all of the team members do not share a common facility (some degree of noncollocation), they may gain access to the same imagery for this meeting via the network and converse via a conference call using a speaker phone as the data are changed in accordance with team agreements or team leader decisions. The telephone arrangement will work for hooking up supplier, or even customer, representatives into these concurrent team meetings provided they are networked with you.

Regular internal team meetings should be held often and in a very informal atmosphere with no requirements for preparation of formal presentation materials. All of the needed materials should be available in the DIG. Without a DIG, the only means you will have for the team to interact concurrently is to hold frequent meetings. Without a DIG, the team data must be changed based on meeting results in a two-step process. With a DIG, the data can be refreshed with new ideas during the meeting with everyone in attendance. This is truly a concurrent engineering environment.

With a DIG, the team information is available to the team at all times. Without a DIG, the team can only gain access to team information in a

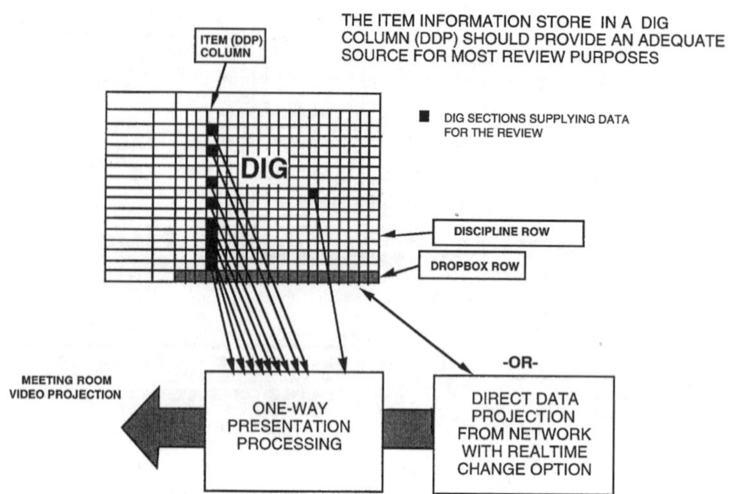

Figure 8.4   The DIG as a meeting presentation source.

meeting because the team information is largely in the minds of its members. A DIG multiplies the team's opportunities for effective concurrent work because everyone's baseline is available at all times.

These techniques can also obviously be extended to formal customer reviews requiring some degree of preparation, as suggested in Figure 8.4, whether all of the participants are congregated in the contractor's facility or separated by some distance and joined through their information and communication systems. The process of creating review data in a computer, only to print paper copies to feed the copy machine producing plastic presentation foils and paper copies is very inefficient. The data can be projected directly from your network with real-time changes during the review for relatively simple matters. Other concerns may require assignment of an action item (entered into the network action item database by someone at a meeting support terminal before the action item discussion is complete) and further study.

Alternatively, the presentation data could be combined into an integrated data set using a presentation tool like Applause or Powerpoint. This approach permits a sequenced step-through of the planned presentation. The network connection in the meeting room also permits the presenter to call up detailed backup data to answer questions about concepts not clearly communicated in the selected presentation materials.

In order to take advantage of these powerful data projection features, meeting rooms must be wired into the network, and the presenter's position in meeting rooms must provide access to computer workstation controls or the presenter must be supported by a computer operator. You can select from machines ranging from an expensive, high-end color projection machine that provides a bright picture in a reasonably well-lighted room to an inexpensive gray-tone overhead projector plate that requires a dimly lighted room. Over time, you should evolve to a condition where you have presenter stations with built-in computer terminals, possibly with the screen flush with the desktop like those used in TV news anchor stations. In addition, provisions should be made for a computer support station, perhaps in the back of the room. You can see a model of a similar equipment configuration in power software presentations like those put on by Oracle. This would be used for action items, minutes taking, and other record purposes during the meeting. A program or company might have one or more workstations installed on carts that can be wheeled into meeting rooms or used in work areas to satisfy this need.

8.3.1.4   The tool-to-tool interface

Initially, your ICDB may not permit interaction directly between the toolbox ring tools. As we said at the outset of this chapter, however, our intent is to use the ICDB concept as a stepping stone toward a true common database. Requirements tools offer an example of how this can be implemented. Figure 8.5 illustrates a requirements tool on the ring that can be updated from a reliability database. The requirements database would have to be structured to assemble the requirements statement from fragments, one of which is a quantity field. For reliability, this field would be updated for all items periodically from the reliability model database maintained by the reliability engineer. Reliability requirements data would always reflect the single authoritative current reliability baseline. This arrangement can be extended to weight, quantitative maintainability figures, and other quantitative parameters.

Other examples of possible tool-to-tool interfaces are:

a. Traceability relationships between databases for statements of work, product requirements, and WBS dictionary.
b. An interface between the company personnel database and all the personnel fields in product development databases identifying people as principal engineers, actionees on action and issue closure items, and team responsibilities.
c. A product system architecture database to which all systems relate their data. Reliability, mass properties, life cycle cost, and maintainability math models all hook to the same definition of the system such that it is possible to gather the complete story for a given item from databases maintained by different disciplines.

Where a tool-to-tool interface exists, it makes unnecessary the use of a DIG for the data treated this way. The utility of a DIG is in providing an organized repository for data that you have not yet figured out how to relate directly in a common database. As a company progresses in its climb toward a true common database, its need for the DIG should decline during the development period, being replaced gradually by the databases joining the tools on the toll ring.

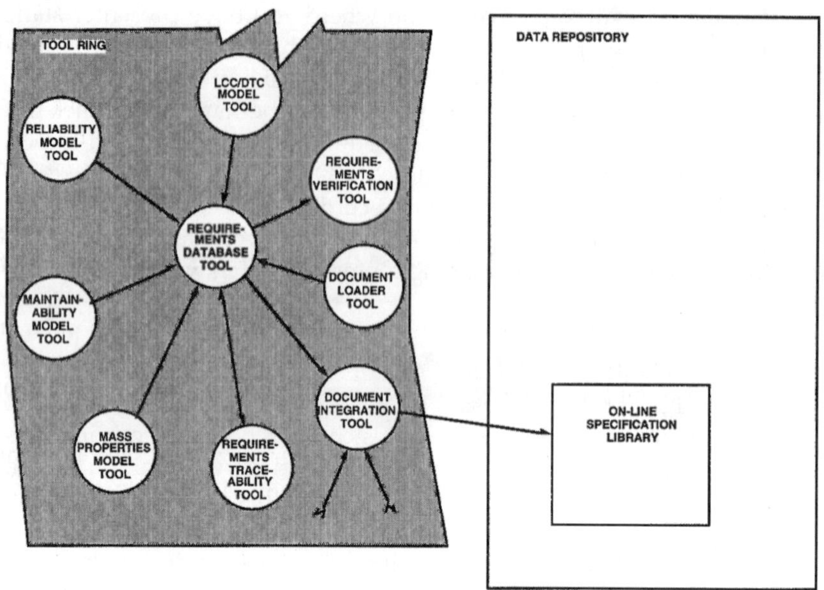

Figure 8.5   Tool-to-tool interfaces.

8.3.1.5   DIG content evolution to the final data repository

As a program matures, the design sketches in the DIG will be converted into engineering drawings located in a protected Computer-Aided Design (CAD) database, and the sketches are no longer needed except as a possible aid in presenting design information at reviews. The manufacturing concepts flow into formal planning data. Procurement data flows into supplier contracts, statements of work, and procurement specifications. Specialty data are transformed into reports conforming to a customer data item description. Requirements database content becomes a pile of specifications. Most of the data in the DIG can be aligned with some terminal form of formal documentation.

Between preliminary design review (PDR) and critical design review (CDR), the content of the DIG should begin flowing out to permanent information residencies. The design concept data in the design subfolder becomes the source of engineering drawings; the requirements subfolder content, possibly phrased in primitive terms, evolves into specifications stored in an on-line specification library; and some analysis data mature into CDRL reports stored in an on-line CDRL library.

From CDR onward, the toolbox ring is connected to these final information repositories, rather than the disappearing DIG, with changes regulated by the strict information configuration management rules applied to those systems.

Clearly companies dealing with DoD, DoE, and NASA will have to convert informal data retained in a ICDB into formal data, but, for a commercial firm, the ICDB content may serve its needs perfectly and provide the company with time-to-market advantages without formal conversion of the data.

8.3.2   Intranet applications

Some companies have created, very inexpensively, internal Internet web pages through which several key data sources are accessible without extensive training in the source applications. The McDonnell Douglas FA-18E/F program is a good example. They applied the comprehensive requirements database RDD-100 from Ascent Logic for the purposes of capturing all of the item requirements appearing in specifications. As every system engineer knows, RDD-100 is at the top of the hill in the hands of a skilled analyst for defining system characteristics, but it is difficult to attain and maintain user skills. The FA-18 program linked the database content via a web page and Interleaf so that anyone could open any specification in read-only mode to find any requirement of interest without needing any knowledge of RDD-100 or Interleaf.

Other source databases can be linked similarly into a web page and accessed by everyone, and with no special training. Meanwhile, the specialist, responsible for each data set can continue to use an effective application with which they are familiar, resulting in the products accessible via the web page. The information components can be updated or replaced as new technology comes on line without great disruption in information access.

8.4   War room or wall

Each IPT, and the PIT for the whole program, should be provided with a wall or room of walls (ideally with stickboard material and white boards applied) upon which they can hang large drawings and materials that help team members visualize team problems and item configuration features that are in progress. Even though a team has access to a DIG or an intranet web page as described above, there are some things that just cannot be easily grasped from the computer screen no matter how ingenious the computer application. A wall full of the right materials can be very effective in triggering new ideas and insights about team problems. There are stories, it is said in Hollywood, that cannot be properly told on the small screen of television, stories that can only be told on the big screen of a movie theater. The same is true in system development where the big screen is a wall full of information.

A powerful combination is formed by integrating DIG projection with the war room by placing a networked computer with projection capability in the war room. This provides a combination of the grand system view on the walls and any selected detailed view of system information via computer projection. The war room should also be equipped with a speaker phone for remote voice access to meetings held in the room. When the DIG data are fully networked, the speaker phone can expand the size of the war room to include the whole facility so remote engineers on a conference call are looking at the same image on their workstation that is being projected in the war room. This makes for a nice compromise to cover those persons who cannot be physically collocated with an IPT. The IPT and PIT should use a war room for regular stand-up meetings and maintain the displayed data to support their changing needs.

8.5 *Virtual teams in your future*

If your enterprise intends to attain or retain a position of leadership in your market, you will have to migrate to a condition of information sharing along the lines discussed in this chapter. The ICDB, an evolving common database, or intranet application will also provide exactly what you need to enable a more flexible teaming arrangement in the form of physically collocated teams. Clearly, future teams will consist of people who are not all physically collocated. Today it is possible to interact effectively in a synergistic way even if physically separated through the use of computer resources designed for that purpose.

Virtual teaming will require a common information depository (our evolving ICDB or intranet web page can serve this need initially) around which our virtual team members may gather for their discussions connected via computers capable of multimedia performance. This will include full motion video of the participants as well as voice and data. These teams will have virtually the same experience as if they were all together in a single war room.

chapter nine

Product system definition

9.1   System development process overview

Program planning for a specific program requires an understanding of the architecture of the system to be created, because it is the things in the system architecture around which teams will be formed. These things in the system should be determined through an organized structured process to avoid omitting needed functionality or including any unnecessary functionality that adds cost without benefit. Therefore, some system development work must take place before a detailed plan of action for the program can be crafted, and that work will have to be planned at a very high level relative to the things in the system. In fact, that planning can very likely take place based on only a program integration team (PIT) rather than a series of integrated product teams (IPT) as well. During that early phase, it will be the responsibility of the PIT to accomplish the front-end decomposition of the customer need and related mission analysis to define what the top-level items are that will become the purpose for identifying IPT on a subsequent phase.

To review, the systems approach entails the following principal steps:

a. Understand the customer's need. This is the ultimate system function because it tells what the customer wants. It defines the problem that must be solved at a fairly high level.
b. Expand the need into a critical mass of information necessary to trigger a more detailed analysis of a system that will satisfy that need. This may entail customer surveys, focus groups, question-and-answer sessions, or mission analysis combined with a high-level functional analysis that exposes needed functionality, system design entailing allocation of the top-level functionality to things placed in a system physical architecture, and definition of requirements for the complete system (commonly captured in a system specification). Through this process, one will determine to what extent and for what elements solutions shall be attempted through hardware and software.
c. Further decompose the need, which represents a complex problem, within the context of an evolving system concept, into a series of

related smaller problems described in terms of a set of requirements that must be satisfied by the solutions to the smaller problems.

d. Prior to design work it is sometimes necessary or desirable to improve one's confidence in their understanding of the requirements or to prove that it is physically possible to produce a design that is compliant. This is commonly accomplished through special tests and analyses, the results of which indicate the strength of the currently available related knowledge and technology base and the current state of readiness to accomplish design work using available knowledge.

e. Application of the creative genius of design engineers and market knowledge of procurement experts with a supporting cast of specialized engineers and analysts to develop solutions to the requirements for lower-level problems. Selections are made from the alternatives studied to identify the preferred design concept that will be compliant with the requirements.

f. Integration, testing, and analysis activities are applied to designs, special test articles, preproduction articles, and initial production articles that prove that the designs actually do satisfy the requirements.

We wish, in this chapter, to focus on step b in the list above in which we come to understand what the system shall consist of. It is the physical things in the system that will require development and around which work will be accomplished by IPT. Conventionally, they are arranged in a hierarchical family tree structure. This chapter tells how one can identify the things that should populate the system architecture. This is also the basis for the work breakdown structure (WBS) important in program planning. In this chapter we will use a construct called an *architecture ID* as well as WBS notation, but in Chapter 12 the proposition will be made that since we have aligned the two in our development work, only one is necessary, and the author will propose that the architecture ID become the WBS.

9.2   System requirements analysis

This whole development process is held together by requirements, especially during the initial development stage. These requirements are statements that define the needed characteristics of a solution to an engineering problem prior to the development of the solution. Requirements are defined as a prerequisite to accomplishing design work because of the nature of complex systems and the need for many people to work together to develop complex systems. All of the requirements for one item are collected into a specification for that item.

9.2.1   Requirements analysis background

In the English-speaking world, requirements are phrased in English sentences that cannot be distinguished structurally from English sentences constructed for any other purpose. It is relatively easy to write requirements once you know what to write them about. A requirement is simply a statement in the chosen language that clearly tells an expectation placed on the design process prior to implementing the creative design process. Requirements are intended to constrain the solution space to solutions that will encourage small problem solutions that work synergistically to satisfy the large problem solution. Requirements are formed from the words and symbols of the chosen language. They include all of the standard language components arranged in the way that a good course in that language specifies. Those who have difficulty writing requirements experience one of three problems: (1) fundamental problems expressing themselves in the language of choice, which can be corrected by studying the language, (2) technical knowledge deficiency, which makes it difficult to understand the technical aspects of the subject about which the requirements are phrased (this can be corrected through study of the related technologies), and (3) difficulty in deciding what to write requirements about.

The solution to the latter problem is the possession, on a personal, company, and program basis, of a complete tool box of effective requirements analysis tools and the knowledge and experience to use the tools effectively. Providing the toolbox is the function of the system requirements analysis process (tasks F4111 of Figure A.3). This toolbox should encourage the identification of product characteristics that must be controlled and selection of numerical values appropriate to those characteristics. So, this toolbox must help us understand what to write requirements about. It gives us a list of characteristics that we should control to encourage synergism within the evolving system definition. If the analyst had a list of characteristics and a way to value them for specific items, language knowledge, and command of the related technologies, he/she would be in a very good position to write a specification.

One approach on what to write requirements about is boilerplate, and this is essentially what specification standards such as MIL-STD-490A provided for many years and what its replacement in MIL-STD-961D, Appendix A is likely to do for many more in military systems. IEEE and other standards bodies provide similar standards for commercial purposes. In this approach, you have a list of paragraph numbers and titles, and you attempt to define how each title relates to the item of interest. This results in a complete requirement statement that becomes the text for the paragraph stating that requirement. The problem with this approach is that there are many kinds of requirements that cannot be specifically identified in such a generic listing. One could create a performance requirements boilerplate that covered every conceivable category with some difficulty only to find that it was more difficult to weed out those categories that did not apply to a specific item than it would have been to have determined the appropriate categories from scratch. This is why one would find no lower level of detail in a specification standard than performance even though 50 pages of performance requirements may evolve during the analysis. Interfaces are another area where boilerplate is not effective at a lower level of detail. Often, environmental and some specialty engineering requirements can be handled partially or completely through standards.

Many design engineers complain that their system engineers fail to flow down the customer requirements to their level of detail in a timely way. Their solution is to begin designing at their level of detail without requirements because their management has imposed a drawing release schedule they dare not violate. These engineers are often right about the failure of their system engineers but wrong to proceed in a vacuum as if they know what they are doing. The most common failure in requirements analysis is the gap between system requirements defined in a system specification given to the contractor and the component level requirements in the performance requirements area. In these organizations in trouble, the goal in the engineering organization is often stated as getting the system performance requirements flowed down to the design level or deriving design requirements from system requirements. Unfortunately, no one in the company seems to know quite how to do that, and avoidable errors creep into the point design solutions in the void that develops.

Flowdown is a valid requirements analysis strategy, as is boilerplate (which the author refers to as one form of cloning), it just is not a completely effective tool across all of the kinds of requirements that must be defined and particularly not effective for performance requirements. Flowdown is composed of two subordinate techniques: allocation (or partitioning) and derivation. Allocation is applied where the child requirement has the same units as the parent requirement, and we partition the parent value into child values in accordance with some mathematical rule. Where the child requirement has different units than the parent requirement, the value may have to be derived from one or more parent requirements through parametric analysis, use of mathematical models, or use of simulation techniques.

The problem in the flowdown strategy is that you must have a set of parent requirements in order to perform requirements analysis. One may ask, where did the parent requirements come from? Parent requirements are derived by applying some form of modeling that leads to the identification of characteristics that the item must have. The model should be extensible with respect to the problem represented by the system being developed such that it can be used to identify needed functionality and be iteratively expanded for lower tiers of the problem. Functional analysis (described later in this chapter) is an example of a means to identify performance requirements. The "flowdown" occurs through the functional analysis activity as an integral part of accomplishing the analysis. Once the physical hierarchy of the product structure is established through functional analysis and allocation, other models can be applied to identify design constraints and appropriate values in interface, environmental, and specialty engineering areas. In the latter two cases, requirements allocation can be very effective in a requirements hierarchy sense.

In addition to the structured analysis, appeal to standards or boilerplate, and flowdown, one could apply one of two other strategies. It is possible to sit down at your word processor and write the great American specification using what the author likes to call free style, unencumbered by all controlling influences. Some would also call this the *ad hoc* approach. If any good content happens to appear in this specification, we will have to thank God because man probably did not put it there using this strategy. Another strategy, often used in combination with the free-style one, is the question and answer game where the analyst tries to think up the right questions to ask customer representatives. This is an important step in the development of the system specification, but you cannot depend on coming up with all of the right questions without aid in the form of structured analysis. Structured analysis is recommended, and it is the only strategy further developed in this chapter.

This is not to say that the other strategies do not have a place at the table. Flowdown or value allocation is very useful within the specialty engineering disciplines, as in the weights statement and reliability math model. Standards properly play a key role in environmental and specialty engineering requirements identification. The Q&A process should be depended on heavily in the early development process in combination with structured analysis to help generate good questions, but the principal tool that ought to be brought to bear is structured analysis.

Our toolbox should help us understand what to write requirements about, what characteristics we should seek to control, what characteristics we need more information on from the customer. Once we have this list of characteristics, we must pair each with an appropriate numerical value and weave words around them to make the meaning clear. We need to quantify our requirements because we will want to be able to determine whether or not a particular design solution satisfies the requirement. This is very hard to do when the requirement is not quantified. As we shall see, this is the reason the author encourages simultaneous definition of requirements and the companion test and analysis, or verification, requirements for proving that the solution satisfies them.

Our system requirements analysis process should clearly tell us what kinds of characteristics we should seek to write requirements about and provide a set of tools to help us add characteristics to our list. We also need the means to quantify them. We seek a list of characteristics that is complete, in that it does not omit any necessary characteristics, and minimized, since we wish to provide the maximum possible solution space for the designer. We would like to have a more specific understanding of how this requirements analysis solution is characterized by completeness and minimization, but there is no easy answer. The best prescription the author can offer is to apply an organized approach that connects specific tools with specific requirements categories and then apply the tools with skill based on knowledge derived in practice or through training.

Figure 9.1 offers an overall taxonomy of every kind of requirement that one would ever have to write in the form of a three-dimensional Venn diagram.

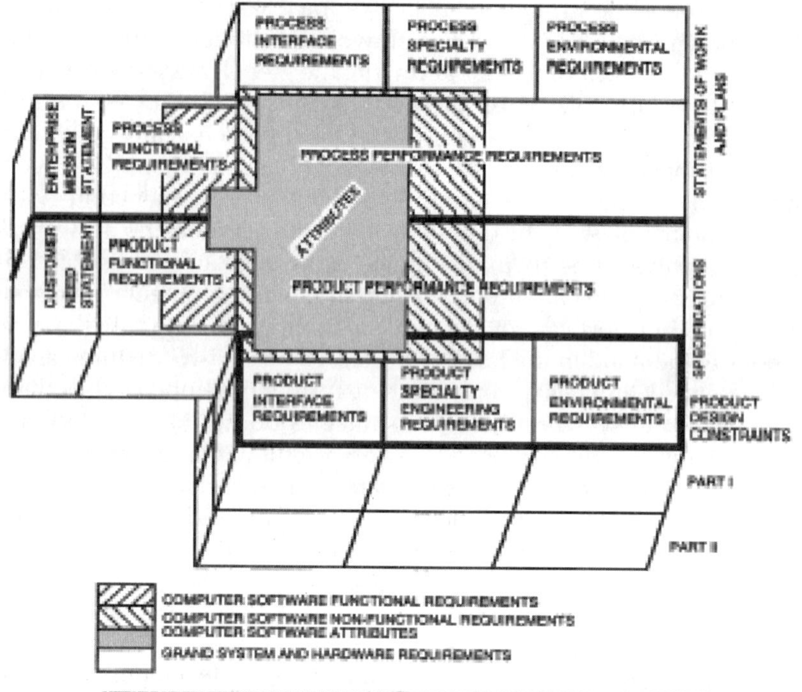

Figure 9.1   Requirements taxonomy.

The top layer, Part I in MIL-STD-490A or the performance specifications in the context of MIL-STD-961D, corresponds to development requirements, often called design-to requirements, that must be clearly understood before design. The lower layer, Part II in 490A context or detailed specifications under 961D, corresponds to product requirements, commonly called build-to requirements. The portion of the construct below the heavy middle line is for product specifications, which are the primary focus of this book. The requirements above this line correspond to process requirements captured in statements of work and plans. Many of the same techniques for specifications discussed in this book apply to process requirements, but most of the book specifically concerns product specifications and proving that the design satisfies their content.

Development requirements can be further categorized as shown in Figure 9.1 as performance requirements and design constraints of three different kinds. Performance requirements tell what an item must do and how well it must do it. Constraints form boundary conditions within which the designer must remain while satisfying the performance requirements. The toolbox offered in this book evolves all performance requirements from the customer need statement, the ultimate system function, through a product functional analysis process that identifies functional requirements. These functional requirements are allocated to things in the evolving system physical model, commonly a hierarchical architecture, and expanded into performance requirements for that item. Functional analysis provides one set of tools. We will identify three other sets for the three kinds of constraints.

As the reader can see from the figure, verification requirements are not illustrated. That is because they are assumed to be paired with the item requirements in our imagination. One could construct another dimension for design and verification requirements, but it would be very hard to envision and draw in two dimensions. This method of verification requirement inclusion corresponds to the fundamental notion that flows through this whole chapter that a design requirement should always be paired with a verification requirement at the time the design requirement is defined. The reason for this is that it results in the definition of much better design requirements when you are forced to tell how you will determine whether or not the resultant design has satisfied that design requirement. If nothing else, it will encourage quantifying requirements because you will find it very difficult to tell how it will be verified without a numerical value against which you can measure product characteristics.

Before moving on to the toolbox for the other categories illustrated in Figure 9.1, we must agree on the right timing for requirements analysis. Some designers reacted to the rebirth of the concurrent design approach with acceptance because they believed it meant that it was finally okay to simultaneously develop requirements and designs. That is not the meaning at all, of course. The systems approach, and the concurrent development approach that has added some new vocabulary to an old idea, seeks to develop all of the appropriate requirements for an item prior to the commitment to design work. We team the many specialists together to first understand the problem phrased in a list of requirements. We then team together to create a solution to that problem in the form of a product design and coordinated process (material, manufacturing, quality, test, and logistics) design. The concurrence relates to the activity within each fundamental step in the systematic development of product accomplished in three serial steps, requirements, design, and verification, not to all of these steps simultaneously. Even in the spiral sequence we rotate between these steps repeatedly.

It is true that we may be forced by circumstances to pursue design solutions prior to fully developing the requirements (identifying the problem), as discussed under development environments above, but requirements before design is the intent. A case in which this is appropriate is when we wish to solve a problem before we have sufficient knowledge about it and it is too hard to define the requirements in our ignorance. A physical model created in a hasty prototyping activity can provide great insights into appropriate requirements that would be very difficult to recognize through a pure structured approach.

9.2.2   *The need and its expansion*

The very first requirement for every system is the customer's need, which is a simple statement describing the customer's expectations for the new, modified, or reengineered system. Unfortunately, the need is seldom preserved once a system specification has been prepared, and thus system engineers fail to appreciate a great truth about their profession — that the development process should be characterized by logical continuity from the need throughout all of the details of the system.

The tools discussed in this section are designed to expand the need statement to fully characterize the problem space. The exposed functionality is allocated to things and translated into performance requirements for those things identified in the physical plane. We then define design constraints appropriate to those things. All of our verification actions will map back to these requirements.

The need statement seldom provides sufficient information by itself to ignite this analytical process, so it may be preceded by efforts to understand the customer's need and the related mission or desired action or condition. In military systems this may take the form of a mission analysis. In commercial product developments, we may seek to understand customer needs through surveys, focus groups, and conversations with selected customer representatives.

This early work is focused on two goals. First, we seek to gain expanded knowledge of the need. In the process, we seek to ignite the functional decomposition process targeted on understanding the problem represented by the need and making decisions about the top-level elements of the solution by allocating top-level functionality to major items in the architecture. Some system engineers refer to this process as requirements analysis, but it is more properly thought of as the beginning of the system development process involving some functional analysis, some requirements analysis, and some synthesis and integration. Second, we seek to validate that it is possible to solve the problem stated in the need. We are capable of thinking of needs that cannot be satisfied. The problem we conceive may be completely impossible to solve or it may only be impossible with the current state of the art. In the latter case, our study may be accomplished in parallel with one or more technology demonstrations that will remove technological boundaries, permitting development to proceed beyond the study.

This early activity should include some form of functional analysis as a precursor to the identification of the principal system items. These top-level allocations may require one or more trade studies to make a knowledge-driven selection of the preferred description of the problem. This process begins with the recognition that the ultimate function is the need statement and that we can create the ultimate functional flow diagram (one of several decomposition techniques) by simply drawing one block for this function F that circumscribes a simplified version of the need statement. The next step is a creative one that cannot be easily prescribed. Given this ultimate function, we seek to identify a list of subordinate functions such that the accomplishment of the related activities in some particular sequence assures that the need is satisfied. We seek to decompose the grand problem represented by the need into a series of smaller, related problems and determine what kinds of resources could be applied to solving these smaller problems so that their synergism solves the greater problem.

If our need is to transport assigned payloads to low Earth orbit and we assign this to the Use System function F7, we may identify such lower-order problems, or functions, as: F71 = Integrate and Prepare For Transport, F72 = Launch, F73 = Transport, and F74 = Release Payload. This particular sequence of lower-tier functions is keyed to the use of some kind of rocket launched from Earth's surface. If we chose to think in terms of shooting payloads into orbit from a huge cannon, our choice of lower-tier functionality would be a little different, as it would be if we elected the Orbital Science Corporation's Pegasus solution of airborne launch.

If our need was the movement of 500,000 cubic yards of naturally deployed soil pèr day, we might think in terms of past point design solutions involving a steam shovel and dump trucks, or we might open up our minds to other alternatives, thus inventing the earth mover (quite some time ago, of course).

So the major functions subordinate to the need are interrelated with the mission scenario, and we cannot think of the two independently. The thought process that is effective here is to brainstorm several mission scenarios and develop corresponding top-level functional flow diagrams (called master flow diagrams or zero level diagrams by some) at the highest level for these scenarios. We may then determine ways of selecting the preferred method of stating the problem by identifying key parameters that can be used to examine each scenario. This is essentially a requirements analysis activity to define quantified figures of merit useful in making selections between alternative scenarios.

As noted above, we may have to run trade studies to select the most effective problem statement as a precursor to allocation of top-level functionality to things. We may also have to accomplish trade studies as a means to derive a solution based on facts for the most appropriate allocation of particular functions. Finally, we may have to trade the design concept for particular things conceived through the allocation process. Trade studies are but organized ways to reach a conclusion on a very difficult question that will not yield to a simple decision based on the known facts. All too often in engineering, as in life in general, we are forced to make decisions based on incomplete knowledge. When forced to do so, we should seek out a framework that encourages a thorough, systematic evaluation of the possibilities, and the trade study process offers that.

The customer need does not have to be focused on a new problem, as suggested in the preceding comments. The customer's need may express an alteration of a problem previously solved through earlier delivery of a system. The early study of this kind of problem may involve determining to what extent the elements of existing systems can be applied and to what extent they must be modified or elements replaced or upgraded. This is not a radically different situation. It only truncates the development responsibilities by including within the system architecture things that do not require new development. Those items that do require new development will respond to the structured approach expressed in this chapter at their own level of difficulty.

This early analytical process may start with no more than an expression of the customer need in a paragraph of text. The products with which we should conclude our early mission and functional analysis work include:

a. A master functional flow diagram (or equivalent diagrammatic treatment) coordinated with a planned mission scenario briefly described in text and simple cartoon-like graphics.
b. An architecture block diagram defining the physical model at the top level.
c. A top-level schematic block diagram or n-square diagram that defines what interfaces must exist between the items illustrated on the architecture block diagram.
d. Design concept sketches for the major items in the system depicted on the architecture block diagram.
e. A record of the decision-making process that led to the selection of the final system design concept.
f. A list of quantified system requirements statements. These may be simply listed on a single sheet of paper or captured in a more formal system or operational requirements document.

So, this mission analysis activity is but the front end of the system development process using all of the tools used throughout the development downstroke. It simply starts the ball rolling from a complete stop and develops the beginning of the system documentation. It also serves as the first step in the requirements validation activity. Through the accomplishment of this work, we either gain confidence that it is possible to satisfy this need or conclude that the need cannot be satisfied with the available resources at our disposal. We may conclude, based on our experience, that we should proceed with development, await successful efforts to acquire the necessary resources, focus on enabling technology development for a period of months or years, or move on to other pursuits that show promise. In the case of commercial products, this decision process may focus on marketing possibilities based on estimates of consumer demand and cost of production and distribution.

9.2.3   *Structured decomposition*

Structured decomposition is a technique for decomposing large complex problems into a series of smaller, related problems. We do this for the reasons discussed earlier. We are interested in an organized or systematic approach to doing this because we wish to make sure we solve the right problem and solve it completely. We wish to avoid, late in the development effort, finding that we failed to account for part of the problem, forcing us to spend additional time and money and bringing into question the validity of our current solution. We wish to avoid avoidable errors because they cost so much in time, money, and credibility. This cost rises sharply the further into the development process we go.

The understanding is that the problems we seek to solve are very complex and that their solution will require many people, each specialized in a particular technical discipline. Further, we understand that we must encourage these specialists to work together to attain a condition of synergism of their knowledge and skill and apply that to the solution of the complex problem. This is not a field of play for rugged individuals except in the leadership of these bands of specialists. They need skilled and forceful leadership from a person who has great knowledge applicable to the development work and who is able to make sound decisions when offered the best evidence available.

During the development of several intercontinental ballistic missile (ICBM) systems by the U.S. Air Force, a very organized process called *functional analysis* came to be used as a means to thoroughly understand the problem, reach a solution that posed the maximum threat to our enemies consistent with the maximum safety for America, and make the best possible choices in the use of available resources. We could argue whether this process was optimum and successful in terms of the money spent and public safety, but we would have difficulty arguing with the results following the demise of the Soviet Union as a threat to our nation's future.

At the time these systems were in development, computer and software technology were also in a state of development. The government evolved a very organized method for accumulating the information upon which development decisions were made, involving computer capture and reporting of this information. Specifications were prepared on the organized systems that contractors were required to respect for preparing this information. Generally, these systems were conceived by people with a paper mindset within an environment of immature computer capability. Paper forms were used based on the Hollerith 80 column card and intended as data entry forms. These completed forms went to a keypunch operator. The computer generated poorly crafted report forms, but it was a beginning and very successful in its final results.

This process included at its heart a tool called *functional flow diagramming* discussed briefly above. The technique uses a simple graphical image created from blocks, lines, and combinatorial flow symbols to model or illustrate needed functionality. It was no random choice of a graphical approach to do this. It has long been well known that we humans can gain a lot more understanding from pictures than we can from text. It is true that a picture is worth 10 to the third words. Imagine for a moment the amount of information we take in from a road map glanced at while driving in traffic and the data rate involved. All of the structured decomposition techniques employ graphical methods that encourage analytical completeness as well as minimizing the time required to achieve the result.

While functional flow diagramming was an early technique useful in association with the most general kind of problem, computer software analysis has contributed many more recent variations better suited to the narrower characteristics of software. U.S. Air Force ICBM programs required adherence to a system requirements analysis standard and delivery of data prepared in accordance with a data item description for functional flow diagramming. While functional flow diagramming is still a very effective technique for grand systems and hardware, it is not as effective for computer software analysis as other techniques developed specifically for it. So our toolbox of analytical techniques should have several tools in it, including functional flow diagramming and several others.

9.2.4 Grand systems and hardware approaches

Functional flow diagramming is the author's preferred approach for grand systems because of its simplicity and generality. Its simplicity is also its greatest fault, as the defenders of more complex modeling techniques would be quick to point out. This process starts with the need as function F, which is expanded into a set of next-tier functions, which are all things that have to happen in a prescribed sequence (serial, parallel, or some combination) to result in function F being accomplished. One draws a block for each lower-tier activity and links them together in a sequence using directed line segments to show sequence. Logical OR and AND symbols are used on the connecting lines to indicate combinatorial possibilities that must be respected. This process continues to expand each function, represented by a block, into lower-tier functions. Figure 9.2 sketches this overall process for discussion.

A function statement begins with an action verb that acts on a noun. The functions exposed in this process are expanded into functional requirements statements that numerically define how well the function must be performed. This step can be accomplished before or after allocation of the function statement to a thing in the physical system architecture, but in any case, the allocation of the function or functional requirement obligates the analyst to write one or more performance requirements based on the allocated function for the thing to which it is allocated. This is the reason for the power of all decomposition techniques. They are exhaustively complete when done well by experienced practitioners. They make it less likely that we will have missed anything compared to an *ad hoc* approach.

This process begins with the need and ends when the lowest tier of all items in the physical architecture in each branch satisfies one of these criteria: (1) the item will be purchased from another entity at that level or (2) the developing organization has confidence that it will surrender to detailed design by a small team within the company.

There are two extreme theories on the pacing of the allocation relative to the functional decomposition work. Some system engineers prefer to remain on the problem plane as long as possible to ensure a complete understanding

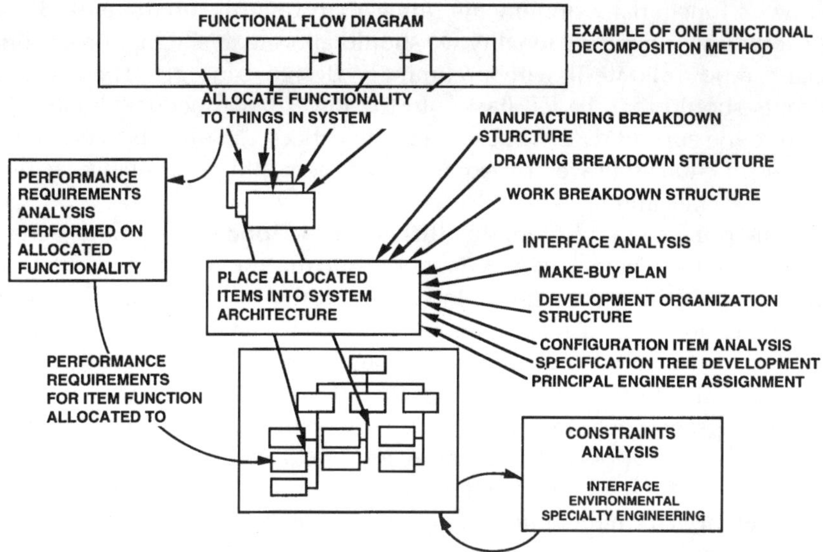

Figure 9.2   Structured decomposition for grand systems and hardware.

of the problem. This may result in a seven-tier functional flow diagram before anything is allocated. In the other extreme, the analyst expands a higher-tier function into a lower-tier diagram and immediately allocates the exposed functionality. This selection is more a matter of art and experience than science, but the author believes a happy medium between the extremes is optimum.

If we accumulate too much functional information before allocation, we run the risk of a disconnect between lower-tier functions and the design concepts associated with the higher-order architecture that results. If, for example, we allocate a master function for Transport Dirt to an earth mover, we may have difficulty allocating lower-tier functionality related to moving the digging and loading device (which, in our high-level design concept, is integrated with the moving device). Allocation accomplished too rapidly can lead to instability in the design concept development process because of continuing changes in the higher-tier functional definition.

The ideal pacing involves progressive allocation. We accumulate exposed functionality to a depth that permits us to thoroughly analyze system performance at a high level, possibly even run simulations or models of system behavior under different situations with different functionality combinations and performance figures of merit values in search of the optimum configuration. Allocation of high-order functionality prior to completing these studies is premature and will generally result in a less than optimum system and many changes that ripple from the analysis process to the architecture synthesis and concept development work. Throughout this period we have to deal with functional requirements rather than raw function statements so, when we do allocate this higher-order functionality, it will be as functional requirements rather than raw function names. Before allocating lower-tier functionality we should allocate this higher-order functionality and validate it with preliminary design concepts. These design concepts should then be fed back into the lower-tier functional analysis to tune it to the current reality. Subsequent allocations can often be made using the raw functions followed by expansion of them into performance requirements after allocation.

Some purists would claim that this is a prescription for point designs in the lower tiers. There is some danger from that, and the team must be encouraged to think through its lower-tier design concepts for innovative alternatives to the *status quo*. The big advantage, however, to progressive tuning of the functional flow diagram though concept feedback is that at the lower tiers the functional flow diagram takes on the characteristics of a process diagram where the blocks map very neatly to the physical situation that the logistics support people must continue to analyze. This prevents the development of a separate logistics process diagram with possible undesirable differences from the functional flow diagram. Once again, we are maintaining process continuity.

The author believes the difference between functional flow and process diagrams is that the former is a sequence of things that must happen, whereas the latter is a model of physical reality. When we are applying the functional flow diagram to problem solving, we do not necessarily know what the physical situation is nor of what items the system shall consist. The blocks of the process diagram represent actual physical situations and resources.

The U.S. Air Force developed a variation of the functional flow or process diagram called the IDEF diagram. IDEF is a compound acronym meaning ICAM Definition where ICAM = Integrated Computer Aided Manufacturing Analysis. In addition to the horizontal inputs and outputs that reflect sequence, these diagrams also have inputs at the top and bottom edges that reflect controlling influences and resources required for the steps. This diagrammatic technique was developed from an earlier SADT diagramming technique for software analysis and applied to the development of contractor manufacturing process analysis. It does permit analysis of a more complex situation than process diagramming, but the diagram developed runs the risk of totally confusing the user with the profusion of lines. Many of the advantages claimed for IDEF can be satisfied through the use of a simpler functional or process flow diagram teamed with a dictionary. These diagrams present a simple view that the eye and mind can use to acquire understanding of complex relationships, and the dictionary presents details related to the blocks that would confuse the eye if included on the diagram. The IDEF technique has evolved into an IDEF0 for process analysis, IDEF1X for relational data analysis, IDEF2 for dynamic analysis, IDEF3 for process description, as well as IDEF4 through 6, only IDEF0 and 1X being well documented at the time of this writing.

Some system engineers, particularly in the avionics field, have found it useful to apply what can be called hierarchical functional analysis. In this technique, the analyst makes a list of the needed lower-tier functionality in support of a parent function. These functions are thought of as subordinate functions in a hierarchy rather than a sequence of functions, as in flow diagramming. They are allocated to things in the evolving architecture generally in a simple one-to-one relationship. The concern with this approach is that it tends to support a leap to point design solutions familiar to the analyst. It can offer a very quick approach in a fairly standardized product line involving modular electronics equipment as a way to encourage completeness of the analysis. This technique also does not support time line analysis as does functional flow diagramming, since there is no sequence notion in the functional hierarchy.

Whatever techniques we use to expose the needed functionality, we have to collect the allocations of that functionality into a hierarchical architecture block diagram reflecting the progressive decomposition of the problem into a synthesis of the preferred solution. The peak of this hierarchy is the block titled system, which is the solution for the problem (function) identified as the need. Subordinate elements, identified through allocation of lower-tier functionality, form branches and tiers beneath the system block. The architecture should be assembled recognizing several overlays to ensure that everyone on the program is recognizing the same architecture: work breakdown structure (finance), manufacturing breakdown structure (assembly sequence), engineering drawing structure, specification tree, configuration or end item identification, make-buy map, and development responsibility.

As the architecture is assembled, the needed interfaces between these items must be examined and defined as a prerequisite to defining their requirements. These interfaces will have been predetermined by the way we have allocated functionality to things and modified as a function of how we have organized the things in the architecture and the design concepts for those things. During the architecture synthesis and initial concept development work, the interfaces must be defined for the physical model using schematic block or n-square diagrams.

Ascent Logic has popularized another technique called *behavioral diagramming* that combines the functional flow diagram arranged in a vertical orientation on paper with a data and interface flow arranged in a horizontal orientation. The strength of this technique is that we are forced to evaluate process and data needs simultaneously rather than as two separate and, possibly disconnected, analyses. The inclusion of architecture synthesis and interface analysis under the functional analysis umbrella does mitigate the differences somewhat. The tool RDD-100 uses this analysis model, leading to the capability to simulate system operation and output the system functionality in several different views, including functional flow, IDEF0, or n-square diagrams.

Behavioral diagramming was suggested by, or independently evolved along the same lines as, a software model called input–process–output (IPO), which was an early attempt to merge computer processing and data analysis, as the object-oriented modeling technique has done more recently.

9.2.5   *Computer software approaches*

A functioning system that is entirely computer software cannot exist, because software requires a machine medium within which to function. Systems that include software will always include hardware, a computing instrument as a minimum, and most often will involve people in some way. Software is to the machine as our thoughts, ideas, and reasoning are to the gray matter that comprises our minds. While some people firmly believe in out-of-body experiences, few would accept a similar situation for software. A particular business entity may be responsible for creating only the software element of a system and, to them, what they are developing could be construed as a system, but their product can never be an operating reality by itself. This is part of the difficulty in the development of software; it has no physical reality. It is no wonder then that we might turn to a graphical and symbolic expression as a means to capture its essence.

We face the same problem in software as we do in hardware in the beginning. We tend to understand our problems first in the broadest sense. We need some way to capture our thoughts about what the software must be capable of and to retain that information while we seek to expand upon the growing knowledge base. We have developed many techniques to accomplish this end over the 40 to 50 years during which software has been a recognized system component.

The earliest software analytical tool was flow diagramming, which lays out a stream of processing steps similar to a functional flow diagram (commonly in a vertical orientation rather than horizontal, probably because of the relative ease of printing them on line printers) where the blocks are very specialized functions called *computer processes*. Few analysts apply flow diagramming today, having surrendered to data flow diagramming (DFD), the Hatley–Pribhai extension of this technique, or object-oriented analysis. Alternative techniques have been developed that focus on the data the computer processes. The reasonable adherents of the process and data orientation schools of software analysis would today accept that both are required, and some have made efforts to bridge this gap. The most recent round of tool development focuses on a merger of these two orientations in what is called object-oriented software development. Earlier attempts included input process output (IPO) which was the basis for behavioral diagramming used by Mack Alfred in the development of RDD, and integration of entity relationship diagramming with structured analysis.

All software analysis tools (and hardware-oriented ones as well) involve some kind of graphical symbols (bubbles or boxes) representing data or process entities connected by lines, generally directed ones. Most of these processes begin with a context diagram formed by a bubble representing the complete software entity connected to a ring of blocks that correspond to external interfaces that provide or receive data. This master bubble corresponds to the need, or ultimate function, in functional analysis, and its allocation to the thing called a system. The most traditional technique was developed principally by Yourdon, DeMarco, and Constantine. It involves expansion of the context diagram bubble into lower-tier processing bubbles that represent subprocesses, just as in functional analysis. These bubbles are connected by lines indicating data that must pass from one to the other. Store symbols are used to indicate a need to temporarily store a data element for subsequent use. These stores are also connected to bubbles by lines to show source and destination of the data. Since the directed lines represent a flow of data between computer processing entities (bubbles), the central diagram in this technique is often referred to as a data flow diagram (DFD).

In all software analysis techniques, there is a connection between the symbols used on the diagrammatic portrayal and the text information that characterizes the requirements for the illustrated processes and data needs. In the traditional line-and-bubble analysis approach, referred to as data flow diagramming, one writes a process specification for each lowest-tier bubble on the complete set of diagrams and provides a line entry in a data dictionary for each line and store on all diagrams. Other diagrams are often used in the process specification to explain the need for controlling influences on the data and the needed data relationships. All of this information taken together becomes the specification for the design work that involves selection of a specific machine upon which the software will run, a language or languages that will be used, and an organization of the exposed functionality into "physical" modules that will subsequently be implemented in the selected language through programming work. A good general reference for process and data-oriented software analysis methods is Yourdon's *Modern Structured Analysis*. Tom DeMarco's *Structured Analysis and System Specification* is another excellent reference for these techniques.

Much of the early software analysis tool work focused on information batch processing because central processors, in the form of large mainframe computers, were in vogue. More recently, distributed processing on networks and software embedded in systems has played a more prominent role, revealing that some of the earlier analysis techniques were limited in their utility to expose the analyst to needed characteristics. Derek Hatley and the late Imtiaz Pirbhai offer an extension of the traditional approach in their *Strategies for Real-Time System Specification* to account for the special difficulties encountered in embedded, real-time software development. They differentiate between data flow needs and control flow needs and provide a very organized environment for allocation of exposed requirements model content to an architecture model. The specification consists of information derived from the analytical work supporting both of these models.

Fred McFadden and Jeffrey Hoffer have written an excellent book on the development of software for relational databases in general and client-server systems specifically, titled *Modern Database Management*. With this title, it is understandable that they would apply a data-oriented approach involving entity-relationship (ER) diagrams and a variation on IDEF1X. The latter is explained well in a Department of Commerce Federal Information Processing

Standards Publication (FIBS PUB) 184. McFadden and Hoffer also explain a merger between IDEF-1X and object-oriented analysis.

The schism between process-oriented analysis and data-oriented analysis, which has been patched together in earlier analysis methods, has been joined together more effectively in object-oriented analysis about which there have been many books written. A series that is useful and readable is by Coad and Yourdon (Volumes 1 and 2, *Object Oriented Analysis* and *Object Oriented Design*, respectively) and Coad and Nicola (Volume 3, *Object Oriented Programming*). Two others are James Rumbaugh et al., *Object Oriented Modeling and Design* and Grady Booch, *Object-Oriented Analysis and Design With Applications*. At the time this book was written, the object-oriented approach was still in the creative phase and had not become standardized, but the company Rational had brought together several of the most gifted object-oriented practitioners from which a unified standard (UML) was likely to evolve. As this book was being written, a fellow system engineer from Kelsey–Hayes, Mr. Steven Ferguson, told the author about an article in *Computer Weekly* (26 Jun 97, p4) that suggested object orientation is "on the way out." Among other things, the article noted that "component-based development is the rightful successor to object-based technology" and that "less than 4% of the code produced using object orientation can be redeployed." If we are lucky, this software modeling process instability will not end until we happen upon a truly effective modeling technique. In the mean time, we will all be driven insane by the proliferation of models.

9.2.6 *Performance requirements analysis*

Performance requirements define what the system or item must do and how well it must do those things. Precursors of performance requirements take the form of function statements or functional requirements (quantified function statements). These should be determined as a result of a structured analysis process that decomposes the customer need as noted above using an appropriate hardware or software technique.

Many organizations find that they fail to develop the requirements needed by the design community in a timely way. They keep repeating the same cycle on each program and fail to understand their problem. This cycle consists of receipt of the customer's requirements or approval of their requirements in a specification created by the contractor followed by a phony war on requirements where the systems people revert to documentation specialists and the design community creates a drawing release schedule in response to management demand for progress. As the design becomes firm, the design people prepare an in-house requirements document that essentially characterizes the preexisting design. Commonly, the managers in these organizations express this problem as, "We have difficulty flowing down system requirements to the designers."

The problem is that the flowdown strategy is only effective for some specialty engineering and environmental design constraints. It is not a good strategy for interface and performance requirements. It is no wonder these companies have difficulty. There is no one magic bullet for requirements analysis. One needs the whole toolbox described in this chapter. Performance requirements are best exposed and defined through the application of a structured process for exposing needed functionality and allocation of the exposed functionality to things in the architecture. You need not stop at the system level in applying this technique. It is useful throughout the hierarchy of the system.

Performance requirements are traceable to (and thus flow from) the process from which they are exposed much more effectively than in a vertical sense through the product architecture. They trace to the problem or functional plane very easily. Constraints are generally traceable within the product or solution plane. This point is lost on many engineers and managers and thus they find themselves repeating failed practices indefinitely.

Given that we have an effective method for identifying valid performance requirements as described under functional analysis above, we must have a way to associate them with quantitative values. In cases where flowdown is effective, within a single requirement category, such as weight, reliability, or cost, a lower-tier requirement value can be determined by allocating the parent item value to all its child items in the product architecture. This process can be followed in each discipline, creating a value model for the discipline. Mass properties and reliability math models are examples. In the case of performance requirements, we commonly do not have this clean relationship, so allocation is seldom effective in the same way.

Often the values for several requirements are linked into a best compromise, and to understand a good combination we must evaluate several combinations and observe the effect on selected system figures of merit like cost, maximum flight speed in an aircraft, automobile operating economy, or land fighting vehicle survivability. This process can best and most quickly be accomplished through a simulation of system performance where we are allowed to control certain independent parameters and observe the effects on dependent variables used to base a selection upon. We select the combination of values of the independent variables that produces the best combination of effects in the dependent variables.

Budget accounts can also be used effectively to help establish sound values for performance requirements. For example, given a need to communicate across 150 miles between Earth's surface and a satellite in low Earth orbit, we may allocate gain (and loss) across this distance, the antenna systems on both ends, connecting cables, receiver, and transmitter. Thus the transmitter power output requirement and the receiver sensitivity requirement are determined through a system-level study of the complete communications link.

This work involved in establishing appropriate values for performance requirements is part of the requirements validation process to the extent that it gives us confidence that the selected values are achievable within the time and money limits established for development.

9.2.7   Design constraints analysis

Design constraints are boundary conditions within which the designer must remain while satisfying performance requirements. All of them can be grouped into the three kinds described below. Performance requirements can be defined prior to the identification of the things to which they are ultimately allocated. Design constraints generally must be defined subsequent to the definition of the item to which they apply. Performance requirements provide the bridge between the problem and solution planes through allocation. Once we have established the architecture, we can apply three kinds of constraints analysis to these items. In the case of each constraint category, we need a special tool set to help us understand in some organized way what characteristics we should seek to control.

9.2.7.1   Interface requirements analysis

Systems consist of things. These things in systems must interact in some way to achieve the need. A collection of things that do not in some way interact is simply a collection of things, not a system. An interface is a relationship between two things in a system. This relationship may be completed through many different media, such as wires, plumbing, a mechanical linkage, or a physical bolt pattern. These interfaces are also characterized by a source and a destination, that is, two terminals each of which is associated with one thing in the system. Our challenge in developing systems is to identify the existence of interfaces and then characterize them, each with a set of requirements mutually agreed upon by those responsible for the two terminals.

Note the unique difference between the requirements for things in the system and interfaces. The things in systems can be clearly assigned to a single person or team for development. Interfaces must have a dual responsibility where the terminals are things with different responsibilities. This complicates the development of systems because the greatest weakness is at the interfaces, and accountability for these interfaces can be avoided. The opportunities for accountability avoidance can be reduced by assignment of teams responsible for development as a function of the product architecture rather than the departments of the functional organization. This results in perfect alignment between the product cross-organizational interfaces (interfaces with different terminal organizational responsibilities) and the development team communication patterns that must take place to develop them. Responsibility and accountability is very clear.

The reader is encouraged to refer to the author's *System Requirements Analysis* or *System Integration* for a thorough discussion of schematic block and n-square diagramming techniques. As a result of having applied these techniques during the architecture synthesis of allocated functionality, the system engineer will have exposed the things about which interface requirements must be written. Once again, the purpose of our tools is to do just this, to help us understand what to write requirements about. The use of organized methods encourages completeness and avoidance of extraneous requirements that increase cost out of proportion to their value.

Once it has been determined what interfaces we must respect, we must determine what technology will be used to implement them, such as electrical wiring, fluid plumbing, or physical contact, for example. Finally, the resultant design in the selected media is constrained by quantified requirements statements appropriate to the technology and media. Each line on a schematic block diagram or marked intersection on an n-square diagram must be translated into one or more interface requirements that must be respected by the persons or teams responsible for the two terminal elements. The development requirements for the two terminal items may be very close to identical, such as a specified data rate, degree of precision, or wire size. The product requirements, however, will often have an opposite nature to them, such as male and female connectors, bolt hole or captive bolt and threaded bore hole, or transmitter and receiver.

9.2.7.2   *Environmental requirements analysis*

One of the most fundamental questions in system development involves the system boundary. We must be able to unequivocally determine whether any particular item is in the system or not in the system. If it is not in the system, it is in the system environment. If an item is in the system environment, it is either important to the system or not. If it is not, we may disregard it in an effort to simplify the system development. If it is important to the system, we must define the relationship to the system as an environmental influence.

We may categorize all system environmental influences in the five following classes:

a. *Natural environment.* Space, time, and the natural elements such as atmospheric pressure, temperature, and so forth. This environment is, of course, a function of the locale and can be very different from that with which we are familiar in our immediate surroundings on Earth, as in the case of Mars or the Moon.

b. *Hostile systems environment.* Systems under the control of others that are operated specifically to counter, degrade, or destroy the system under consideration.

c. *Noncooperative environment.* Systems that are not operated for the purpose of degrading the system under consideration but have that effect unintentionally.

d. *Cooperative systems environment.* Systems not part of the system under consideration that interact in some planned way. Generally, these influences are actually addressed as interfaces between the systems rather than environmental conditions because there is a person from the other system with whom we may cooperate to control the influences.

e. *Induced environment.* Composed of influences that would not exist but for the presence of the system. These influences are commonly initiated by energy sources within the system that interact with the natural environment to produce new environmental effects.

As noted above, cooperative environmental influences can be more successfully treated as system interfaces. Hostile and noncooperative influences can be characterized through the identification of threats to system success, and the results joined with the natural environmental effects. The induced environment is best understood through seeking out system energy sources and determining if those sources will interact with the natural environment in ways that could be detrimental to the system.

The natural environment is defined in standards for every conceivable parameter for Earth, space, and some other bodies. The challenge to the system engineer is to isolate those parameters that are important and those that are not and then select parameter ranges that are reasonable for the parameters that will have an impact on our system under development. The union of the results of all of these analyses form the system environmental requirements. It is not adequate to stop at this point in the analysis, however.

Systems are composed of many things that we can arrange in a family hierarchy. Items in this hierarchy that are physically integrated in at least some portions of their system operational use, such as an aircraft in an aircraft system or a tank in a ground combat system, can be referred to as end items. We will find that these end items in operational use will have to be used in one or more environments influenced by the system environment but, in some cases, modified by elements of the system. For example, an aircraft will have to be used on the ground, in the air through a wide range of speed and altitude, and in hangars. These are different environments that we can characterize as subsets of the system environment definition. The best way to do this is to first define the system process in terms of some finite number of physical analogs of system operation. We may then map the system end items to these process steps at some level. Next, we must determine the natural environmental subsets influencing each process step. This forms a set of environmental vectors in three-dimensional space composed of a process set, architecture set, and environmental set. In those cases where a particular end item is used in more than one process, we will have to apply some rule for determining the aggregate effect of the environments in those different process steps. The rule may be worst case or some other one. This technique is called *environmental use profiling.*

The final step in environmental requirements analysis involves definition of component environmental requirements. These components are installed in end items. The end items can be partitioned into zones of common environmental parameters as a function of the end item structure and energy sources that will change natural environmental influences. We define the zone environments and then map the components into those zones. In the process, we may find that we have to provide an environmental control system in one or more of the zones to reduce the design difficulty of some components. We may also conclude, given that we must have at least one such system, that we should relocate one or more components into the space thus controlled. So the environmental requirements for a component are predetermined by the zone in which it is located, and we may derive its environmental requirements by copying (cloning) the zone requirements. Component environmental requirements may have to be adjusted for shipping and other noninstalled situations.

In summary, this set of three tools (standards, environmental use profiling, and end item zoning) may be used to fully characterize the environmental requirements for all system items from the system down through its components. In all cases, we must be careful to phrase these requirements in terms that recognize our inability to control the ultimate natural environment.

The discussion above focuses on the effects of the environment on our system. We must also consider the effects of our system on the natural environment. This is most often referred to as environmental impact analysis. It has not always been so, but today we must be sensitive to a bidirectional nature in our environmental specification. Our efforts in the distant past were very small compared to the tremendous forces of nature. Today, the scope of our access, control, and application of energy and toxic substances is substantial, and potential damage to local and regional natural environments is a real concern. Environmental impact requirements are commonly defined in laws and regulations. We can determine in what ways the environment will be exposed to system-generated stress by evaluating all system energy sources, toxic substances used, and any exhaust products. The author prefers to classify this topic under the banner of environmental requirements analysis, but it could be thought of as a specific specialty engineering discipline.

These effects must be considered both for the life of the system during its use and, perhaps more importantly, for its disposition following its useful life. There can be no better example of how difficult and dangerous this can be than the case of the nuclear weapons cleanup process seriously begun in the 1990s after the end of the nuclear confrontation between the USSR and the United States, which lasted 40 years. Throughout the confrontation, both sides were so concerned with survival and the urgency of development that little thought was given to the problems that they may be causing for the future. By the time the problem could no longer be ignored, it was so substantial that a solution was more difficult than the complex scientific and technical problems that had to be solved to create it. One can become very unpopular expressing interest in system disposition during heady system development times, especially with one's proposal or program manager, but customers will be increasingly concerned with this matter as the problems we attempt to solve become more general and their scope more pervasive.

Computer software environmental requirements are limited. The natural environment impinges on the computer hardware within which the software functions but does not in any direct way influence the software itself in the absence of hardware deficiencies that cause computer circuits to fail due to environmental stresses. The context diagram used to expose the boundary between the software and its environment is the best place to start in determining the appropriateness of any environmental influences directly on the software. One such requirements category of interest commonly is the user environment describing how the user will interact with the software. These influences could alternatively be covered under the heading of interface requirements. One must first determine whether the operator is inside or outside the system, however, and this is no small question.

The principal element in the software environment is in the form of cooperative systems exposed through interface analysis. What other software and hardware systems will supply information or use information created by the system in question? The noncooperative system component may exist in very complex software systems, but it is hard to characterize in terms of an example. A very common and increasingly important environmental factor is the hostile environment. This matter will more commonly be addressed as the specialty engineering discipline called *security engineering*, however, than as an environmental consideration.

9.2.7.3 *Specialty engineering requirements analysis*

The evolution of the systems approach to development of systems to solve complex problems has its roots in the specialization of the engineering field into a wide range of very specialized disciplines for the very good reasons noted earlier. Our challenge in system engineering is to weld these many specialists together into the equivalent of one all-knowing mind and applying that knowledge base effectively to the definition of appropriate requirements followed by development of responsive and compliant designs and assessment of those designs for compliance with the requirements as part of the verification activity.

There are many specialized disciplines recognized including: reliability, maintainability, logistics engineering, availability, supportability, survivability and vulnerability, guidance analysis, producibility, system safety, human engineering, system security, aerodynamics, stress, structural dynamics, thermodynamics, and transportability. For any specific development activity, some or all of these disciplines will be needed to supplement the knowledge pool provided by the more general design staff.

Specialty engineers apply two general methods in their requirements analysis efforts. Some of these disciplines use mathematical models of the system, as in reliability and maintainability models of failure rates and remove-and-replace or total repair time. The values in these system level models are extracted from the model into item specifications. Commonly these models are built in three layers. First, the system value is allocated to progressively lower levels to establish design goals. Next, the specialty engineers assess the design against the allocations and establish predictions. Finally, the specialists establish actual values based on testing results and customer field use of the product.

Another technique applied is an appeal to authority in the form of customer-defined standards and specifications. A requirement using this technique will typically call for a particular parameter to be in accordance with the standard. One of these standards may include a hundred requirements, and they all flow into the program specification through reference to the document unless it is tailored to reflect precisely what is required on the specific program. Specialty engineers must, therefore, be thoroughly knowledgeable about the content of these standards; familiar with their company's product line, development processes, and customer application of that product; and acquainted with the basis for tailoring standards for equivalence to the company processes and preferred design techniques.

9.3   Architecture synthesis

Return now to Figure 9.2 and note the development of the architecture as a result of the allocation of functionality to things. In building the physical architecture model we should take into consideration the integration of several different views. One of these is the finance perspective, often referred to as a work breakdown structure (WBS). It is important that the system engineer and the finance people come to a common perspective on the architecture. In the past in the development of systems for DoD, the finance community was responsible for the WBS and system engineering for the top-down development of the architecture. These were seldom totally consistent because the WBS on DoD programs was determined by an appendix of a military standard (MIL-STD-881) before the functional analysis had finished to offer a functionally derived architecture. DoD has now moved to a position that system engineering should be responsible for defining the WBS.

So we should be able to establish a single notation that serves both as a WBS and architecture identification. The author would prefer the architecture identification system suggested in earlier chapters involving a base 60 system of numbers and letters (upper case less O and lower case less l). If it is necessary because of customer requirements to maintain separate identification systems, then we would have to map the two together, leading to some inefficiency.

Large customers have gotten themselves into the habit of using a WBS that covers two different classes of things. DoD, NASA, and other large customers recognize product and process components in the WBS. The product WBS is simply an overlay of the architecture (identical or mapped). In this configuration, only the product-oriented component can be defined through functional analysis of the product. The remainder is commonly defined from a boilerplate used on past programs. The process-oriented component is associated with things that are not easy to connect up with the product. The author maintains that this is a mistake that causes unnecessary complexity in program planning and management. The premise is that it is possible to hook all cost and schedule parameters to product.

It is especially easy to do this when we accept the IPT approach. The program planning process is simplified more when we use the architecture IDs as a WBS for finance purposes. If there must be financial component complexity, let it fall entirely within the finance world and their computers. Leave the human part of the process as simple as possible.

9.3.1   Product-only WBS

If we recognize that work can be accomplished against the whole system as well as any element of the system, we shall have a workable model that will enable an efficient transform between generic enterprise work planning and the work needed for a specific program. The DoD customer has, in the past, been so concerned with the need to control cost and schedule that they have forced contractors to apply a planning model that denies them an efficient map or transform between their identity and the needed program work. Therefore, these contractors have generally failed to develop an identity through common written practices to be applied on all programs, sound internal education programs (on-the-job training and mentoring, as well as more formal methods) focused on that generic process, and efficient methods for building program planning structures based on their generic process definition.

Figure 9.3 illustrates the proposed relationships between product items, functional departments, generic work commonly required on the enterprise's programs, and time. The understanding is that the enterprise is managed through a matrix with lean functional departments responsible for providing programs with qualified people skilled in standard practices found to be effective in the company's product line and customer base and good tools matched to the practices and people skills. Each of these functional departments has an approved charter of work it is responsible for performing on programs and which provides the basis for personnel acquisition and training.

The front vertical plane in Figure 9.3 illustrates the aggregate functional department charter by showing the correlation between all functional departments and all possible work they may be called upon to do on programs. The darkened blocks indicate this correlation. The cross-hatched block in each column indicates the department that should supply the persons to lead that particular effort on programs in the product development teams assembled about the architecture. The indicated departments must prepare practices supportive of the practice defined by the task lead department (cross-hatched block). The task lead department should also be primary in providing tools to do the work for that task on all programs and for providing any training in that activity.

The reader should note that when an enterprise fully projectizes its workforce it looses the agent that can accomplish the generic work useful in providing the foundation for all programs. Each program becomes a new process factory. Therefore, there can be no central process, there can be no

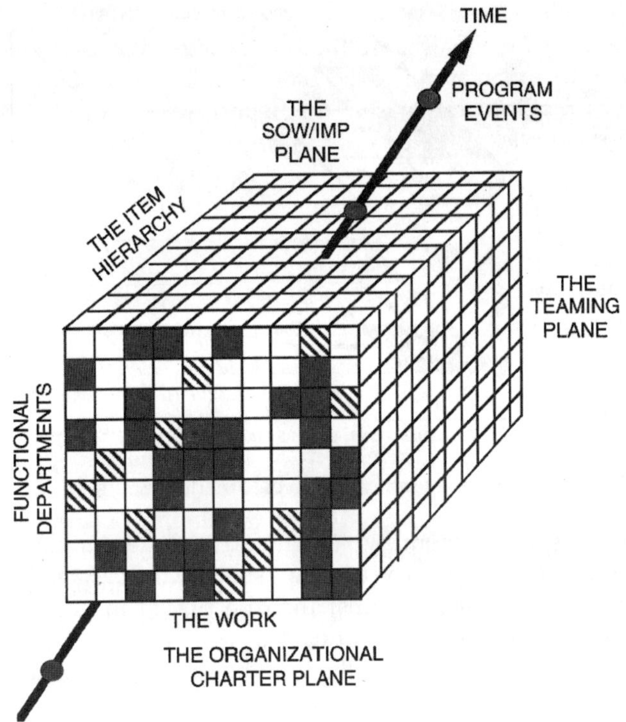

Figure 9.3   Product-only WBS enables good planning.

continuous improvement, there can be no progress, only a never-ending struggle, at best, at the lower fringes of excellence.

In the suggested arrangement, the WBS and the architecture become one since there is no process component of the WBS. When planning a program composed of items identified on the item hierarchy (architecture or WBS) axis in Figure 9.3, we must select the work required from our generic work axis to form the program planning data, which may consist of a general identification of the work at a high level called a *statement of work* (SOW) and a more detailed expansion of that work definition called an *integrated master plan* (IMP). This is referred to in Figure 9.3 as the SOW/IMP plane. This work selection leads to a clear identification of the functional depart ments from which personnel should be selected and the practices that must be applied on the program. Once we have decided the architecture nodes around which we will form cross-functional teams, the teaming plane tells us to which teams these people from the functional departments should be assigned.

The SOW that results from the first level of correlation between the product architecture and generic work must then be expanded into a more detailed plan referred to here as an IMP. For each SOW paragraph, we need to identify in a subparagraph in the IMP the work to be accomplished by

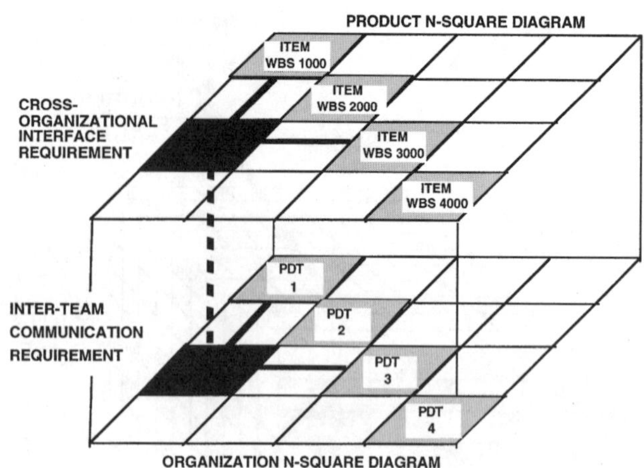

Figure 9.4   Interface development benefits.

each contributing department, which should include the cross-functional, co-process integration work to be accomplished by the task lead department. All of the work thus selected must then be placed in a time context by mapping it to major events terminating that work and corresponding to the conduct of major reviews forming the foundation of the program integrated master schedule (IMS). We see the program plan as a series of nodes arranged in a series-parallel network mapped to trios in our matrix illustrated in Figure 9.3. These nodes, in turn, are traceable to a combination of contributing functional departments, product architecture or WBS, and generic work elements.

In that we should form the work teams focused on the product architecture or WBS, each team inherits a contiguous block of budget in the WBS and work in the corresponding SOW and IMP section, a clearly correlated integrated master schedule (IMS) expands upon the IMP in the time axis, and a specification at the top level of the hierarchy for which they are responsible is needed. The cross-organizational interfaces, which lead to most program development problems, become exactly correlated with communication patterns that must occur between the teams to develop those interfaces as shown in Figure 9.4. Thus, people are clearly accountable for doing this work well.

This product-only WBS has been discussed in terms of a physical product, but most products also require deliverable services such as training, system test and evaluation work, site activation work, and other services. Figure 9.5 suggests that these services can be framed in the context of physical entities, some of which may be delivered and others not. The customer training program may entail the services of teaching the courses developed initially or for the whole life cycle. Indeed, the whole system could be operated by the contractor throughout its life, and this would clearly be a service, but these services can be related to the things in the system. The

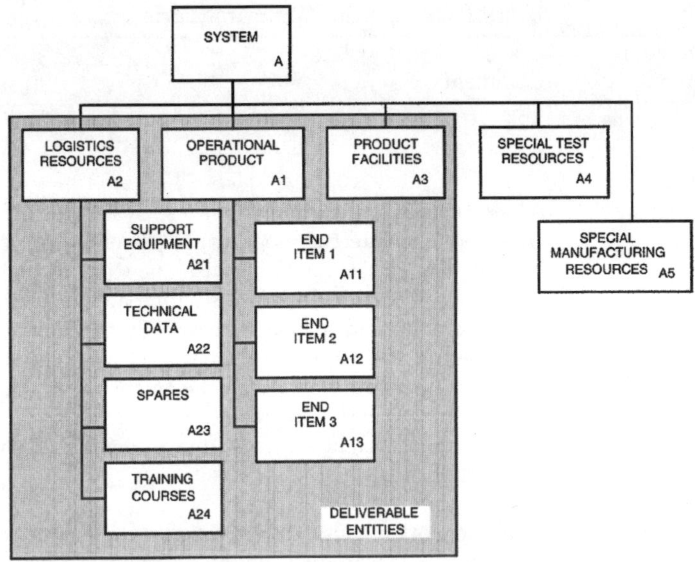

Figure 9.5 The product-only WBS.

author distinguishes between the services that would be performed on the delivered system and those required to create the system. The latter should all be coordinated with the physical product.

How about a contract that entails only services? The WBS for such a contract can still be formed of services to be delivered, but there remains a distinction between the work that must be done to prepare for performing those services and the performance of those services. The former can be oriented about an architecture of services being designed. The latter constitutes the product that is delivered. Table 9.1 lists several often troublesome traditional WBS elements and correlates them with the product-only blocks of Figure 9.5 to show that this can be done. Note that some of these entities would not normally be delivered, such as the tooling and special test equipment, but the customer may very well own this equipment in that it was developed on the contract.

It is especially important to note that the program management and system engineering services work need not be treated as a service. It can be associated with the system level of the product if nothing else, but by distributing these elements through the architecture where they belong, we distribute the cost and schedule parameters associated with these skills into the teams, PIT and IPT. Also, we support the notion that system engineering is a process, rather than an organization or profession that should be applied by everyone throughout the system architecture hierarchy.

So at this point, the enterprise has prepared itself by developing an identity composed of an organizational charter plane showing the map between common tasks that must be applied on all programs and their

Table 9.1 Nonproduct WBS Transform

Nonprime item work element	Comment
Training	A24
Peculiar Support Equipment	A21
Systems Test and Evaluation	A1 elements as appropriate to the test. System testing may be mapped to architecture item A or A1 depending on the scope of the planned testing. Otherwise, the work maps to the activities associated with items within architecture A1.
Systems Engineering/ Program Management	Elements of system A as appropriate. System engineering work accomplished at the system level would map to A. Deliverable operational product element work would map to A1. This same pattern would be extended down to team management and the systems work accomplished within the context of those teams.
Data	A22
Operational Site Activation	A3 as the work relates to facilitization and preparation of sites from which the product will be operated and maintained.
Common Support Equipment	A21
Industrial Facilities	A4, A5
Initial Spares and Initial Repair Parts	A24

functional department structure. Each functional department has developed written practices telling how to do the tasks mapped to them coordinated with the designated task lead department's practice and tools available to do the work. This how-to information could be in detailed text format in practices manuals or could be provided in training materials traceable to the content of an abbreviated practice. This generic identity data is continuously improved based on lessons learned from ongoing programs and is the basis for in-house training.

When a program comes into the house, the work is planned by first performing a functional decomposition and analysis on the customer need, which yields a preferred architecture synthesized from the allocated functionality and the aggregate views of engineering (hardware and software influences), manufacturing, procurement and material, finance, logistics, and quality within the context of a program integration team (PIT). The product architecture becomes the WBS (base 60 identification suggested here can be substituted with any other system), which is the basis for development of the program statement of work (SOW). The SOW is created by the PIT by linking high-level tasks with the WBS elements as discussed earlier. WBS prefixes could be used to cover the most complicated case where one prefix may be for recurring and the other for nonrecurring. Figure 9.6 illustrates this whole environment, which will be further developed in Chapters 10 through 12.

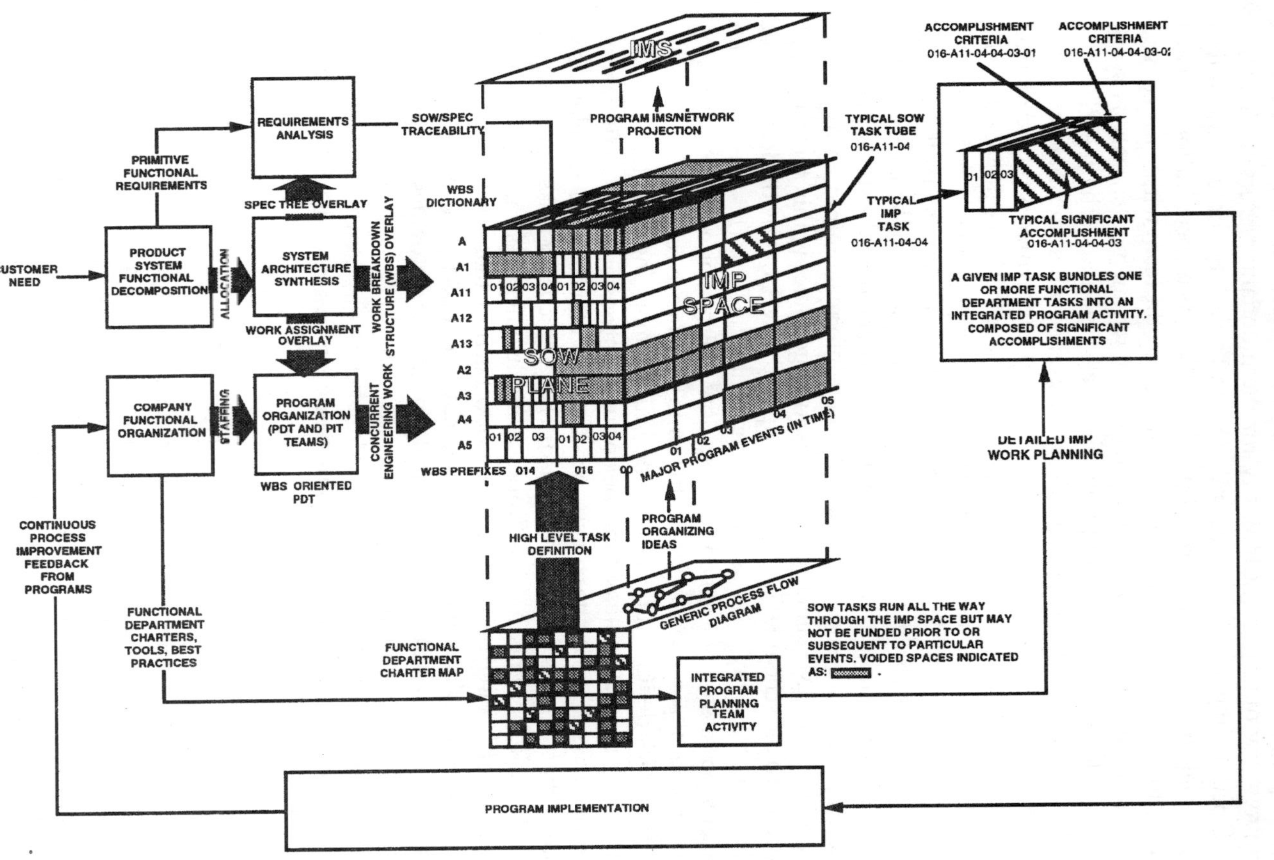

Figure 9.6 The grand planning environment using product-only WBS.

The WBS elements also become the basis for formation of the integrated product teams (IPT), each of which receives a contiguous block of WBS, SOW, and related budget in a simple fashion that will lead to a need for relatively low management intensity for the mundane preserving management focus on the remaining real problems and risks. The continuing requirements analysis process performed by the PIT yields a specification for each team defining the design problem to be solved by the team within the context of the planning data and budget. Lower-tier requirements analysis work may be delegated to the appropriate teams and monitored by the PIT.

An IMP is formed by expansion of the SOW with additional detail that tells what has to be done and by whom. How-to information is provided to all programs by reference in the IMP to generic how-to practices provided by the functional departments rather than through narratives unique to each program. All IMP tasks at some level are linked to major program events through an IMP task-coding technique to be taken up in subsequent chapters. These events are used to define planned milestones coordinated with major program reviews. Scheduling specialists craft the raw information into an integrated master schedule (IMS) linking the events to target dates within the context of an event-driven schedule. Teams expand upon the IMS to provide details about their planned activity so as to be consistent with the IMS. This process can be aided by possession of a generic process flow diagram or network that places generic tasks at some level of indenture into a sequence context.

The work definition flows into the IMP by mapping functional department charters to SOW tasks and performing integration and optimization on the details to craft a least-cost, most-effective process for developing a product that satisfies the customer need. These functional tasks mapped into the IMP tasks become the significant accomplishments phrased in terms of the result of having performed that work. Each of these tasks includes at least one accomplishment criteria whereby one can tell when the task is finished. The aggregate of all of this planning work provides the program with a sound foundation.

Figure 9.6 uses USAF Integrated Management System work cataloging methods that will be further explained in Chapters 11 and 12. The author will offer an alternative in Chapter 13.

9.4   *A general theory of structured analysis*

Earlier chapter content expanded upon one possible structured analysis approach the author refers to as traditional SRA, used on many military programs and applying functional flow diagramming as a key element in the model. It happens that this is but one of many possible modeling approaches. A more general exposition of these methods entails an N-faceted approach to a problem space. Figure 9.7 illustrates the general situation, showing a three-faceted approach of which there are several. Where N = 0, it corresponds to a nonstructured approach, also referred to as *ad hoc*. In this case, we wander without structure through the problem space looking for meaning.

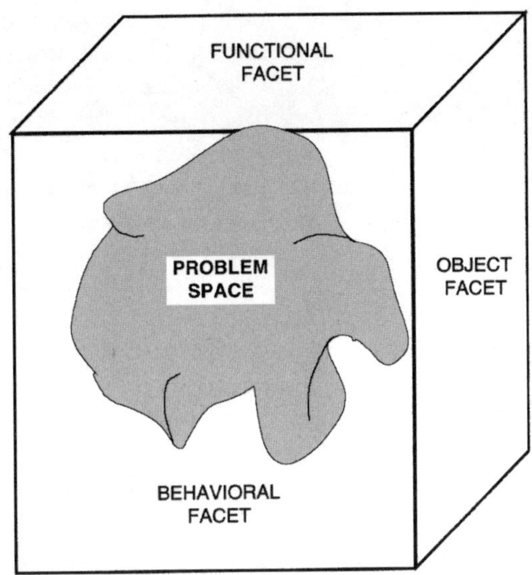

Figure 9.7   Three-faceted structured analysis approaches.

Table 9.2   Three-Faceted Approaches

3-Faceted Structures	Functional Facet	Object Facet	Behavioral Facet
Traditional SRA	**Functional Flow Diagram**	Architecture Block Diagram	Schematic Block Diagram, Time Line
Modern Structured analysis	**Data Flow Diagram**	Software Modules	P Spec and State Diagram
Object oriented analysis	Data Flow Diagram	**Object Diagram**	State Diagram

In three-faceted approaches, we apply particular modeling structures to capture the results of our analysis of the problem space. We stare at the problem space through these facets, commonly initiating the search for knowledge about the space using one of the three facets as the lead perspective, printed in bold in Table 9.2, that lists three particular three-faceted approaches. The author has encountered some organizations and analysts who use a single-faceted approach, using state diagrams where the product line is heavily into dynamic time-dependent behavior.

The first line in Table 9.2 corresponds to the approach described earlier in this chapter. So N can be any number from 0 to 3. It may be possible to construct more complex faceted approaches, but we should keep in mind that there are benefits in simplicity. In fact, we are faced with a continuum between very simple models that communicate powerfully via vision to our brain and very complex modeling structures that offer a very rich modeling environment that we have difficulty understanding visually.

chapter ten

Interface development in a new world

The other chapters of this book are focused on the relationships between the elements of the functional organization that will supply the enterprise with the talent and resources with which to create product systems and the teams that will be formed around the product architecture on programs. This chapter focuses on the relationships between the elements in the product system through interfaces and the teams that develop the elements of the product system. There should exist on a program a well-developed correlation between these team relationships and the interfaces between the products for which the teams are responsible.

Interface development is the most difficult part of the system development process because interfaces have two terminals, and all the more difficult when one party is not responsible for both terminals. Chapter 9 addressed the need to define the requirements for these interfaces. In this chapter we will explore the management and documentation aspects appropriate for these interfaces. Interface development has always been a difficult management problem, but rapid changes in ownership of companies common in the 1990s made it even more difficult. This chapter extends common interface development techniques, which are explained in some detail, to the full range of organizational relationships in both static and dynamic business relationship.

10.1   Interface fundamentals

10.1.1   Interface defined

A system is a collection of things that interact to satisfy a common objective, goal, or need. The things may be hardware, software, facilities, or people accomplishing procedures. An interface is a relationship between things in a system, between systems, and between systems and their environment. Figure 10.1 illustrates this arrangement at the top end of the system and interface concepts. The system is composed of things (not illustrated) internal

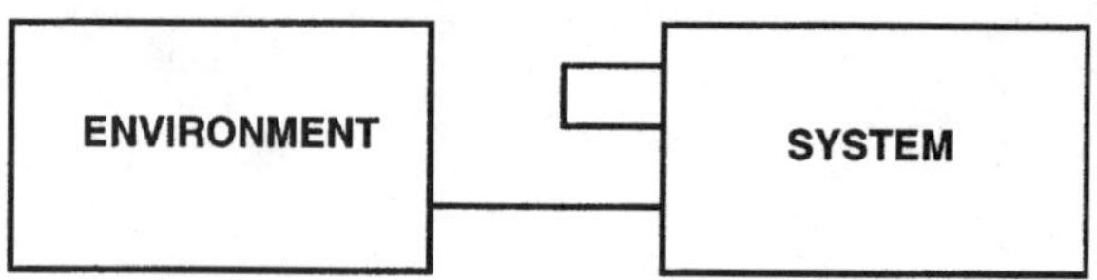

Figure 10.1   Ultimate system abstraction.

to the system which interact via the interface loop (all of the internal interface) attached to the system block, and the system interacts with its environment via the interface shown connecting the two blocks.

Everything in the universe that is not in the system is in the system environment. Some of these things are related to the system in some way, while many are not and can be ignored.

The environment consists of space, time, and natural forces applied to things throughout the universe, as well as other systems linked by one of the following relationships: cooperative, noncooperative, and hostile. Cooperative systems can best be defined in terms of interfaces between the two systems because there is someone on each end of the interface to take responsibility for his/her end, and the relationship can be negotiated. Noncooperative and hostile relationships have to be treated as environmental influences because the relationship cannot be negotiated by persons responsible for the ends.

10.1.2   Interface characteristics

All interfaces have three fundamental aspects: two ends, or terminals, and a media through which the relationship is consummated and maintained. The ends touch two things that are in the system in the case of an internal interface or one thing in the system and one element of the system environment in the case of an external interface.

10.1.3   Classes of interface media

There are many ways to partition all interfaces into subsets relative to the kinds of media employed as indicated in the following outline.

Electrical	Mechanical
Power	Structural or physical
Control	Mechanism or linkage
Radio or radar	Human manipulative
Signal bus	Fluid
Electrical noise	Pneumatic (air or other gases)
Optical	Liquids (nonhydraulic)
Visible light	Hydraulics
Infrared (IR)	Acoustic
Ultraviolet (UV)	Sound in air
	Sound in water

The technology associated with these different kinds of interfaces determines the way we would define the requirements for a particular interface in terms of volts, flow rate, forces applied, or intensity. Some organizations prefer to partition all interfaces into the following subsets: functional, physical, and environmental.

10.1.4 Functional vs. physical plane identification of interfaces

The interfaces that exist between the things in a system are predetermined by the way we allocate functionality to the things in the system. Some people claim that they can foresee this pattern on the functional plane prior to allocation of functionality through the use of functional n-square diagrams (more about these diagrams in a moment), but the author maintains that this only works when one uses a hierarchical functional depiction and is not fruitful when applying a sequence-oriented functional portrayal. The n-square relationships between a sequence of functions is sequential in nature rather than predictive of physical interfaces that will be required. Most people, including the author, have to identify the interfaces within the context of the physical plane and the product architecture based on the way functionality was allocated.

10.2 Interface depiction

Interface relations between things in the system can be exposed to humans through the use of schematic block diagrams or n-square diagrams. Tabular lists could be used featuring pairs of architecture columns to identify the interfaces uniquely, but these are not easy for humans to use commonly. Such a listing is a good idea in the form of an interface dictionary to supplement a diagrammatic portrayal but not as the sole means of formally identifying interfaces.

10.2.1 Schematic block diagrams

Using this technique, the things from the system architecture are illustrated as blocks connected by lines which signify the interfaces between them. Nondirected lines can be used at the high level where the relationship may entail many individual interfaces from one or more media classes. At the individual interface level, we can use directed lines associated with particular media classes.

Figure 10.2 illustrates one of these diagrams at the detail level where directed lines make sense. We define the way the things shown as blocks in Figure 10.2 are related by defining the interfaces joining them. Items U, V, and Y interact with the system environment, whereas the other items are related only internally.

The things are identified by names and their architecture IDs, and the interfaces are identified by interface IDs. This diagram should be accompanied

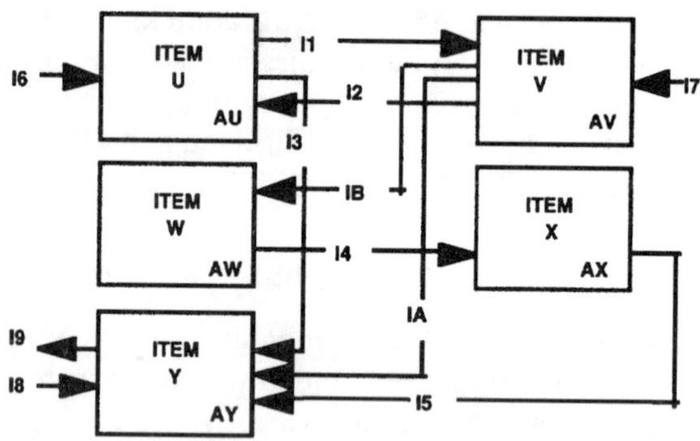

Figure 10.2 Schematic block diagram example.

ENV	I6	I7			I9
	AU	I1			I3
	I2	AV	IB		IA
			AW	I4	
				AX	I5
I8					AY

Figure 10.3 N-square diagram example.

by an architecture diagram and two dictionaries, an architecture and interface dictionary where details can be given and responsibilities defined. The complete set of diagrams and dictionaries should be captured in a system definition or description report and maintained during early program development work.

10.2.2 N-square diagrams

Figure 10.3 illustrates exactly the same architecture and set of relationships as shown in Figure 10.2 but uses the n-square diagramming technique. The things in the system are listed down the diagonal along with the external (ENV). Interface IDs are inscribed in the intersections. This diagram should include a directionality arrow as noted in the upper right-hand corner of the diagram. Interface I9, for example, sends something from the environment to item Y, whereas interface I2 sends something from item V to item U.

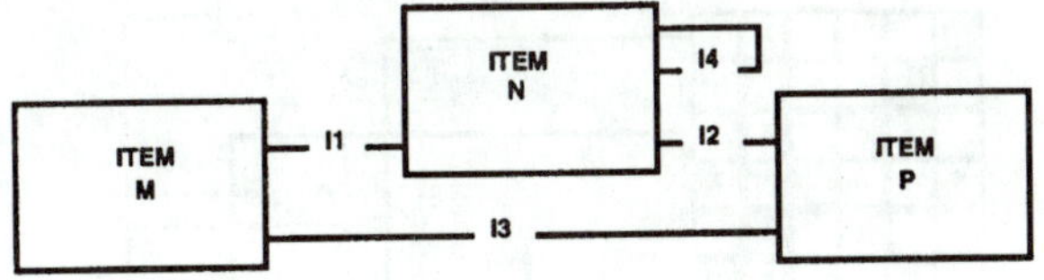

Figure 10.4   Interface viewpoints.

10.3   Terminal responsibility

10.3.1   Three faces of interface

The interfaces that appear in systems are looked upon differently by different persons as a function of the number of terminals of an interface that touch the item for which they are responsible. There are obviously three possibilities: one end, both ends, or neither end touch the architecture item in question. Let us look at these responsibilities in the context of the principal engineer for item N in Figure 10.4.

Note that both terminals of interface I4 touch item N. Clearly, we should expect that this interface can be developed by the principal engineer responsible for Item N. This is an example of item N Principal innerface. Neither end of interface I2 touches item N so we can place this interface in item N Principal's outerface set, and we should not bother this principal about this interface. Both of the remaining two interfaces in Figure 10.4 have one terminal that touches item N and another that touches a different item. These are all in item N Principal's crossface and examples of true interfaces. The true nature of product interfaces can only be exposed in the context of the organizational responsibilities for these interfaces.

10.3.2   Cross-organizational interface

The most difficult problems in the development of systems to satisfy complex needs commonly reside at the interfaces we have described as crossfaces. The author would prefer to call these *cross-organizational interfaces*, where the responsibilities fall upon teams or organizations rather than individual principal engineers. It is very important that the system engineer be able to sort out which interfaces fall into this set, because it corresponds with the most serious problems and risks that can occur on programs.

The easiest way to uncover these interfaces is with a schematic block diagram overlaid with responsibility outlines. Wherever the interface lines cross an organizational responsibility line, it identifies a cross-organizational interface. This same effect can be seen when using compound n-square diagrams as shown in Figure 10.5. Given that each square represents the items under the responsibility of a cross-functional development team at

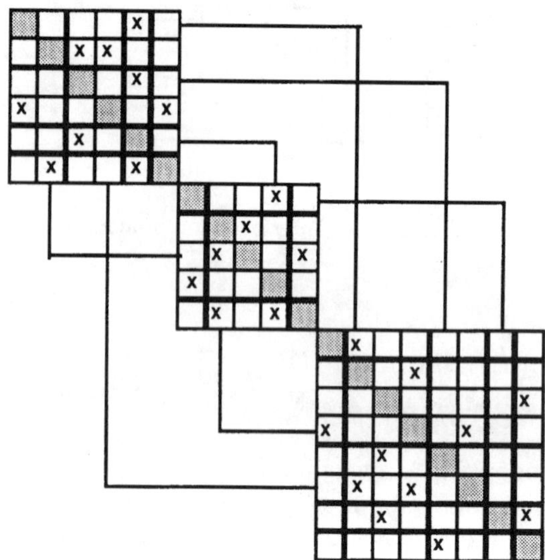

Figure 10.5   Cross-organizational interface.

some level, the interfaces denoted by the lines external to the team squares are cross-organizational interfaces. The interfaces depicted by Xs in each team n-square diagram are innerfaces relative to the team, and we should be able to expect that each team can handle those responsibilities.

10.4   Interface documentation

10.4.1   Requirements documentation

The development of product designs should be based on predefined requirements captured in specifications focused on the items in the product architecture. The specifications generally have a one-to-one correlation with these items. We have several choices when capturing the requirements for the interfaces between these items. First, we could capture the requirements for each item in the corresponding specification pair. Alternatively, we could create a document that corresponds not to the items but to the interfaces between them, an interface control document (ICD). Where an ICD is used, one should be careful not to double-book the requirements definition. If the ICD is to be used throughout the life of the product, the corresponding specification pair should reference the ICD and not include the requirements. If the ICD is only used as a basis for reaching agreements on the interface requirements to put in the terminal specification pair, they should be deleted from the evolving ICD as agreements are reached and the requirements transferred to the specification pair until such time as the ICD is a void when it disappears.

10.4.2   Design documentation

Engineering drawings are commonly used to capture the design solution to the problem defined in specifications. Drawings can, of course, be used to capture the design of interfaces. These drawings could be released independently and/or included in what is sometimes called a Part II ICD, which defines the requirements for acceptance of the interface terminals. The ICD used during development, in this case, would be called a Part I ICD.

10.5   Traditional management of interface development

The traditional view of interface management is appropriate to a relatively stable world where ownership of the enterprises involved in a development effort remain essentially unchanged over the life of the development effort. The management activities can be collected into three subsets as a function of the contractual relationships between the parties.

10.5.1   Internal interface management

An internal interface must be developed by the organization responsible. The same enterprise is responsible for both interface terminals. The program should assign interface responsibility to the teams responsible for development of the things in the system as a function of the way the terminals touch these items as discussed above in the context of innerface and crossface. All innerface should be the responsibility of the one team whose item both terminals touch. The teams should be held accountable for the crossface in pairs, but one of them could be made responsible for coordinating the interface development based on a receiver or sender rule.

This arrangement can be structured through several team levels on a large program with each team responsible for all items and interfaces at their top item level as well as all interfaces subordinate to that level flowing down to their lower-tier teams held accountable for the subsets through as many layers of teams as necessary.

10.5.2   Supplier interface management

Development work accomplished by suppliers should be controlled through contractual relationships coordinated with an item specification and statement of work. The interface requirements should be clearly defined in the procurement specification and controlled through the contract. This is a case of a contractual relationship between the parties.

10.5.3   Associate interface management

It can easily happen on government contracts that the two parties to an interface do not have a contractual relationship between them. This happens

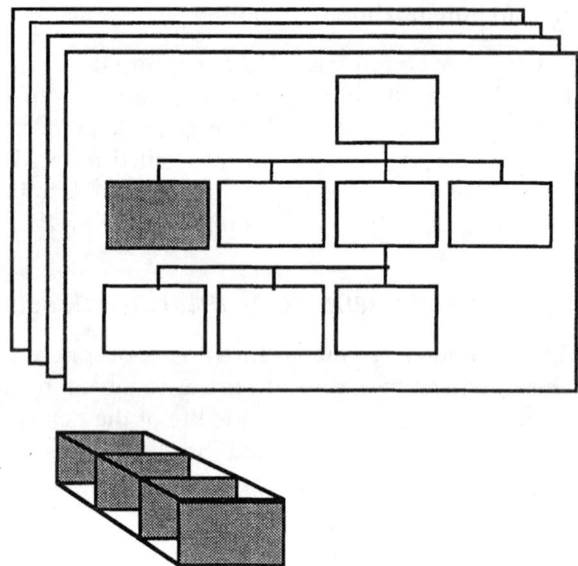

Figure 10.6   Cross program relationships.

when two contractors are under contract to the same customer. This is often called an *associate contractor relationship*. The common government customer should include interface management clauses in the contracts for each of the contractors, forcing them to cooperate in the development of a compatible interface defined in an ICD.

One of these contractors, or a third contractor also involved in the program, will be assigned the lead interface responsibility, developing an interface management plan and draft ICD. Both of these documents should be offered to the parties to the interface being controlled for review and approval. These associate relationships also often involve management through a joint team staffed by representatives from the contractors and the common customer co-chaired by the customer and the lead interface contractor. This interface control working group (ICWG) acts to adjudicate differences of opinion on these interfaces and is the final authority for the definition of the interface.

10.5.4   *Cross-program interface management*

Some companies and programs become involved in additional interface development problems because of the use of items in multiple applications. The gray item in the product architecture illustrated in Figure 10.6 must be compatible with the internal interfaces corresponding to that application. However, if this same item must be used in three other applications as well, the interface development problem becomes significantly more difficult.

In addition to an integration activity within each program to ensure that internal program interfaces are compatible, it is necessary to implement a cross-program integration activity to adjudicate the different demands of the several programs for these interfaces, with the goal being to develop a single interface that is used in the same way in all applications. Where this is not possible, maximum commonality and minimum uniqueness is desirable.

This effect can be realized in a large company characterized by vertical integration of the product line. One division may provide all of the other divisions with engines or transmissions. To the extent that these units can be identical for similar loads and performance requirements, the aggregate enterprise will benefit from a cost perspective.

This same effect can be realized as well when the final assembler wishes to purchase a common design for several products, as in the use of a common engine for several lines of automobiles. The general solution for this problem is an additional integration agent corresponding to the common item with representatives from the several programs involved.

10.6   Management of interfaces within the context of ownership dynamics

10.6.1   A new reality

In years past, companies tended to remain fairly stable relative to their ownership and employees tended to remain employed by the same company for a considerable period of time. Neither of these is a characteristic of aerospace today and may not be for an indefinite period. These factors introduce a new dynamic into the development of systems and, in particular, the development of interfaces within these systems.

The development of the things that compose systems will always be a clear one-on-one arrangement that can adjust to any of these changes fairly easily. However, since interface development involves a two-to-one relationship and the organizational relationships between the parties responsible for the things can be in a state of flux and because the development of cross-organizational interfaces has always been the most difficult part of developing systems, this dynamic suggests for a potential need for a methodological change.

10.6.2   The use of cross-functional teams

No matter the development situation, one should apply cross-functional teams to the development of systems that are organized as a function of the product architecture. Teams simplify shifts in relationships because there are clear relationships between the organizations, no matter how they may change relative to contracts or ownership, and the product elements between which the interfaces connect. Organization of programs using the functional

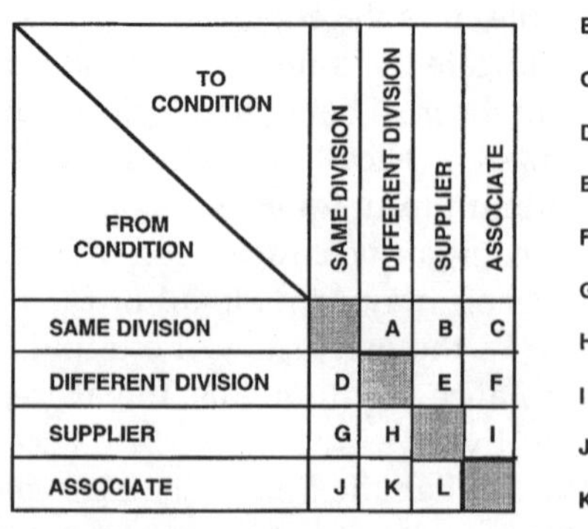

A   START TREATING LIKE A SUPPLIER
B   START TREATING LIKE A SUPPLIER
C   START TREATING LIKE AN ASSOCIATE
D   INTEGRATE INTO INTERNAL TEAM STRUCTURE
E   CONTINUE TREATING LIKE A SUPPLIER
F   START TREATING LIKE AN ASSOCIATE
G   INTEGRATE INTO INTERNAL TEAM STRUCTURE
H   CONTINUE TREATING LIKE A SUPPLIER
I   START TREATING LIKE AN ASSOCIATE
J   INTEGRATE INTO INTERNAL TEAM STRUCTURE
K   START TREATING LIKE A SUPPLIER
L   START TREATING LIKE A SUPPLIER

Figure 10.7   Organizational relationship changes.

department structure will lead to increasingly difficult management challenges because the interface relationships between the things for which the organizations are responsible are invariably complex and difficult to manage.

10.6.3   Team relationship dynamics

For any one business situation, we have covered the several kinds of relationships that might exist between two organizational entities during the development of a system. These techniques are effective when implemented in the context of the existence of an overall system agent that clearly assigns interface responsibilities and is able to monitor and audit the performance of all the organizational elements working the program relative to their interface responsibilities.

In addition, all of the organizational entities in the program should accept the responsibility for integration and optimization of the work and system elements subordinate to their immediate responsibility. Thus, at all levels one finds effective integration efforts being applied across the cross-organizational interfaces.

Another characteristic of a sound program from this perspective is the existence of a comprehensive model of system interface, either in the form of a schematic block diagram or n-square diagram where it is easy to see interface responsibilities because of the use of a responsibility overlay.

The relatively new element that creeps into the interface management process is instability in the business relationships between the organizations responsible for the cross-organizational interfaces. Figure 10.7 identifies all of the ways the organizational relationships can change between two organizations during the development of a system due to movement from one business relationship to another. Four kinds of business relationships have been included, as described earlier, and the matrix intersections are uniquely keyed to a description of the appropriate response on a program for each one.

Because two of the relationships are supplier oriented (supplier and other division), you might expect that there are only three kinds of responses to all of these possible changes. The key to flexibility here is a clear definition of all of the interfaces all of the time, no matter what the relationship between the various business entities. Whether two teams are in-house teams, supplier/buyer, or associate, they should have a clear understanding between them of what the interfaces are between the items for which they are responsible, and they should be managing those interfaces for compatibility.

10.6.4   Tools and documentation flexibility

What may change as we move from one business relationship to another is the formality of the documentation of the interfaces unless we look into more effective ways to capture the interface requirements than are currently used. Current requirements database tools are really focused on the things in systems and do not adequately address the interfaces between them. So one action that would greatly improve interface development amidst business relationship dynamics would be to have available some effective interface management tools.

The ideal tool would allow us to capture the interface requirements relative to the interface model captured in the tool and the item requirements about the item model captured in the tool. This could be done in a fairly simple way by linking the requirements to either an architecture ID or an interface ID (see Figure 10.2). Because the tool would also know what the terminal architecture IDs were for any interface, it should be easy for the tool to drop the appropriate interface requirements into the specifications if they were to be documented, using the specification pair pattern, or to drop them into an ICD otherwise. Supported by this kind of tool suite, the business relationships between the enterprises could shift with considerable freedom, and a stable model of the interfaces would remain under control.

10.7   International effects

The development of systems to satisfy complex needs is very difficult in the best of cases, but it becomes much more difficult when program teams include people who speak different first languages, where the teams are separated by considerable distance as can be the case in international programs, and when item common interfaces across multiple programs are involved.

A program applying these arrangements must select a principal language and require that all participants master that language and provide for the best possible communications across the distances involved. Many product managers find that they must first travel to the distant place and meet the people involved. Thereafter, they can interact effectively via telephone, email, fax, and videoconferencing. Without that personal contact, communication is much less effective. Development under these conditions can be made even more difficult when a demanding schedule must be maintained. Difficulty of communications on programs should carry with it additional schedule slack rather than less.

10.8 Summary

The standard interface development techniques that have been used for decades are still useful, but contractors must be sensitive to the dynamics of business ownership as a result of downsizing, instability of ownership, and the special difficulties of international cooperation. These enterprise and program relationships will still yield to standard interface development techniques, but they may have to be applied in a more flexible fashion than in earlier times.

Key elements in the management of evolving product interfaces during development are the use of cross-functional, product-oriented teams and clear definition of interface responsibilities. All interfaces should be carefully developed based on these clear rules of responsibility.

chapter eleven

Generic program process development

11.1   Aggregate planning goals

In the aggregate, our enterprise must create a work planning environment that satisfies three important goals:

a. Clearly state the responsibilities of the functional departments in terms of providing coordinated quality resources to programs (people, tools, and practices) that are continuously improved in a way coordinated within the enterprise.
b. Clearly state the work necessary to perform on a program so as to encourage understanding of the work at all levels and minimize the management difficulty, with the result that management intensity can be placed on smoothing current discontinuities, and identification and mitigation of future potential risks.
c. Minimize the program planning transform complexity so everyone can contribute from a position of knowledge in pouring generic planning data into the program framework to create the minimum-cost, most effective program plan that results in customer best value and enterprise profit.

In order to satisfy these goals, we should clearly identify the entities of interest for functional departments responsible for the generic data and the programs and name them for easiest understanding by the humans responsible in each case. If there must be complexity, let it be hidden from the people in computer databases as through mappings between one class of data and another. Let the humans refer to things in ways with which they are familiar and in the simplest possible way.

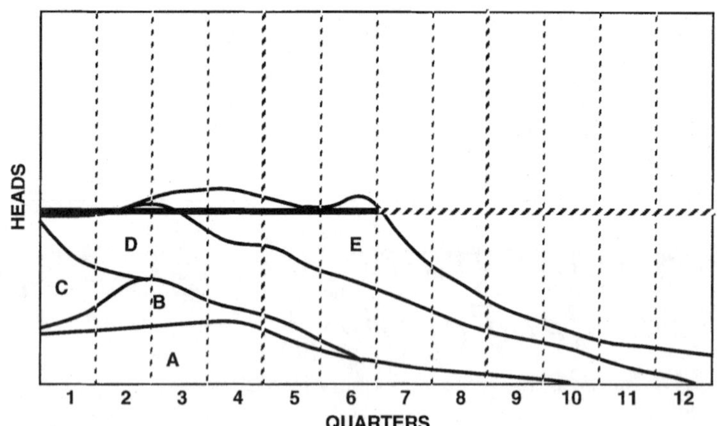

Figure 11.1   Aggregate functional department personnel demand curve.

11.2   Generic planning data structure

On the functional side of the divide, we are interested in generic tasks (the functions of our generic process diagram) that must be accomplished on many things on every program. We then allocate this functionality to functional departments and integrate the sum of those allocations to any one department to craft a set of department charters so that the departments are clear what their responsibilities are.

These functional departments are responsible for acquiring and training qualified people to do the work identified in their charter, so they have to know what the aggregate program demand in the future is going to be for these people in order to permit them to match that demand. Thus, the program budgets must reflect back into the functional departments in some intelligent way.

In Figure 11.1, it is possible for the manager of the functional department that is presented with this information to plan on a particular head level for the next six quarters, recognizing that this kind of chart always decays toward zero the farther out you go in time. What is shown is that the aggregate budget from programs A through E piles up to create an aggregate demand of the number of heads corresponding to the heavy horizontal line in the figure. This is the number of people the department can support for the near term. If the department has fewer than this number, then the current workforce will be over worked or be required to work overtime to maintain program schedules. If the functional department has more people than this number, it may have to reassign personnel or, in the worst case, lay off or furlough some people until the demand climbs back up. Finally, the functional departments have crafted process knowledge in process manuals, such as a system engineering manual, and coordinated with the contents of those manuals access to how-to knowledge that people can use to find out how to do the work.

11.3   Program planning structure

11.3.1   Reviewing program management needs

Program management needs some way to identify work in terms of what has to be done, who is responsible for that work, a way to determine what portion if any of a particular work activity has been accomplished at any particular time, and where the program is relative to use of budget and time. We will coordinate all work with the product architecture as noted in the last chapter, using either the term WBS or architecture ID synonymously, and link teams to these same entities making a clear connection between work and responsibility. In order to satisfy the other needs, we shall need some way of interrelating time and goals to the work.

Our generic process and planning data will have to be applied to programs in a variety of ways as a function of where we are in a given program. A generic task may have a different flavor early in a program vs. later in the same program. Two concepts used to characterize work in a timing sense are *program phasing* and *program events*. Phases are major pieces of programs contiguous in time. Events are instantaneous occurrences along the program timeline corresponding to milestones representing important program achievements or goals.

11.3.2   Program phases

Customers of large and complex programs, such as the Department of Defense (DoD) and NASA, seek to control risk and pace the expenditure of funds by establishing a structure known as phases. A first phase will have very modest goals to define the system functionality and concept, the related mission, top-level system architecture, and demonstrate that the solution posed is feasible with risks identified and controllable. A second phase may involve defining requirements for all of the things in the system and implementing designs, manufacturing and quality planning, and testing to show that the right product has been crafted for the requirements. A third phase would involve manufacture of the product system, customer acceptance of the product, shipment or deployment of the system, and setting up initial operations.

Different customer agencies have developed different ways to break the overall system life cycle into phases. Figure 11.2 illustrates a three-phase program in the interest of simplicity and coordinates with those phases the implementation of the enterprise generic process. In the first phase the enterprise is developing concepts, applying some of the tasks on the front end of its generic process. In the next phase, some of the work from the first phase must be repeated for two reasons. First, we become smarter as we go, finding that as we know more about our proposed solution we would have done it differently if we could have foreseen the problems we have run into. So we change prior decisions, ripple the results, and go on from that point. Second,

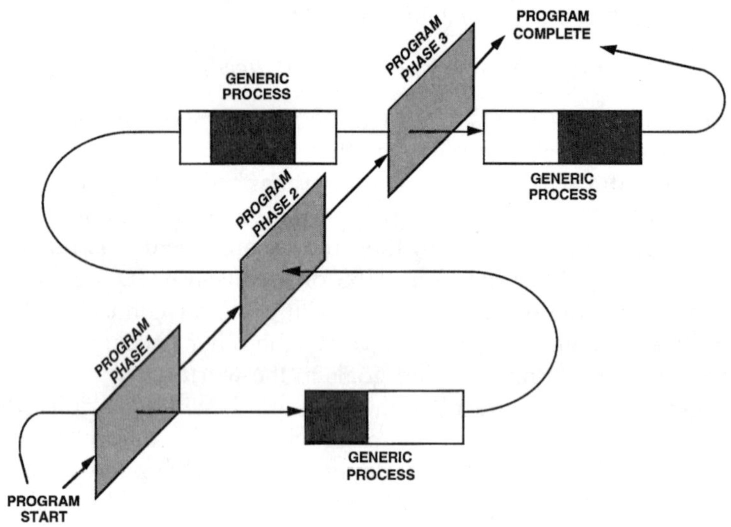

Figure 11.2   Phasing and process correlation.

in the second phase, we are expanding our definition of the product system downward in detail using the same decomposition and definition techniques from our generic process that we applied to higher-tier elements. So generic process repetition, or as some people call it, *iteration*, occurs.

Customers of large and complex systems insist on phasing structures because it helps them control risk. The overall problem they seek to solve through a contract with a developer is often too complex to clearly see the preferred solution in all of its details at once. If a contract were let for development and manufacture of the whole system based on early studies, it would likely be the wrong system that was procured because we simply are not well enough informed about the possibilities and risks at that time. By breaking the whole program into phases or chunks, the customer and contractor are able to lay out clear and specific goals for a collection of work coordinated with a schedule and an amount of money. At the end of that work period, it is possible to clearly establish whether or not the goals were met, one of which should have been to clearly define a set of objectives for the next phase. The next phase gets us closer to the overall program goals — a product system in the field performing as required by the customer — and the program continues either until that objective is attained, or it has been proven that the system is not feasible, or that the current contractor is not the right party to implement it.

The enterprise, in applying its generic process to the phases, does repeat some activities from phase to phase, but over the whole program period, they execute all of the steps in the process moving from early activities to later activities.

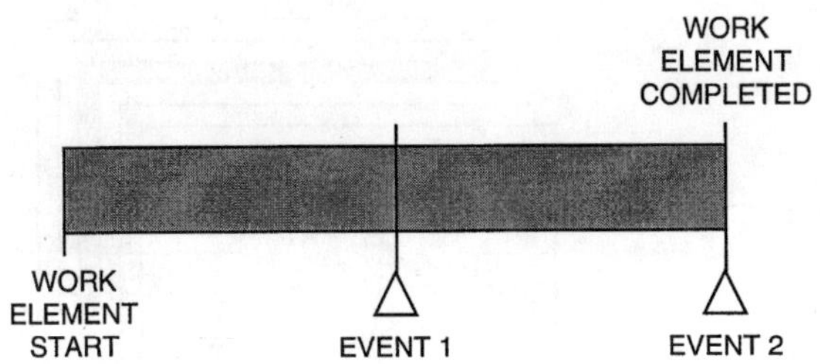

Figure 11.3 Program events.

In planning a specific program, we have to be careful to select work from our generic process that is appropriate to the phase. It would not be appropriate to include detailed design work in an early program phase, for example. Nor would it be appropriate to include mission analysis in a later program phase. If our generic process includes the major reviews that our customer respects, such as system requirements review (SRR), system design review (SDR), preliminary design review (PDR), and critical design review (CDR), and these major reviews act to terminate the program phases, we can use those major reviews in our generic process as boundaries for the process steps that should be selected for a particular phase. The reader should note that the generic process diagram in Appendix A does include these major reviews, which are clearly fed by prerequisite work.

11.3.3   Program events

Each phase may have spaced within it one or more milestones, also referred to as an *event*. These are instantaneous happenings corresponding to some planned program activity having been achieved or toward which we are working. Every program work element at every level should terminate in an event, and it may be appropriate for us to identify intermediate events as well within a particular work element. In that a phase is an element of work, it too should terminate with an event and may have intermediate events.

Events provide opportunities to estimate our program progress. It should be very clear how one determines when any work element has been completed, when we can claim the terminating event or milestone, so every task should have a terminating event and an associated completion criteria by which any reasonable person could judge that the task is complete and about which all reasonable people could agree. In Chapters 12 and 13 we will use events as one of the program work element identifiers. As noted in Figure 11.3, these events may be terminal (event 2) or intermediate (event 1) types.

We can describe the work illustrated in Figure 11.3 by referring to these event numbers. For example, if this complete work element was identified by the designation 1200-02, we could identify the part of the work between

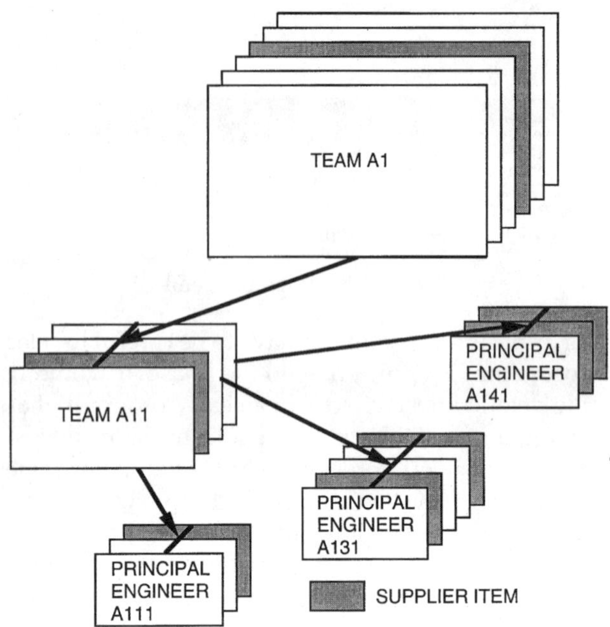

Figure 11.4   Team responsibility structure.

the beginning and Event 1 as 1200-02-01 and the latter component as 1200-02-02, where the latter symbols by convention mean the work element ending at the event defined by the last two numbers (01 or 02 in this case). By convention, this work element begins at the previous event or the beginning (Event 0) as appropriate.

11.3.4   *Teaming phenomena*

When developing systems characterized by great complexity and scope, we find it is not efficient to group everyone together. It is more efficient to decompose the grand system problem into a series of smaller problems, another example of chunking complexity for compatibility with human intellectual capacity. Our system decomposition process and IPT teaming concept are driven by this phenomena. There will be system-level work required to accomplish the decomposition activity driving the team identification, and each of the teams we establish will have to repeat some of this same work to identify subordinate items. This same process can continue through several levels. Figure 11.4 illustrates this expanding process in the context of Team A1, which will be responsible for item A1 of a system A composed of six major elements, each under the responsibility of an IPT.

Each of these IPTs will accomplish essentially the same work to define a system element and decompose it to lower levels of indenture. If the system is sufficiently complex, it may be appropriate for IPT A1 to act as an integration team for four subordinate teams (A11, A12, A13, and A14). These teams could involve development of several lower-tier elements. For example, team A11 might be responsible for developing items A111, A112, and A113 through principal engineers drawing support from specialists on the A11 IPT. In each of the teams above the principal engineer level, there is both IPT work and integration work we have placed in the context of a PIT.

The system will very likely involve procurement of items at all levels of the system architecture. The gray blocks in Figure 11.4 indicate procurement items. Item A4 is being procured, and it is at a very high level in the system. This could be an associate item, customer furnished, of a major subcontract item. It may make sense to have a representative of the supplier involved as a member of our PIT to encourage compatibility of interfaces. We might also establish an IPT supplier team A4. It is important to note that, because of the way the program manages the teams, it makes little difference whether they are internal or supplier in nature. In either case, the team will have responsibility for a specific architecture item which has associated with it work breakdown structure (WBS), planning and scheduling data, specification, and budget components.

Identification of an item as a supplier item significantly affects the continuing decomposition of the subordinate product architecture. The subsequent work performed in the company after this identification of a particular item will follow one of three paths, generally as a function of the contractual relationship between the company and the supplier.

a. Customer furnished equipment should not require any development work other than the influences it may have on the product under development. The customer is essentially the responsible team providing the item.
b. A supplier item may be procured to a specification and statement of work requiring considerable in-house coordination with the supplier. The supplier must design, manufacturer, and test the item to comply with the specification, and the process applied by the supplier must be compliant with statement of work content.
c. A supplier item may be procured to a set of engineering drawings and a statement of work. In this case, the supplier must manufacture an item that has already been designed by the buyer.
d. A supplier item may be purchased "as is" off the shelf and either used in that configuration or altered in accordance with an altered item drawing.

11.4   Making the transform between generic and program models

We have crafted a fine definition of our process and our identity in prior chapters that we have convinced ourselves will work in a real-world program (well, at least the author is convinced). There are two program characteristics

Table 11.1   USAF Model Structure

Identifier	Example	Meaning
WBS Prefix	013	A prefix can be used to collect work into different categories of interest like nonrecurring and recurring.
WBS Number	1100	Work Breakdown Structure identification.
SOW	05	A paragraph in the SOW under 013-1100 identifying a fairly high-level element of work.
Event Number	04	A major program milestone.
Significant Accomplishment	03	One of one or more work elements contributing to work element 013-1100-05-04.
Accomplishment Criteria	01	Identifies one or more ways one can tell when the work is accomplished in element 013-1100-05-04-03.

that we have to handle before we should all accept the model as constructed. These realities are customer interest in phasing structures and the effects of teaming as recommended throughout this book. Finally, in this chapter we will construct two methods of cataloging work on programs. The first, more fully discussed in Chapter 12, is the USAF Integrated Management System approach, and the second is a method developed by the author and presented in Chapter 13.

The U.S. Air Force developed a marvelous work identification system in the early 1980s and applied it to several programs, the F22 Fighter Aircraft Program among them. They did not include a way to link it to an in-house generic work identification system, but the author did so in earlier books (*System Integration* and *System Engineering Planning and Enterprise Identity*). Some of this work is repeated in Chapter 12. Some problems with this linkage caused the author to pursue other alternatives over a period of four years, and the conclusions from that search are included in Chapter 13. You may come up with a third method more in keeping with the way your company captures identifying data and the finance system used to identify program work units.

11.4.1   The USAF integrated management model

This model identifies six entities for each element of work on a program. The number strings formed by these six identifiers uniquely identify every piece of work in a plan called an Integrated Master Plan (IMP) and program schedule called an Integrated Master Schedule (IMS). Table 11.1 offers an example of the structure.

A complete identification sample would then be 031-1100-05-04-03-01. These identifications can be loaded into a database with fields as noted in Table 11.1 and assigned in contiguous blocks to IPT in that the teams will

Table 11.2   JOG System Engineering Model Structure

Identifier	Example	Meaning
Program	P05	The fifth program in-house using this method.
Architecture/WBS	A11	The architecture ID as explained in Chapter 10.
Program Event	E4	Major program milestone as in the USAF model.
Generic Function	F11221	Generic work identifier from the Table A.1 process model.
Functional Department	D262	The Requirements Analysis department will provide the personnel to accomplish this work on the program.

be assigned as overlays of the WBS that this model is based upon. In previous books, the author has shown how you can link up the significant accomplishments to your internal generic model of work, but it does require an inefficient and redundant identification system. Chapter 11 covers that linkage that does not appear to violate any aspect of the USAF system.

Under this system, all work related to product entities is defined in the IMP in the numerical sequence noted above, but other work that is difficult to map to product entities is described in narratives in the IMP.

11.4.2   JOG System Engineering model

The author has developed a method of identifying work using five identifiers as listed and explained in Table 11.2. The model is named after the consulting company operated by the author. The elements of this system are consistent with the content of this book and involve a minimum of mapping between different identification systems. It does eliminate the SOW tradition, which some DoD programs have also done. As in the USAF model, the IMP would be created from alphanumeric ordered work identification strings like the one described in Table 11.2.

A full identifier then would be PO5-A11-E4-F11221-D262. The added program identifier permits all program planning data to be contained in a single database if desired with each record program linked by this field. Within the context of a single program this identifier could be dropped, of course, simplifying the work identification. The prefix letters could also be dropped but are applied in this book to avoid confusion.

The reader will note that the identifier does link to the functional department providing the personnel to do the work on the program and this linkage offers the functional department managers the ability to skim all of the records in an enterprise database linked to his/her department, generating a labor demand curve looking into the future. Thus, the manager knows what the personnel demands are going to be for his/her discipline on all enterprise programs. In some cases, the functional manager may be able to work with one or more programs through the PIT to adjust work scheduling to smooth the demand peaks. In other cases, it may be necessary to hire more staff in a timely way or consider ways to trim staff during some periods.

Table 11.3 Model Correlation Matrix

USAF identifier	Example	JOGSE identifier	Example
NA		Program	P05
WBS Prefix	013	NA	
WBS Number	1100	Architecture/WBS	A11
SOW	05	Generic Function	F11221
Event Number	04	Program Event	E4
Significant Accomplishment	03	NA	
Accomplishment Criteria	01	NA	
NA		Functional Department	D262

The system also links to the generic work definition as in F11221 (Transform Allocated Functions Into Performance Requirements) around which the enterprise can have constructed estimating models based on past performance. Every program will not have a degree of uniqueness that inhibits development of these vital statistics.

11.4.3   Correlation between models

Table 11.3 shows how the two work identification systems align as well as how they do not. The JOG System Engineering model is oriented toward the enterprise perspective, whereas the USAF model is primarily focused from the acquisition agent's perspective. The few differences could be reconciled fairly easily so that an enterprise could apply either model. We could add a prefix identification between the Program identifier and the Architecture identifier of the JOG System Engineering model, removing one of the differences. The significant accomplishment of the USAF model is actually satisfied by the Functional Department field in the JOGSE model, if one followed the author's application of the significant accomplishments identifier to link the program work to functional departments, as discussed in Chapter 12. The JOGSE model does not include an accomplishment criteria in the work identifier, but the database using the identifier as a key field set should include an accomplishment criteria field as well as a task description, name, and other data.

chapter twelve

Program work definition and integration using the USAF integrated management system

12.1 The ultimate requirement and program beginnings

The customer's ultimate requirement is his need, a simple sentence or paragraph that succinctly describes the needed product in terms of its effect on other systems, the environment, or both. This need may be stated by a Department of Defense (DoD) customer as a new threat that must be defeated, an ordnance transportation and/or delivery demand, or an undersea war fighting capability, for example. It may be phrased by a commercial concern with appreciation for their hoped-for customers. On the surface it would appear that the fundamental difference is that the commercial company must understand its customer base and their needs well enough to know what will sell, whereas the military contractor is told what is needed by the customer in a request for proposal (RFP). Actually, every healthy DoD contractor is in the same boat with the commercial companies and must be constantly examining marketing possibilities, developing new ideas that their customer base may find useful, and studying their customer's current situation and potential needs. Also, the DoD contractor increasingly must work to understand all of the stakeholders for a given system, possibly even extending to the population at large.

So the phrase "customer need" need not scare away those interested in applying this process to commercial practices. In commercial business, it just may be more difficult for the company to find out what the customer's needs are. Success may require good market research, good intuition, lucky guesses, rapid development, and a very fine pencil for cost and schedule control. Whether the system is of a military or commercial nature, no one must respond to customer needs, however they become aware of them, with an ineffective, *ad hoc*, hit-and-miss process. The planning process included in this chapter can be applied to government or commercial development programs. The first step in this organized process is to understand clearly the customer's need.

In the structured, top-down development model encouraged in this book, this need is expanded into a system operational requirements document, system requirements document, or system specification through requirements analysis, system modeling and simulation, mission and operations analysis, environmental and environmental impact analyses, and logistics and basing analyses early in program phasing. In addition to the system requirements identified in these early phases, the system architecture is also defined based on the functionality needed to satisfy the need, and it is used to fashion a product work breakdown structure (WBS), a hierarchical arrangement of product material and services cost elements. A DoD program may use some derivative of canceled MIL-STD-881 as a guide for the WBS, but the evolving product architecture based on allocation of needed functionality should be used as the principal input for the WBS. The customer's needed functionality must drive the system architecture, and therefore the WBS, and not the finance community and their inflexible cost-tracking computer implementations.

The program planning work must begin in the proposal period, or early marketing efforts in a commercial situation, and continue during program execution. Where a proposal is required, the management volume or section should reflect much of the content of the program plan. In some cases, the contractor will be required to provide a formal program plan, system engineering management plan (SEMP), and/or integrated master plan (IMP) with the proposal in draft, preliminary, or final form. In the commercial situation, a new product is a good reason to frame a new development plan or update an old one for new conditions, however simply it may be stated. Granted, there are some commercial product lines richly dependent on serendipity exposed through basic scientific research, an uninhibited creative mind, or good intuition where a degree of nonstructure may be conducive to success. Even in these situations, once the product possibility reaches the conscious mind, its development and distribution can fit into the pattern described here with some tailoring.

At the time this book was written the world was unwinding from several decades of fierce competition between the East and West involving an intense system development rivalry to develop the unstoppable weapon system and to counter the other side's systems. This competition energized a search for technology that sometimes resulted in new systems requiring a clean sheet of paper approach, with several facets pushing the state of the art simultaneously. The systems approach matured from this clean sheet of paper environment, but it can be applied effectively on programs entailing mild to massive reengineering or modification of existing systems to accommodate new or changed conditions.

In these cases the system engineering products described in this book may or may not have been created when the system was new, or they may have long since been lost or disposed of if they were. So even if a world-class system engineering job had been done on the original system, the results of that work might not be available for use in changing that system.

In these situations, what can be made clear is the boundaries of the system in terms of its architecture. We can then determine what new functionality is needed and which of the elements of the system architecture must be modified, deleted, added, or replaced to achieve the new goals set for the changed system. Once this is known, we can apply essentially the same system engineering process to the development of the affected items as we would apply to the elements of a new system. It may be a case of working from the middle out rather than from the top down, but it can be done in an orderly process following the tenets of this book.

Several years before this book was written, U.S. Air Force customers had been required to follow MIL-STD-499A in the implementation of a system engineering program. This standard required contractors to develop a System Engineering Management Plan (SEMP) telling how they intended to perform the technical and management activities required by the contract. Some contractors wrote these plans as extensions of the proposal in the same sales language used there rather than as a plan for their own guidance in program execution. This pattern of behavior was a mistake. Whether a customer requires a management plan or not, your company needs one, and it should be written by your company for your company based on a marriage of your customer's needs and your capabilities. This is a fairly complex integration process, linking customer needs and contractor capabilities, composed of many specialized component parts.

In this chapter we will explore the use of an expansion of a U.S. Air Force initiative referred to as *integrated management system*. It is a variation of the planning process described in MIL-STD-499B (rejected by Department of Defense prior to approval) that encourages the development of an Integrated Master Plan connected to generic planning data rather than a SEMP. The reader interested in commercial markets should not become disenchanted because we are going to use a U.S. Air Force model. The fact is that the Air Force has really seized on an excellent planning and management model applicable to almost any development situation.

Unfortunately, the Air Force did not coordinate its model with contractor generic planning structures. Why should they have done so, and how could they have foreseen a structure that all of their contractors would be comfortable with? To complete the picture from the contractor's perspective, we will glue onto the Air Force model an internal contractor generic planning capability and continuous process improvement module. We will also merge this with the planning requirements of the integrated management system.

In applying the resulting planning system, the contractor or commercial company first must understand themselves, their capabilities, and their best practices, leading to good customer product value. Next, for each program, they must accomplish a transform between this knowledge of their capability and an expression of the program plan in terms of the customer's need. Then this plan must be well executed. Finally, you must take advantage of lessons learned from each program execution to continuously improve your generic self. This provides an integrated environment within which to attain and maintain a world-class capability in your product line. System integration plays a central role in realizing this goal because we must accomplish our work through the integration of the work products of many specialized people both in program planning and execution of that planning.

12.2   Program plan tree

We would all doubtless accept that a specification tree is necessary on a large program to introduce order into the product requirements development effort. What is not as universally accepted is a similar tree for program plans that capture the requirements for the development and production process. All too often programs are implemented allowing planning documentation to be autonomously prepared by the several functional departments (engineering, manufacturing, finance, etc.) contributing work to a program. These plans may be generic company plans or procedure manuals applied to the program or specifically written for the program. Many companies become caught in the trap of trying to be totally responsive to every customer's initially stated process requirements to such an extreme that they redesign themselves for each customer in terms of these plans. A forward-looking company will apply a continuous process improvement concept to its generic procedures in combination with a rigorous customer procedures tailoring effort on each program to take advantage of the practice-practice-practice notion that world-class athletes use on the road to greatness.

Autonomous, functional department planning, disconnected from program requirements, is the target of this chapter. Program process plans should be architected just as the product systems requirements and components should be. Plans should exhibit traceability from the top-level plan down through the lower-tier plans. A structured, top-down planning process will ensure mutual consistency of all of the program plans with a minimum of surprises during program execution. In preparing program plans, it should not be necessary to come up with a new program design for each proposal and program. It should not be necessary to redesign your company for each customer. We need to find out how to apply the practice-practice-practice technique to our work through careful program planning for specific programs and apply continuous process improvement to our methods with a long-term view. Company personnel should be applying the same proven process, incrementally improved in time, to each program and in the process become expert in their specialized disciplines and be able to accomplish needed work at less cost and shorter time frames. This should lead to reduced customer cost and a more competitive position and/or bigger profits.

Very little of the business that a company tries to gain through proposal or marketing efforts involves radically new initiatives. Most of our energy is applied to prospects that are close to our historical product line. Therefore, most of a company's procedures and plans should apply in any new program. This is especially true if your company already applies an energetic continuous process improvement program.

Given that we are organized in a matrix structure as discussed in Chapter 7, our functional departments should have procedures covering how they perform their function on programs as members of cross-functional teams. The program must knit the functional methods into a coherent process appropriate for the particular product system under development as appropriate to the development phase. In this process, programs should not be allowed to substitute alternative processes creatively without acceptance by functional management because the functional departments should be deploying the very best methods they have developed over time based on continuous improvements fed by lessons learned from prior program experiences.

There are, however, two sound reasons for permitting programs to deviate from the current best practices. First, it is through program implementation that improvements can be developed and tested. A particular program may be asked to experiment with a particular technique in defining interfaces, for example. Perhaps the company's history is to use schematic block diagrams and the proposition is that program XYZ will use N-square diagrams instead. The second reason is that the customer may have a valid need for a job to be done differently than our current best practices cover. Perhaps the company uses a particular computer program and related procedures to capture logistics support analysis data. The customer may have a big investment in capturing the data in a different computer data structure and have a perfectly valid reason for needing different data than your system will address. You will simply have to adjust your practices to the customer's in this case or run the risk of loosing the competition. In the process of doing so, you may find improvements that can be woven into your preferred practice, but the suggestion is that this should be the exception, and the rule should be to follow internal procedures while incrementally improving them.

Given that we all accept that programs should be conducted in accordance with prepared plans for each activity, what plans are needed? The program plan can be very simple, giving the overall schedule in very broad terms, stating the customer's need, ground rules and policy, top-level program organization and responsibilities, and reference to other documentation. Figure 12.1 is based on the integrated management system but includes a SEMP and subordinate plans to satisfy the IMP requirement for narrative material covering certain planning areas. The Figure 12.1 plan tree illustrates one way to satisfy the requirements of the integrated management approach. The indicated plans provide requirements for the process that will result in the product system. They should be mutually consistent, and this can be demonstrated by establishing traceability between the plans in the patterns suggested by the plan tree.

These plans, in combination with program schedules, tell the humans populating the program what to do, when to do it, and who should do it. Where do these plans come from? What are the right plans to write? Who should prepare them? As mentioned above, we can allow our functional

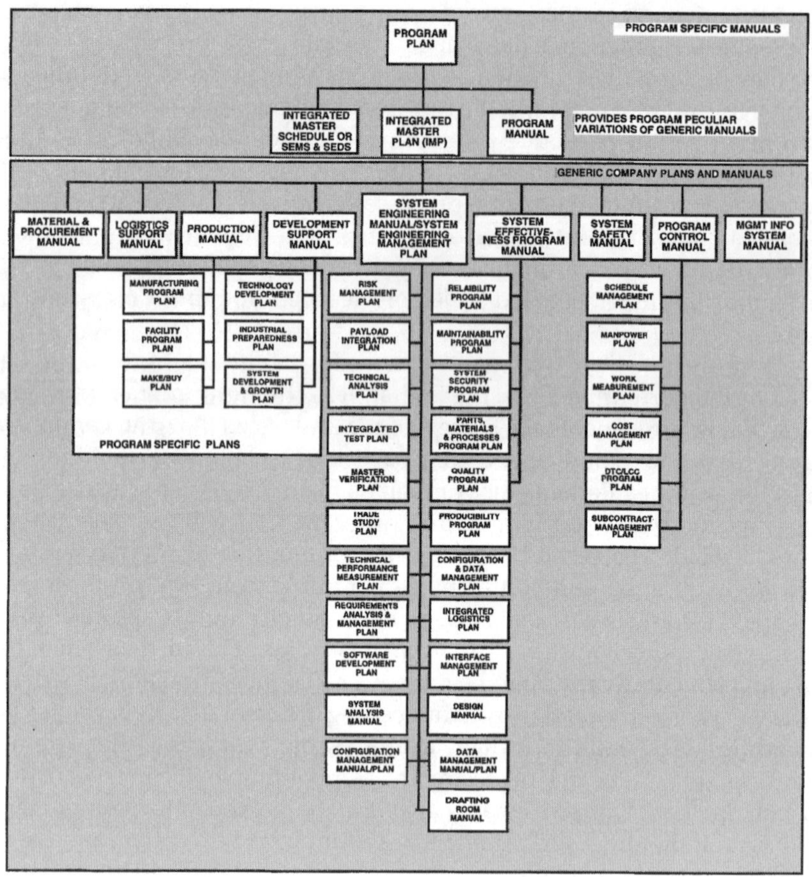

Figure 12.1 Program plan tree example.

departments to autonomously develop program plans and then try to fit them together during program execution. This is a bottom up or grass roots approach to planning. Alternatively, we could approach program implementation planning as we should approach product system development — systematically from the top down.

The fundamental difference between past planning processes and the integrated management system is that the integrated management system defines the work to be accomplished very rigorously linked to the program statement of work and WBS by a common work identification coding system. In the approach suggested in this chapter, generic functional department manuals become detailed narrative descriptions of how the work defined in the IMP will be accomplished. Also, the integrated management system calls for an integrated master schedule (IMS) rather than the system engineering master schedule (SEMS) called for in MIL-STD-499B. These are essentially the same except in name. The advance made by the integrated management system is to absolutely link all work defined in the IMP with the schedules in the IMS using a common work identification system. The remainder of the chapter focuses on this approach.

If you are forced by a DoD customer's requirements to plan a program using an approach excluding the integrated management system, the IMP is simply removed from Figure 12.1, and the tree collapses to the program plan, and the IMS may be renamed the SEMS. You may conclude from this that the IMP is an extraneous layer of planning, but you are encouraged to finish the chapter before making that conclusion a permanent part of your belief system.

In the integrated management system approach, you would not normally prepare all of the plans subordinate to the IMP. The content of all of those plans could be incorporated as part of the IMP in narrative form. They are separated here for two reasons. First, a company may have to respond to a program not involving an IMP/IMS approach, so the book should offer people in that situation this alternative. Second, the principal approach offered in this chapter recognizes a planning approach that makes best use of the limited time available during proposals (by using the generic planning data as narratives referenced in the IMP rather than physically including them there); focuses proposal work on how these data can be blended into a program-specific IMP; and encourages company use of standard procedures, incrementally improved over time, that provide customers best value. You cannot develop the latter unless you are allowed to practice-practice-practice. If you redesign your company for each new contract (new IMP narratives, for example), you will never realize a single identity, and your workforce will not be able to benefit from repetition.

Regardless of whether your common customer base can be expected to call the integrated management system in your RFPs, you are encouraged to develop the planning approach offered in this book. You should, of course, sound out your current and potential customers on this planning approach and get their feedback on acceptability. Generally, you will find them to be delighted with the thoroughness of your approach.

Please understand that the intent is that all of the generic plans subordinate to the IMP in Figure 12.1 are intended to be available as a function of ongoing functional department planning that defines generically how the various tasks in department charters are to be accomplished. On a given proposal and subsequent program work, that basis should already be in place. The proposal work should focus on fashioning a program plan, IMP, and IMS appropriate to the customer's needs, drawing on your standard planning data as narratives referenced in the IMP, and crafting any program-peculiar planning data required. If you have 10 programs in house, for example, each of them would have its own program plan and IMP, but each IMP would reference the same set of functional manuals that tell the people on the programs how to do their tasks.

It is true that some customers will be uncomfortable with your generic planning in that you can change it without their approval after a contract has been awarded. Some customers will want review authority over your internal plans, and that can be a nightmare when you have multiple customers with very different interests. In such cases you may have to run a copy of a generic plan and reidentify it for specific use on a program. This program-specific copy could thereafter be changed with customer review without causing chaos in your internal documentation. This is a slippery slope that will force you to deviate to some extent from the practice-practice-practice notion. If all of your customers take this tack, then you may not have achieved very much with a "standard" process. You must understand your customer base and ensure that your standard process will not offend them. You cannot construct your generic process in a vacuum.

The party line in your proposals and in conversations with your customers should be that you are motivated to continuously develop generic procedures that will provide all of your customers with the best possible value. In cases where the needs to satisfy two or more customers are in conflict, you should consult with those customers as part of the process improvement activity. You should also offer them access to your generic internal planning data on the basis that it is proprietary. Let them act as one of the pathways through which you can detect incompatibility using management by exception. One excellent way to do this is to provide internal read-only access to all planning data via computer network throughout your facility and extend this access to your customer base either at their facilities or only at your own.

In any case, the specific work required for each customer's program would be clearly defined in their program-unique and customer-approved program plan, WBS, SOW, IMP, and IMS. These documents would reference the generic planning data that only tells how to do these tasks. The principal obstacle in implementing this approach is, of course, that everyone has not done a fine job of documenting their practices. In the author's opinion, this is not a valid basis for rejection of the offered planning method, nor does it represent an impossible barrier. It means that you must begin developing these generic practices now and keep improving them over time. Your competition may yet give you time to improve your performance, because they are very likely in every bit as much difficulty as you are in this respect.

In the case where a company is organized in a projectized fashion, or only deals with a single product line, you still have to acquire your personnel from the same source as everyone else, from the human race. We are all knowledge limited and are forced to specialize. So even if you have no matrix and only a project structure in a company with one or more projects, you will have to accomplish program work using specialists whose work patterns should be standardized, one program to another. This standardization can be captured in the kinds of documents shown in Figure 12.1 under the IMP.

12.3 *Know thyself through generic program planning data*

In the ideal situation, our company will have been involved in a continuous process improvement program for some time, and we will have developed a generic set of good practices coordinated with our tools, personnel knowledge, and skills base. This data should include: (1) a generic process flow diagram that hooks these tasks into relative time (Appendix A of this book), (2) a functional department charter listing all the tasks for which each functional department is responsible for maintaining and improving company technology and capability (Appendix B of this book), (3) a task planning sheet for each charter task (to be explained later), and (4) an integrated enterprise System Engineering Manual or generic System Engineering Management Plan (SEM/SEMP).

The generic process flow diagram provides a framework into which all of the charter task descriptions fit as pieces into a puzzle. In this model, the process diagram is a simple process flow diagram. It could be expressed as a PERT, CPM, or other network diagram; V diagram; generic schedule diagram; or Gantt chart. The time axis could be in some nominal time measure or the complete period of execution equated to 100%. We can picture this diagram as a rubber sheet being stretched or compressed to satisfy a particular customer's needs with some paths being deleted, others possibly added, as a function of the contract.

Figure 12.2 offers an example of a generic planning sheet for one of the tasks illustrated on the enterprise process flow diagram. It happens to be a reliability task. Much of this information will have to be changed as it is applied to a specific program, in particular the intensity of task application, but it is helpful to have this information as a beginning point. If you have these sheets available as a generic input to the program planning process, you can save time and introduce your company's standards for process quality into the program planning process commonly accomplished during the proposal development period. If you have no enduring generic planning data now, these forms could be filled in initially as part of your next proposal task estimate and later iteratively improved upon in each subsequent proposal process using the incremental continuous process improvement notion.

Some of the charter tasks expressed on the generic process diagram will entail personnel from only one department working alone. Many tasks will require cooperative effort on the part of personnel from two or more departments, depending on how you are organized and the business in which you are involved. Therefore, the atomic structure of generic task definition includes a task number and a department (or specialty) number. The generic task described in Figure 12.2 would be referred to as F4111471-2164. This identification number appears quite long due to the number of process layers included.

With generic planning data in hand, we can apply this data in a Tinker Toy or Erector set fashion to particular customer needs expressed in their request for proposal. What we need is an efficient transform process between generic planning data and specific program planning data. If you look at the planning data prepared in many companies, it includes a mix of functional department and product-oriented data. Our transform process must map the generic planning data into a program and product context understandable to

FUNCT. TASK ID	F41114471
DEPARTMENT	216-4, RAM Engineering
TASK NAME	Reliability Allocations (Task 202)
PROCEDURE REF	Company Procedure 42-8; MIL-STD-756B (tailored); *Reliability Engineering*, ARINC, Prentice Hall, Chapter 6.
OBJECTIVE	Define a quantitative reliability figure for each system item.
DESCRIPTION	The analyst studies the planned design concepts provided by the IPT designer(s) and allocates the system or lower-tier failure rate to subelements from top to bottom in each architectural layer and branch based on available historical data or predictions determined by the item composition.
SIGNIFICANT ACCOMPLISHMENT	Reliability figures assigned to all system items to a depth defined for the program.
COMPLETION CRITERIA	The task is complete when every configuration item, procured component, and other items as defined on the program have been assigned a reliability figure and the complete set of data has been checked for internal consistency, customer requirements compliance, and been approved by the PIT.
EVENTS	Model structure complete at SDR, allocations loaded by PDR, and predictions loaded by CDR.
RESOURCES	
TOOLS	Five networked IBM compatible computers each running RAMEASY, Microsoft Word 5.0, and Microsoft Excel for first two quarters. Thereafter, two machines are adequate. Network server memory space 50mb.
PERSONNEL	At peak work load, six experienced reliability analysts skilled in failure rate prediction and reliability math and familiar with equipment used in missile systems.
FACILITIES	Each analyst requires a standard company engineer's workstation. Personnel will be assigned to teams, with two assigned to PIT and four to product teams, one per team during the first two quarters of the program. Thereafter, two analysts on PIT only.
INPUTS	1. Design concept information from item designer 2. System architecture definition
OUTPUTS	1. Reliability Model complete with a quantitative reliability number for each configuration item and procured item. Data users include: IPT as a source of reliability requirements, Availability analyst as a source of reliability data for computing availability

BUDGET EST PER QUARTER	1	2	3	4	5	6	7	8	9
HEADS	2	5	5	5	2	2	2	2	1
MAN-HOURS	1008	2520	2520	2520	1008	1008	1008	1008	504

Figure 12.2   Sample functional task definition form.
```
```

our customers and valuable to us as a basis for managing the program through product-oriented integrated product development teams.

We have assumed throughout this discussion that the company is organized into a matrix structure. You might not be organized in a matrix, but you will still need to focus on the many specialties the technology base associated with your product line demands. The advantage of the matrix is that the functional organization can provide process continuity and a continuous improvement focus, while the program organization runs the programs using resources derived from the functional departments. The danger occurs when the functional organization is allowed to enter the program work supervision path.

12.4   Integrated management system overview

During the late 1980s and early 1990s, the U.S. Air Force, working with several contractors on several major programs including the F-22 Fighter Aircraft Program, evolved an extremely important acquisition management methodology called *integrated management system*. One of the beautiful aspects of this Air Force integrated planning initiative is that it strips away a buildup of past confusion and shines a brilliant light on the important things in the program planning process. The fundamental notion is that we should first understand the requirements for the product system and then apply a structured approach to design a program that will produce a system compliant with those product requirements. The program design is captured in a system specification, work breakdown structure dictionary, statement of work, a list of deliverable data, a set of program plans, and program scheduling data. The contents of the plans are requirements for performance of the process that will produce the product defined by a customer system specification. There should be no work planned or performed on a program that does not contribute in a perfectly clear way to satisfying the customer's product needs expressed in their system specification. This shockingly simple concept is, sadly, not always realized in practice.

This methodology recognizes six fundamental documents that collectively contain the requirements for the product system and the program through which the product will be created. Figure 12.3 illustrates the traceability relationship between these documents from the ultimate requirement, the need. The system specification should be driven by the customer's need, and it should recognize a particular top-level architecture that is expanded into a work breakdown structure (WBS) dictionary for the purpose of structuring the program for management purposes. A statement of work (SOW) defines the work that must be accomplished for the product system and each WBS element of the system. The required work must then be planned in an integrated master plan (IMP) that fits the work elements into a framework of major program events. Finally, the planned work hooked to major program events is scheduled in time in the integrated master schedule (IMS). The system specification, SOW, and IMP all represent the top ends of trees of specifications, statements of work (including supplier SOWs), and plans, respectively.

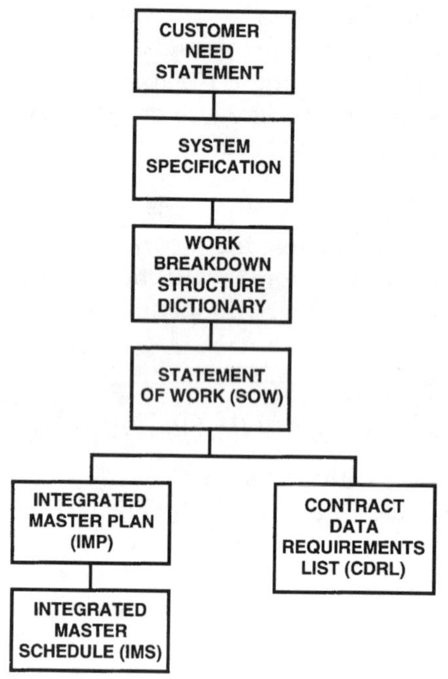

Figure 12.3   Program planning document stream.

Some large system customers, like DoD, use a device called a *contract data requirements list* (CDRL) to define their requirements for formal delivery of data about the product and the process for creating it. This document lists each item of data that must be delivered and tells what format it must be in (referencing a data item description, or DID), when it must be delivered, to whom it must be delivered, and how many copies are required. These items are commonly tied to paragraphs in the statement of work and WBS numbers for management purposes. Some of these information products will be part of the delivered product, such as technical orders that customer personnel will use to understand how to operate and maintain the product system. Other data items, such as cost schedule control system reports, meeting minutes, and schedule updates, will be focused on reporting progress in the product system development and manufacturing effort.

12.5   Generating the six primary documents

12.5.1   The system specification

Depending on the program phase we are discussing, the system specification may be available to us or may not yet exist. Preparing the system specification may be one of the program tasks if the contract involves a very early program phase. More likely, the contractor will receive with a request for proposal some kind of requirements document: such as a system requirements document, operational requirements document, or draft system specification from the customer. This input will have to be completed or transformed into a system specification to the mutual satisfaction of the customer and contractor. Commonly, this transform happens during the proposal process with a system requirements review (SRR) scheduled very early in the program execution. Refer to the book *System Requirements Analysis* by the same author for a description of a process that will result in a quality system specification. It often happens that the customer's original need has become lost by the time they are ready to let a contract for the development of their system. Sometimes a need statement is never phrased, but even when it is prepared, it is not uncommon for it to have been so thoroughly digested in the process of developing the system requirements that it simply passes from view. Some people would maintain that once you have the system requirements defined, you do not need one anyway. The author disagrees. The need is the ultimate requirement and, if it truly represents the customer's need, it offers useful guidance throughout the development process.

If you cannot find a need statement in materials provided by the customer, you should ask them for it. If they cannot or will not produce one, write one and get customer acceptance. This may be very difficult, especially if the customer has many faces, including one or more users and a procurement agency. However, just because it is difficult, you should not conclude that it is not needed. Quite the contrary, the harder it is to phrase and gain acceptance of the customer's need, the greater the need to press on to a conclusion. A lot of good work will result that will eliminate many false starts during later program execution. This same policy should be pursued in the definition of system requirements — the harder it is to gain acceptance, the harder you should work to understand the customer's needs and achieve agreement.

All of the requirements in all of the lower-tier product requirements documents should be traceable from and to this document. All of the requirements in the system specification should be traceable to the customer's need through a logical process of functional decomposition, allocation, and requirements analysis.

12.5.2   *The work breakdown structure (WBS) dictionary*

The WBS has, in the past in the DoD market, all too often been crafted by finance people in industry and government based on one of several MIL-STD-881 appendices reflecting different kinds of systems. Even though MIL-STD-881 encouraged that its appendices be used as guides only, the mindless way it was applied often had a chilling effect on the application of a sound system functional decomposition approach to the development of product systems. Despite the evolution of a conflicting functional system architecture on such a program, the inflexible computer cost management tools used by the customer and contractor finance communities often inhibited alignment of the WBS to the needed functionally derived architecture. Under the Air

<pre>
                    Missile System
                      Air Vehicle
                          Propulsion (Stages 1…n, as required)
                          Payload
                          Airframe
                          Reentry System
                          Guidance & Control Equipment
                          Ordnance Initiation Set
                          Airborne Test Equipment
                          Airborne Training Equipment
                          Auxiliary Equipment
                          Integration, Assembly, Test, and Checkout
                      Command and Launch Equipment
                      System Engineering/Management
                      Systems Test and Evaluation
                      Training
                      Data
                      Peculiar Support Equipment
                      Operational/Site Activation
                      Industrial Facilities
                      Initial Spares and Repair Parts
</pre>

Figure 12.4   Example of a work breakdown structure.

Force initiative, the government has encouraged that the WBS be a responsibility of the system engineering community rather than the finance function. MIL-STD-881 has also been terminated.

The WBS dictionary provides a hierarchical organization of product material and services needed to satisfy the customer's need. The WBS must span the complete system, allowing everything in the system to be placed in some WBS category. The content of the WBS must reflect what is needed to satisfy product system functionality and should not be chosen rigidly and arbitrarily based on some financial model. Figure 12.4 lists the partial content of Appendix C of MIL-STD-881B for missile systems. Only the Air Vehicle, a 2nd-level WBS element, is expanded to the 3rd level.

MIL-STD-881 did not prescribe a numerical coding system for the structure, but one is always assigned. For example, in Figure 12.4 the Air Vehicle might be assigned WBS 1000 and the other 2nd-level items 2000, 3000, and so on. WBS 0000 would be assigned to the whole system in this case. If the system included an Air Vehicle and other things, like a launch site and a final assembly factory, WBS 0000 would be assigned to the complete system and each of these elements, sometimes called segments, would be assigned thousand level WBS codes. Some programs are so complex that prefixes are assigned to permit cost accumulations in different useful patterns. The WBS 012-1000 might be assigned for nonrecurring development of the Air Vehicle (1000), while 014-1000 is assigned to recurring manufacturing of the Air Vehicle. Costs can be accumulated in WBS 1000 across all WBS prefixes for the complete Air Vehicle cost and within prefix 012 for all development cost.

Lower-tier WBS identification is accomplished by using different numbers in the hundreds and/or tens place of the four-digit WBS code. For example, WBS 1100 might be assigned to the first-stage propulsion system, 1200 to the second-stage propulsion system, and 1210 to the rocket engine of the second-stage propulsion system. We have chosen a four-digit code so far, but there is nothing to prevent the selection of a five- or six-digit code. In a very complex system, a decimal expansion could be added, such as 1210.05 for an engine turbo pump. In this fashion, the WBS can identify everything in the product system in a very organized fashion to any level desired. It is not necessary to apply a unique WBS number to each product item throughout the system hierarchy, however. The WBS is a management tool and should give managers insight into the system structure. The WBS is actually (or should be) an overlay on the functionally derived system architecture which should be expanded to the component item level (valves, black boxes, mechanical assemblies, etc.). Not all of this detail is needed for management of the system development. Ideally, the WBS numbering should be the same as the architecture identification numbering, eliminating one unique numbering system.

12.5.3   The statement of work

The next step in the programmatic requirements development process is to determine what work must be performed to develop, design, manufacture, test, and deploy every product element depicted in the WBS. Every bit of work we perform on a development contract should be included in the statement of work (SOW) at some level of detail and should be traceable to the product requirements in the system specification. The SOW, therefore, will tell what work must be accomplished for each product WBS element at some level. Only then can the program work be said to flow from the system requirements.

In the past on DoD programs, the SOW has often been prepared by someone in the customer's program office copying customer-created boilerplate SOW material from a similar past program SOW into the new program SOW. The integrated system management approach calls for the contractor to write the SOW based on a WBS derived from the same work that produced the definition of the system contained in system specification. The contractor must decide what work must be done within the context of his plant, personnel base, and product history in order to create a product system that satisfies the provided requirements and to organize the work around the breakdown in the WBS.

Two competing contractors may very well offer the customer two very different statements of work with their proposals because they have different plants, product histories and experiences, and personnel mixes. Neither may necessarily be better than the other, only different for these reasons.

Some customer program offices applying the IMP/IMS approach do not call for a SOW, preferring to pass directly from the WBS to the IMP. The author prefers to recognize the SOW as a progressive planning step, but once it is created, it is simply an indenture within the IMP so it could be set aside as a separate document.

If we have properly identified program work, it should be possible to establish traceability between the paragraphs of the SOW and the paragraphs of the system specification. Where data item description (DID) CMAN 80008A is called by a DoD customer to define the system specification format, the product-oriented SOW paragraphs can be traced to paragraphs under system specification paragraph 3.7, which captures requirements for major items in the system architecture or WBS. Paragraphs under 3.7 should have been initially conceived from an orderly functional decomposition of the customer need and the WBS developed from that same analytical process.

We can picture all of the supplier statements of work strung out from the system SOW in a tree structure branching from the SOW element of Figure 12.3. Work traceability should exist between the process requirements we accept in the applicable documents referenced in the statement of work with our customer and the process requirements we lay upon our suppliers in procurement SOWs. Otherwise, some elements of the complete product delivered by us (containing supplier elements) may not be compliant with our customer's requirements. The Air Force planning initiative brings into focus the need for traceability, not only through the specification tree but through the SOW tree as well, and between a SOW and its companion specification. This means that the complete product and process definition for a system can be unfolded from the customer's need in a structured top-down development effort.

Commonly, the system level SOW is written to cover all of the work in several WBS indentures. We prepare supplier statements of work for major suppliers, but the system SOW commonly provides the only work definition coverage for the prime contractor. The integrated product team (IPT), or concurrent engineering, paradigm suggests an interesting alternative to this arrangement. We could prepare the system level SOW to cover only the system level work under the responsibility of the PIT and write internal SOWs for each major system element identified in the WBS and to which an IPT will be assigned. These internal SOWs would each define the work that must be accomplished by one of the teams. The product element would be covered by a specification defining the product requirements, and the team SOW would define the process or work requirements that must be accomplished to satisfy the corresponding product requirements.

This arrangement results in product requirements and work definition documents aligned perfectly with the product system elements and the development responsibility definition and should result in great precision in management of the program. In order to be successful in this approach, you have to have thoroughly studied the customer's need and requirements and decomposed their need into a stable architecture that can be assigned to product teams. Instability in this whole structure can result in a tremendous amount of parasitic work.

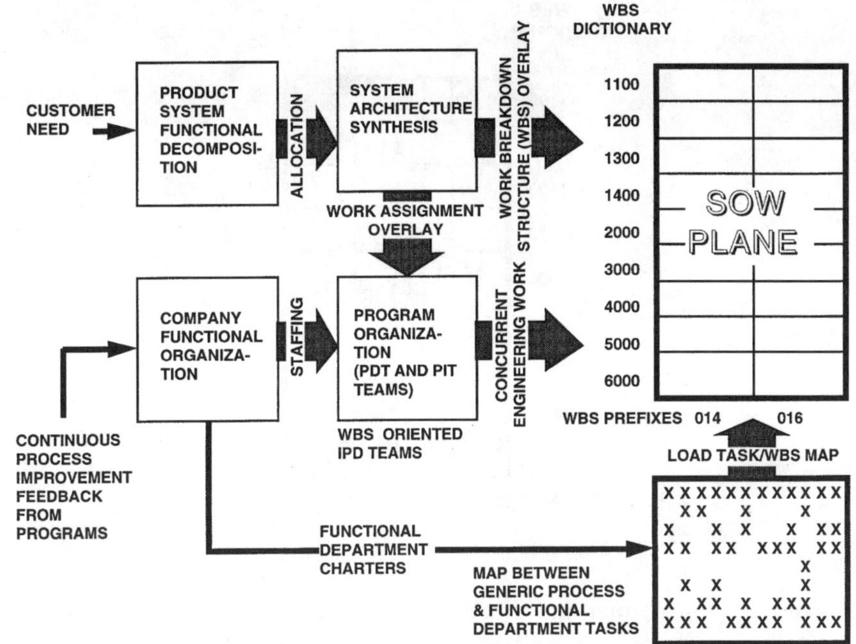

Figure 12.5   Initial SOW task loading.

The SOW, as the name implies, identifies work that must be accomplished at a high level to provide the items identified in the WBS. Now, how do we identify or describe the work that must be done in detail? A three-step process is suggested.

The first step in transforming generic planning data into the specific program plan for a given program is to map the generic tasks to the WBS. Figure 12.5 illustrates this process. It can be accomplished in a top-down or bottom-up fashion and be accomplished at any level of management desired either by the program team or functional management. Two alternatives for these three planning process components are listed below under headings in the form Development Direction/Planning Level/Planner, and the reader can imagine other possible combinations:

a. *Bottom-up/department chief level/chief.* Each functional department chief, or their representative, lists under each WBS, their department tasks that must be performed.
b. *Top-down/upper management level/program planning team.* A program planning team, composed of people from functional departments, accomplishes the map between the functional department charters and the WBS. The planners shop the functional charters for tasks that satisfy the customer's needs expressed by the WBS dictionary.

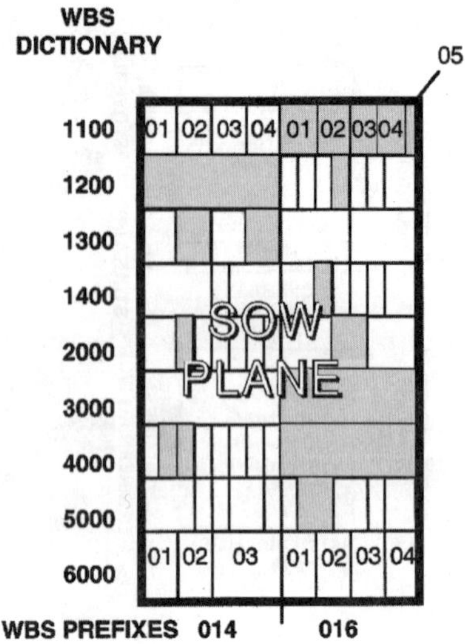

Figure 12.6   The SOW task plane.

Step 2 is to organize the functional task inputs into a program context within the WBS-driven SOW hierarchy. In the author's preferred scenario, no matter how we orchestrated the first step above, the program should form an integrated planning team composed of people from the functional departments assigned to work on the proposal or program initiation work. Ideally, these same people will later be assigned to work on the actual program from those departments.

The planning work for each WBS should be led by the person who will be responsible during program execution for that WBS, if possible. In the case of product-oriented WBS elements, at some level this lead person should become an IPT leader. The planning team members for a given WBS must study the functional charter task sheets (see the sample in Figure 12.2) assigned to their WBS and organize, integrate, or synthesize them into a specific set of major program tasks that must be accomplished to satisfy the work element goal or purpose. The aggregate results of this work forms the SOW plane of the IMP Work Space shown in Figure 12.6. We might conclude, as a result of this planning exercise, that all of the work that must be done for WBS 014-1200 (prefixes not necessarily used in a specific situation) can be achieved within the context of major program SOW tasks 014-1200-01 through 014-1200-07, 016-1200-01, and 016-1200-02, for example.

Some of the detailed functional task sheet inputs may be gold plated, even unnecessary empire-building attempts. Some may be understated or even missing. They must be assembled into the fabric of the program by people experienced in the company's product line, organizational structure, functional department charters, and the needs of the customer. This is an application of ALL CROSS integration seeking out a condition of minimized completeness.

We have to assure ourselves that we have covered all of the necessary work and have not introduced any unnecessary work. For each planning sheet offered, we have to ask, "What would happen if we did not do this work or did it at a lower level of intensity?" For each WBS element we have to assemble the complete stack of planning sheets (physically or in our computer) and ask, "Have we covered everything that needs to be done to satisfy this work element?" There is, of course, no substitute for successful program planning experience in this integration process, no matter what kind of whiz-bang computer system you are fortunate enough to have.

Prior to kicking off this SOW task planning exercise, the WBS should be well developed and made available to everyone involved in the work planning exercise. It is a great source of confusion when the WBS is being changed twice a day in the middle of the work planning activity. Of course, the WBS may have to change based on new knowledge developed in this process, but the more stability that can be introduced into the WBS prior to beginning the more detailed SOW planning work the better.

Very little of the planning information shown in Figure 12.2 need be included in the SOW, but it is needed for the overall planning process, including budget determination (cost estimating), detailed program planning, and program facilitization, staffing, and equipping. Planning sheets like that shown in Figure 12.2 received from the detailed task planners form the atomic structure of the detailed planning activity that will result in the IMP and IMS. These inputs must first be organized within the WBS, as discussed above, and woven into an understandable presentation in the SOW. Figure 12.7 includes a sample SOW fragment framed in the integrated system management style organized not by paragraph numbers but by an expanding work ID number formed from the WBS prefix (if used), WBS number, and SOW Task number.

These SOW tasks are fairly generic in nature and are appropriate for a wide range of programs. Note that in the two alternative mapping processes we discussed above you could have used a top-down list like this created by the program prior to the functional department loading of the SOW tasks. The alternative is that you allow the functional departments to load their tasks into the WBS numbers and integrate the result to gain insight into the SOW tasks by combining and unifying. The author prefers the former method, but no matter how you gained insight into them, they are the program tasks to which we have mapped all of the functional charter tasks.

Twenty different departments may have mapped a charter task to SOW task 014-1200-01. A system engineering department could have indicated they will coordinate the requirements development process for the item/team. Fifteen different specialty engineering departments (reliability, maintainability, mass properties, design-to-cost, etc.) may have indicated they must develop a particular requirement for the item. Functional charter task F41114471-2164, displayed in Figure 12.2, could be one of these. A manufacturing engineering organization and logistics department will both cooperate in development of requirements that are consistent with manufacturing

014	Nonrecurring work.
014-1200	Nonrecurring upper-stage development work.
014-1200-01	Define upper-stage requirements. Conduct analyses to define appropriate requirements for the upper-stage based on system requirements and the evolving system concept.
014-1200-02	Review and approve upper-stage requirements. Subject the requirements to formal review and approval prior to design authorization.
014-1200-03	Design upper stage. Develop a preliminary and detailed design fully compliant with item requirements.
014-1200-04	Provide design test and analysis support. Perform analyses and tests for the purpose of validating the adequacy of the design concept.
014-1200-05	Review and approve design. Subject the design in preliminary and detailed stages of maturity to a formal review by qualified specialists and approve same with possible redirection.
014-1200-06	Verify design by analysis. Analyze upper-stage design for compliance with requirements.
014-1200-07	Verify design by test. Conduct testing to verify that the design complies with item requirements.
014-1200-08	Identify material sources. Define the source of each item of material needed to manufacture the item (make/buy).
014-1200-09	Develop manufacturing process. Define facilitization, tooling and test equipment, and personnel needs.
014-1200-10	Develop inspection process. Define inspection steps necessary in coordination with manufacturing process definition.
016-1200	Recurring upper-stage work.
016-1200-01	Acquire and distribute material. Execute material acquisition plans, stock received material, and make available to the production process.
016-1200-02	Manufacture, assemble, and test upper stage. Perform all necessary steps to assemble materials defined in engineering drawings in accordance with manufacturing planning and inspection data. Test upper stage in accordance with approved written instructions for the purpose of assuring item readiness for acceptance by the customer.
016-1200-02	Deliver upper stage. Package and ship item to launch site.

Figure 12.7   Sample SOW fragment.

and system operation needs. A test and evaluation department will have to define verification requirements. Finally, an operations analysis department may have to perform a special simulation to define the most cost-effective guidance accuracy allocation for the item. In the case of each line item in this list, we should find one functional department identified with principal responsibility (the bold Xs in Table 12.1).

In the process of combining all of these functional department charter tasks into SOW task 014-1200-01, we have also defined the specialists who must be brought together in a team environment to concurrently develop the requirements for the item. The WBS manager or IPT leader who performed the task integration process will have derived knowledge of his/her

Table 12.1   Task Responsibility Matrix Fragment

SOW task	Task name	Stat	Team			Functional dept						
			1	2	3	A	B	C	D	E	F	G
014-1200-01	Define upper-stage requirements	C	X					X	X		X	
014-1200-02	Review and approve upper-stage requirements	C	X				X					
014-1200-03	Design upper stage	A	X				X		X	X	X	X
014-1200-04	Provide design test and analysis support	A	X									
014-1200-05	Review and approve design		X				X					
014-1200-06	Verify design by analysis		X					X		X		
014-1200-07	Verify design by test		X				X					X
014-1200-08	Identify material sources		X								X	
014-1200-09	Develop manufacturing process		X					X				
014-1200-10	Develop inspection process		X									X

team composition as a function of the planning experience. Table 12.1 illustrates a fragment of a SOW task responsibility matrix that we could now begin to build. The matrix will be useful in controlling program implementation. Each SOW task is identified by number and name (using a subset of those identified in Figure 12.7), followed by the task status (STAT) column, IPT responsibilities definition (left blank because we discuss it next), and needed specialized functional department participation. Given that we have progressed in the program to the point that requirements are complete and approved, the "C" in the Stat column would convey that condition. The "A" in the design row would indicate that the team may perform design work and supporting analysis and test work.

12.5.4   Integrated master plan and schedule

Now that we know how the product system and development process will be organized (WBS) and what work must be done (SOW content expanded by the detailed planning sheets mapped to the SOW tasks), it is necessary to determine who will be responsible (not in the functional organization sense but within the program IPT structure), how it will be done, and when it will be done. We could establish a tree of plans to define these things like that illustrated in Figure 12.1, topped by a program plan flowing down to a system engineering management plan (SEMP) and subplans such as: manufacturing plan, procurement plan, quality assurance plan, etc.

The Air Force integrated management system initiative calls for an integrated master plan (IMP), and that is the pattern we will follow here. This does not mean that a program team should overlook all the other plans shown in Figure 12.1. Many of those are still appropriate as IMP subplans or integral IMP narrative content. The IMP can replace the program plan and SEMP but may not provide the detail needed by manufacturing, quality, and many other functions. The integrated management system initiative recognizes that all IMP content does not fit neatly into the highly organized structure explained here. It also calls for narrative sections where appropriate. The content of the functional lower-tier plans, if prepared, can provide the narrative data encouraged in the IMP.

The author proposes that the enterprise create a generic system engineering manual (SEM) and that the IMP for each program reference this document rather than replicate its content in IMP narratives contained in each program IMP, possibly with differences. Other functional department plans can have the same relationship to other IMP narratives. The author's idea of the ideal situation is that the contractors should have a set of functional department plans or manuals that describe how their chapter tasks are to be accomplished. One of these plans would be a generic SEM/SEMP that would tell how the company accomplishes the system engineering process on all contracts. This generic SEM should be mapped to the content of the preferred customer system engineering standard with possible tailoring clearly defined.

Following this pattern for a given contract/program, the company would write a WBS dictionary/SOW/IMP/IMS document set and in the IMP, where the customer requires narrative materials, refer to the lower-tier functional plans for detailed coverage. This arrangement allows company personnel to follow the same practices on all programs developing and maintaining skills in a single process. Continuous process improvement based on program lessons learned can then be applied to a sound base to move toward the very best possible capability constrained by available resources.

12.5.4.1   Program events definition

The IMP expands on the SOW, telling how and by whom the work will be accomplished, and it provides a means to determine successful work completion. For each SOW task, at some level of WBS indenture, the IMP must contain planning data that identifies a series of major events through which the development of that product WBS will be accomplished. These events are selected with effective program management in mind. We do not want so many events that we are spending all of our time reporting. Nor do we want so few that we do not know what is going on. This is like selecting test points in a system design that allow rapid identification, isolation, and correction of problems. Too many test points cost money unnecessarily and are tedious to use in isolating faults, while too few test points lead to ambiguity in fault isolation. These same extremes apply to program health monitoring systems for management.

The principal event discriminator will be defined by a customer or by your marketing department's determination of a potential date that commercial competitors might be ready to market a competing product. In DoD this is called an *initial operating capability* (IOC) marked by a specific date in the future. This date, or its commercial equivalent, defines a point in time when the program must have completed testing and have produced enough product articles to equip the customer for what they have defined as IOC. This might correspond to two aircraft squadrons, one armored battalion equipped, or production of the first day's assembly line run of new cars.

Your scheduling experts will then have to fit all program activity in between the program beginning time and IOC, with some quality, manufacturing, and logistics events extending beyond IOC. A generic process diagram, drawn on a rubber sheet in our imagination, can be useful as noted in Figure 12.8. The major events on the generic diagram can be pushed and pulled into alignment with the realities uncovered by the scheduling experts.

Depending on the customer, they will insist on a particular stream of major reviews, possibly those included in Table 12.2. Other major events, or milestones, unique to your product line can be added based on the kind of

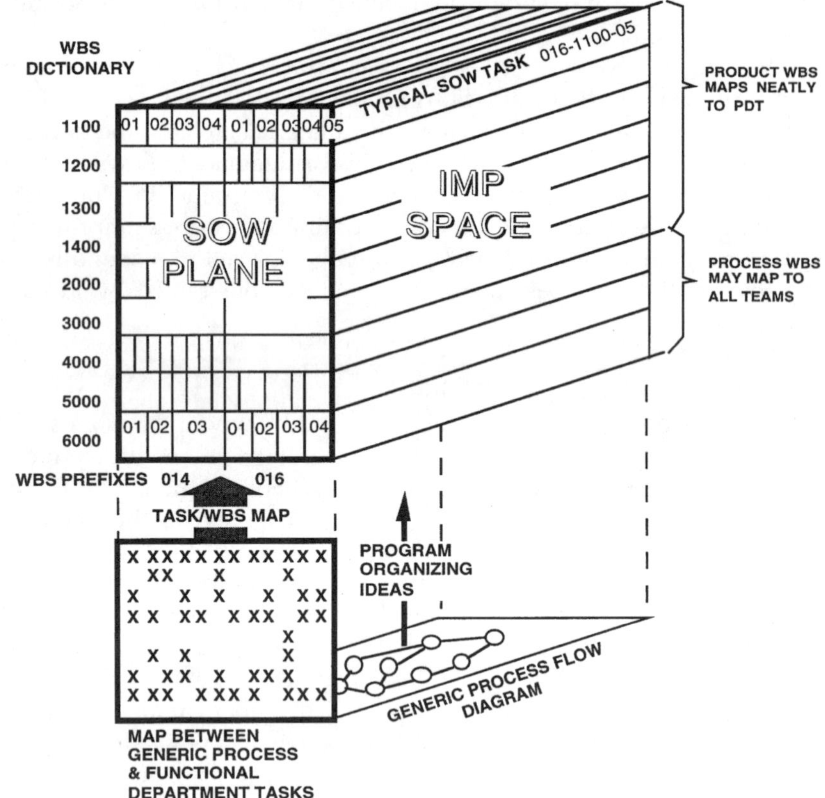

Figure 12.8   Projection of the generic process diagram.

product, maturity of the product development process, and degree of intensity with which customer or company management wishes to manage the program. The list may include only a single program phase or recognize the full sweep of the development, deployment, and operational steps in the development portion of the system life cycle.

The list of program events in time must now be applied to the IMP space as shown on Figure 12.9. The rubber sheet generic process diagram now can be stretched and compressed to conform with the program-specific events and be used as an aid in transforming the rough SOW task planning results into the fine structure provided by IMP task planning.

All of the SOW tasks now need to be laid into this time axis. The team responsible for each SOW task should expand the planning detail for their activities by linking up each SOW task with the events. This breaks up each SOW task into a number of segments like the sample illustrated in Figure 12.9. Some of the periods of time between events will correspond to voids for some SOW tasks as indicated by shading in Figure 12.9, meaning no work is planned for those periods and tasks.

Table 12.2 Major Program Event List

ID	Acro	Event name	Event description
01	SRR	System Requirements Review	Joint understanding of requirements reached.
02	SDR	System Design Review	System requirements validated with risks identified and mitigated to an acceptable level.
03	IRR	Item Requirements Review	Approval of item requirements as a prerequisite to item design work.
04	PDR	Item Preliminary Design Review	Acceptance of readiness to undertake detailed design.
05	PDR	System Preliminary Design Review	All item PDR complete and acceptance of detailed design entry.
06	CDR	Item Critical Design Review	Design acceptance and authority to initiate manufacturing of test articles or low rate production.
07	CDR	System Critical Design Review	All item CDRs complete.
08	FCA	Functional Configuration Audit	Proof that design will satisfy requirements.
09	PCA	Physical Configuration Audit	Proof that product will comply with design, quality, and manufacturing planning.
10	IOC	Initial Operating Capability	Enough resources produced and delivered to provide a complete operating capability at some service level.

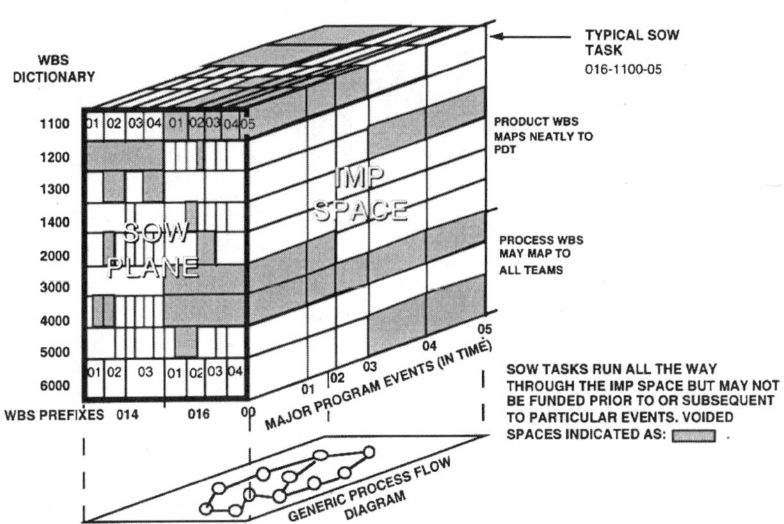

Figure 12.9 IMP space partitioning by event.

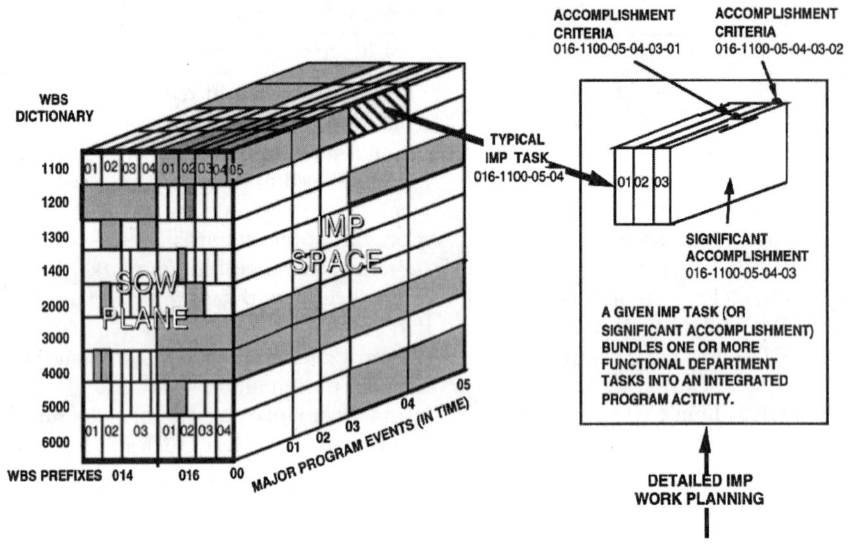

Figure 12.10 IMP task definition.

12.5.4.2 *Final work definition steps*

The work components formed by mapping generic functional tasks into the WBS dictionary, partitioning and combining the work mapped to specific WBS numbers into major SOW tasks, and the time delimitation of those SOW tasks into events provide us with manageable increments of work that are associated with specific WBS numbers, most of which will map to IPT in an unambiguous fashion if we have organized the teams properly to reflect the product structure. To enhance the management potential for these tasks, the integrated system management initiative calls upon the planner to identify one or more significant accomplishments for each of these IMP tasks, and for each accomplishment to identify a measurable accomplishment criteria through which we can objectively determine when the accomplishment can be claimed.

Figure 12.10 illustrates these final IMP task definition parameters. An IMP task is a work increment between two events within a given SOW task. It is composed of one or more subtasks, each of which is characterized by a significant accomplishment statement. Each significant accomplishment must have identified for it one or more unambiguous accomplishment criteria.

Table 12.3 illustrates one simple string in this planning process using a dash-delimited numbering system. Paragraph 3.7.1.2 of the system specification calls for a space launch vehicle upper stage to accomplish some previously defined functionality. WBS 014-1200 was selected to identify upper-stage development work. The SOW has expanded on WBS 014-1200 to include several work tasks needed to satisfy the WBS 014-1200 requirement. Based on an analysis of the evolving IMS (or SEMS), we selected

Table 12.3 Single Planning String Example

Document	Reference	Planning content
SYSTEM SPEC	3.7.1.2	Launch vehicle upper stage
WBS	014-1200	Launch vehicle upper-stage development
SOW TASK	014-1200-01	Accomplish design work needed to define the upper stage configuration.
IMP EVENT	014-1200-01-06	Item critical design review
SIGNIFICANT ACCOMPLISHMENT	014-1200-01-06-01	Design complete
ACCOMPLISHMENT CRITERIA	014-1200-01-06-01-01	95% of drawings released
IMS EVENT/DATE	014-1200-01-06	20 January 1994

completion of design as a critical event through which to manage the program such that we could review design progress before we committed to manufacture of the flight test vehicles. Traditionally in the DoD environment, this event is called the Critical Design Review (CDR). We will accept that the launch vehicle design is complete if 95% of all planned engineering drawings have been formally released at that time. Scheduling, with the concurrence of program management, determines that 20 January 1994 would be the ideal time at which to hold the CDR.

This is only one planning string. It may require thousands of these branching strings to complete the whole planning database for a large program. It quickly becomes obvious that computer technology could be helpful just to capture and organize all of this information. At the time this book was being written, several of the computer tool companies with requirements database tools were beginning to see a potential market for their application to programmatic as well as product requirements development.

All of these documents have a similar structure of a paragraph, WBS, SOW, or IMP task number; title; and text. These tools could be used to capture the specifications, WBS dictionary, SOW, and plans, and include traceability across the current valleys between them as well as generate the documents, or parts thereof, on demand.

When you have linked up all of the work in strings like the one illustrated in Table 10.3, using manual or computerized methods, you should have a mutually consistent network of management data that provides implementation guidance for people doing the work and management test points for effective program health monitoring.

12.5.4.3   Final IMS development

As we are developing the IMP content, we do not obligate ourselves to define scheduling constraints, only to identify the work that must be done. Once the work is defined, scheduling must arrange the tasks corresponding to the significant accomplishments in time, constrained by the events previously located in time. In the process, we will find that some things cannot be accomplished as planned in the available time, requiring adjustments in event timing and the times between events. This may require several iterations before all of the inconsistencies shake out and the complete set of planning data is ready for use.

12.5.4.4 *Planning process summary*

Figure 12.11 combines all of the steps described over the past few pages into a single diagram to better illustrate the intended continuity between the steps outlined. Our approach has been to associate generic functional department work descriptions with WBS numbers defined for the particular product system derived from an understanding of the needed system functionality.

These functional tasks were then combined by the responsible teams into coherent SOW tasks oriented about the product WBS structure. We next determined some major milestones to use in managing program progress and projected these events into a space created by expanding the SOW plane into the time dimension. Finally, we further organized the work content of each IMP task defined by a SOW task and a terminal event into a series of specific significant accomplishments, each with predefined accomplishment criteria.

12.5.5 *Contract data requirements list*

The contract data requirements list (CDRL) identifies every item of data the customer expects to have formally delivered. This should be driven by the work that will be accomplished as defined by the SOW. Every task listed in the SOW that will produce a data item of value to the customer in managing the program or understanding the technical or administrative flow of events, should be referenced there in terms of its CDRL identification, and all of these items collected to form the CDRL. The CDRL items should be determined from the SOW tasks and, once identified, should be fed back into the SOW for reference.

Customers sometimes mindlessly boilerplate the CDRL as they do the SOW rather than deciding in an organized fashion what, of the information the contractor must produce to develop the product and manage the development, they need for the same purposes. Customers realize that formally delivered data costs contractors more than data generated only for internal consumption, so there is a naturally limiting cost boundary at work. The customer should be able to acquire data developed on the program anyway whether the item is on the CDRL or not. This data follows the contractor's format and content definition rather than the customer's, as in a CDRL, and requires a special request on the customer's part to obtain it.

The contractor should maintain an organized list of all program data and make it accessible to all contractor team members in a program library. A DoD customer calls this list (exclusive of CDRL items) a *data accession list*

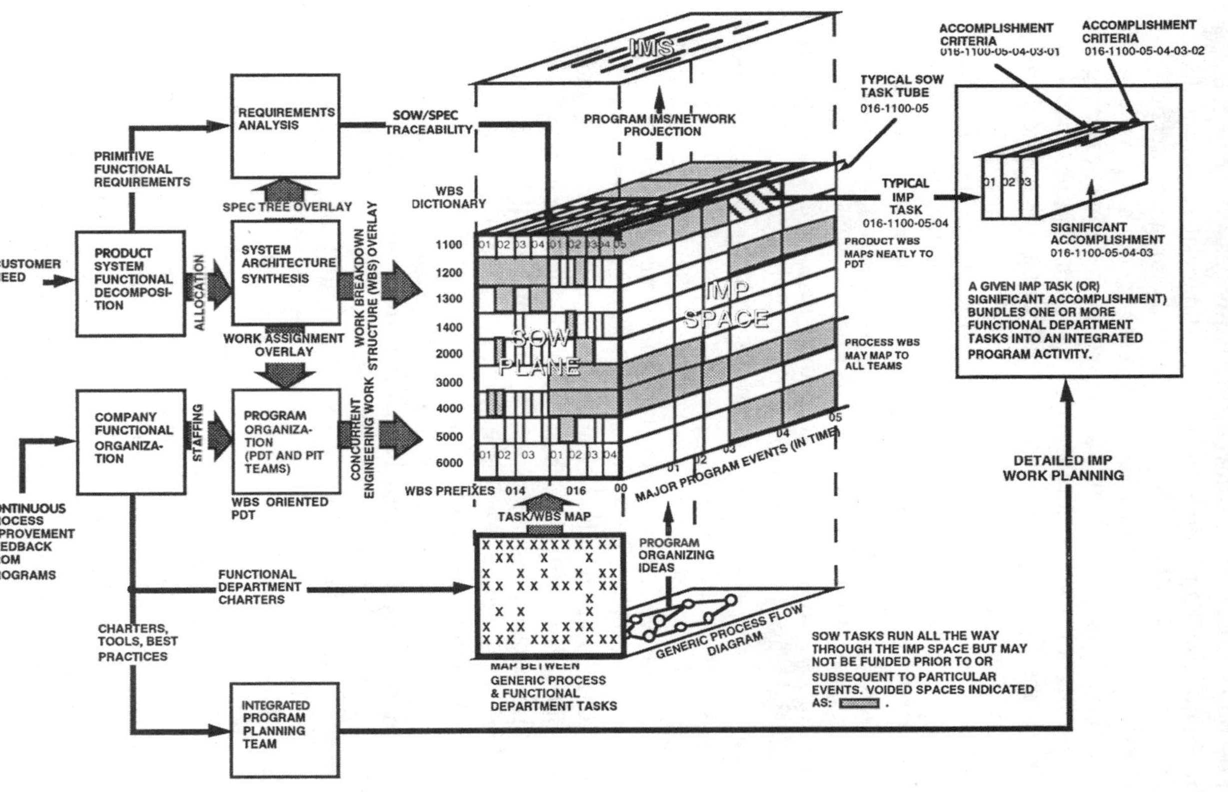

Figure 12.11 Overall program planning scenario.

(DAL). This list names all of the data the customer can acquire by specifically requesting it and paying a separate fee under the terms of the contract. The contractor benefits by having this same data available for internal use rather than it becoming lost in desk drawers and waste baskets.

The CDRL must include schedule and financial reporting data that allows the customer to determine the health of the program. In addition, the customer should wish to receive technical performance measurement (TPM) data that offers insight into how the technical development is progressing in terms of meeting a small number of key system requirements. Together these three items inform the customer about evolving program cost, program schedule, and product performance.

The customer will normally wish to make the system specification, WBS, SOW, CDRL, and IMP contractual documents requiring a formal process (contract or engineering change proposal) to change. The IMS, however, should be a CDRL item because it will have to change as a function of showing progress, if for no other reason. All other CDRL items should be selected with care from the work identified in the SOW to provide the customer insight, at reasonable cost, into the evolving product requirements and design synthesis and into the product development, testing, and production processes.

Ideally, the DAL and CDRL items should be in electronic media stored on a computer network server and accessible from all workstations in the contractor's facility. The contractor can then also quickly respond to a customer request to electronically deliver any item listed in the DAL and in the program library. Very little data is produced today by means other than a computer. Therefore, most items that a customer would be interested in receiving as a CDRL exist in electronic media within the contractor's business. Yet many customers continue to insist on delivery of tons of paper documents that will fill ever-expanding storage space. Alert customers will begin to require delivery of CDRL items in electronic media. This will change the nature of CDRL delivery. Instead of periodic delivery of paper copies of updated documents, the contractor will be required to refresh a customer database at a particular interval or at particular milestones.

The DAL should exist as an electronic data delivery conduit from the program on-line library to a customer. A customer could be charged a periodic rate plus a fee for each call to gain access to this library section. There are some fine precedents for this kind of service in the form of on-line databases open to public access, such as CompuServe. This kind of automated process would result in important savings for large customers like the government. This arrangement will require special provisions in the contract and good discipline on the part of customer personnel to avoid over-running cost targets.

The company that is able to integrate all of this information product into a central program library has a distinct advantage in concurrent engineering because concurrent engineering is, or at least requires, effective communication of ideas. The people working on the project will also never have to worry about using out-of-date paper copies of documents. Incidentally, the facilities people in both customer and contractor ranks will also benefit from fewer file cabinets.

12.6 Work responsibility

The IMP must relate program work to work execution responsibility. First we must decide how we will organize to execute the program. In Chapters 3 and 7 we discussed the organizational structure preferred by the author, namely matrix management characterized by: (1) programmatic integrated product development teams and (2) functional departments that provide qualified people as well as proven tools and procedures to programs. Of course, other forms of organizational structures can successfully execute a program. For small companies with little product line differentiation, forms other than matrix may even be an advantage. This book attempts to focus on a situation in which a matrix is advantageous for the reasons suggested in Chapters 3 and 8.

If we are to follow this pattern, our program work must not be assigned to functional department directors, managers, and chiefs, but rather to responsible IPT. In the process, if we are not careful, we may develop a team structure that is completely unworkable in execution. The team structure must be aligned with the WBS (product architecture) in order to preclude conflicts in assignment of budgets and tasking to teams.

The product teams must be aligned with the product structure reflected in the WBS because the budget will be aligned with the WBS in order to satisfy the cost/schedule control system (C/SCS) criteria required by a DoD customer or an equivalent entity by a non-DoD customer. In 1991 the criteria for DoD contracts was relocated from DoD Instruction 7000.2 to newly released DoD Instruction 5000.2, but remained unchanged in content. The C/SCS criteria requires us to manage budget through intersections in a matrix of WBS and functional organizations.

We wish the teams to align with the WBS so the budget for all team tasks can be simply assigned. The larger the number of crossovers that exist between WBS, teams, and tasks, the more complex the program will be to manage. At the same time, the WBS must align with the functional architecture allocated from needed system functionality This means that the WBS must track the evolving functionally derived architecture reflected in the system specification. This combination will result in the fewest possible crossovers between organizational interfaces and product interfaces, referred to as cross-organizational interfaces, which lead to program problems.

This combination also results in the simplest possible integration task. System integration is a difficult task no matter how expansive. We can do a better job at it if there is less of it to do. Coordination of the organization of the work, product, and performing organization leads in this direction. Figure 12.12 illustrates this point by showing a perfect correlation between a product N-square diagram and an organizational N-square diagram. An

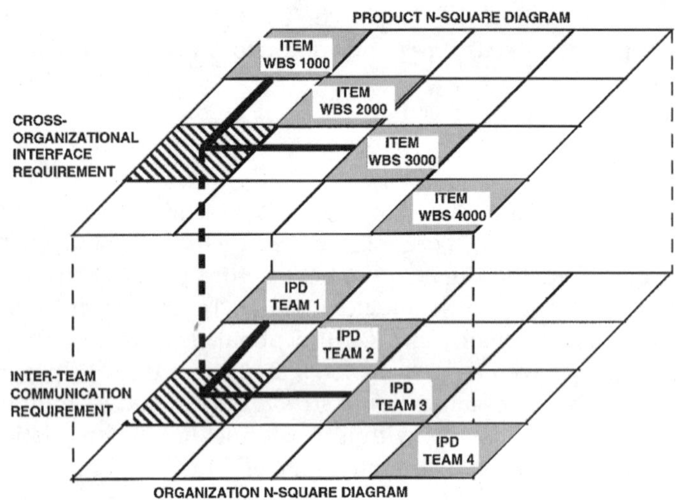

Figure 12.12   Product and teaming N-square alignment.

N-square diagram displays the relationships between N items by noting these relationships in the square matrix intersections. The N items are listed down the diagonal.

On the product N-square diagram we see a requirement for an interface between two specific items identified on the diagonal. On the organization N-square diagram we see a need for the two teams corresponding to the product items joined by this interface to communicate about this interface. If our IPT is organized about the same structure that the product system uses, the team communication patterns will match the product item cross-organizational interfaces perfectly. It is precisely these cross-organizational interfaces that traditionally lead to development problems. If there is a complex relationship between product development responsibility and product composition, there will be interface development problems leading to unnecessary cost and schedule impacts.

Either the SOW or the IMP can cross-reference the work to the program organization team structure. Since the SOW lists all tasks, you might conclude that it would be a better place to locate the task responsibility matrix than the IMP. The IMP selectively expands on SOW content, linking work to specific events and the accomplishments corresponding to those events for management purposes. The author encourages that this matrix be placed in the IMP because the SOW tends to be a simple task list for people with green eyeshades, while the IMP is a plan for us humans. As the product and team structures mature, we must check for emergence of misalignments between the WBS and the team structure, and work toward nulling them out.

Every task identified in the SOW should have some kind of procedural coverage that tells how that task will be performed. This may be in the form of a customer standard (such as MIL-STD-961D), internal company procedures, or commercial standards (such as the ANSI series). This can easily be

Table 12.4   Task/Responsibility/Procedure Matrix

WBS task	Responsible team							Procedural reference
	1	2	3	4	5	6	7	
1000.01	X							MIL-STD-1422 (Tailored)
1000.02		X						Company Practice 128.23E
1000.03			X					MIL-STD-490A
1000.04	X							Company Practice 153.5
1000.05						X		Program Manual 14.24B

defined in a task/procedures matrix or integrated with responsibility assignment into a SOW task/responsibility/procedures matrix, such as the fragment illustrated in Table 12.4.

12.7   Who plans the program?

We have left one thing unanswered up to this point. As a result of our wonderful planning work, we know what must be done, who shall do it, when it shall be done, and how it shall be done. But, who should have accomplished this planning work in the first place? Should we let the functional departments engage in bottom-up, grass roots planning integrated by the program? Should the program or proposal team do all of the planning?

The author believes that few proposal teams or program staffs at the time this book was written accomplished their program planning activities in a purposeful integrated fashion, including defining traceability throughout the product and process requirements stream as suggested in Figure 12.2. If the systems approach (or concurrent development) method is useful in developing product systems, why should we not apply it to the program planning process? We, the program team, are, after all, a system. Let's try it.

First, who should participate? Table 12.5 lists the planning documents discussed above and correlates them with some generic principal functional organization responsibilities. Your company may be organized differently, but these organizational entities are probably fairly widely recognized. While these documents should be developed in approximately the order listed (from top to bottom), we should not permit autonomous work on any of them. There will be insufficient time to sequentially develop these documents during a proposal preparation period of 30 to 60 days. More important, their content must be mutually consistent. Figure 12.13 offers a rough schedule showing the relative timing suggested.

The recommended integrated planning approach involves forming a program planning team with membership of representatives from each functional department noted in Table 12.5. Someone identified by the proposal or program manager should lead the team. All members should be in close proximity, encouraging easy conversation and a close working relationship. They must have available to them good telephone and computer data com-

Table 12.5   Program Integrated Planning Team Responsibilities

Document	Functional responsibility
SYSTEM SPECIFICATION	Systems engineering
WORK BREAKDOWN STRUCTURE	Systems engineering & finance
STATEMENT OF WORK	Systems engineering & IPT Managers or candidates
INTEGRATED MASTER PLAN	Systems engineering
INTEGRATED MASTER SCHEDULE	Scheduling
CONTRACT DATA RQMT LIST	Data management

DOCUMENT	PROGRAM PLANNING PERIOD
SYSTEM SPECIFICATION	<---------->
WORK BREAKDOWN STRUCTURE	<----->
STATEMENT OF WORK	<------>
INTEGRATED MASTER PLAN	<---------->
INTEGRATED MASTER SCHEDULE	<------------->
CONTRACT DATA RQMT LIST	<----->

Figure 12.13   Program planning timeline.

munications facilities as well as adequate wall space for posting information for integrated viewing.

The suggested integrated planning approach is not rocket science any more than is the concurrent or integrated product development approach. It simply requires clear definition of responsibilities, cooperation between the parties, excellent interpersonal and communication skills, and a shared appreciation for the discipline of traceability. The system architecture identified in the system specification must be respected in the WBS. All work must be listed in the SOW and linked to the WBS. All events, accomplishments, and criteria in the IMP must link to the tasks defined in the SOW. All tasks appearing on the IMS must correspond with the SOW and IMP tasks, and all events on the IMS must correspond to those respected in the IMP.

Throughout the planning process, the selected IPT managers must contribute to WBS and SOW development and maintain vigilance for crossovers between WBS, team definition, and work responsibilities. They must also work to develop a cost estimate in a proposal situation. Also in a proposal situation, this team will have to coordinate its work with the management volume writing team.

This planning activity probably cannot be accomplished in a straight-line fashion. While preparing the IMP, we will get insights into changes in the SOW that might ripple through other documents. All team members must have available to them the full content of all the evolving documents and be familiar with that content through an almost continuous conversation between team members and access to the actual text on their computer screen and a stickup on a wall in close proximity.

Team members must respect the hierarchy in Figure 12.2, and the content of the documents should be developed in the sequence illustrated there. Each component of each document expands into lower-tier document details. This pattern repeats through the hierarchy. As this expansion is developed, the traceability links should be captured. The information developed by the team members assigned to each document must be constantly checked for traceability and consistency by someone responsible for planning integration.

Some readers may think that the material in paragraph 12.5 should have been placed prior to the planning material in paragraphs 12.3 and 12.4. At the same time, you have to be familiar with the Air Force initiative before appreciating the opportunities for accomplishing the planning work in an integrated fashion. The author agonized over this dilemma for some time until concluding that the reader should first understand the planning relationships outlined in the Air Force initiative, then be exposed to the integrated planning team concept not explicitly included in the initiative. You should now scan the previous material in this chapter with the perspective of the integrated planning team concept in mind.

12.8 *A generic SEM/SEMP for you*

Some people greeted the workup to release of MIL-STD-499B with screams of, "The sky is falling." You would think that, after many years of contract performance under its predecessor 499A, all DoD contractors would have put in place the fine system engineering activities that they had been describing in the many SEMPs they had written over the years and submitted with proposals. The SEMPs were never contractual and, sadly, often were never opened by the contractor subsequent to winning the contract. So, the adverse reaction to 499B was often based on not then having in place, despite the fine stories told in SEMPs past, an effective systems approach and a concern for how to acquire an effective process in a reasonable time. Many companies dispatched people to the several MIL-STD-499B short courses only to have them return with confirmation that they were in very big trouble.

Whether your company must respond to one standard or another or has no constraints placed on its process, you really should have in place an effective systems approach, because it will result in a product with better value for your customer and more business for your company. If your competitors are able to put in place an effective systems approach and you cannot, you will have great difficulty matching their product cost, schedule, and quality.

As noted earlier, you will find a copy of a complete generic integrated System Engineering Manual/System Engineering Management Plan (SEM/SEMP) attached to the author's prior book, *System Engineering Planning and Enterprise Identity*. This document can be used as a basis for your company SEM/SEMP. The computer disk supplied with the book contains this document in editable format in Microsoft Word.

It does not follow, of course, that by acquiring a good SEM/SEMP that you will become an excellent systems house overnight, though the process of writing one can be quite an education in the right direction. This document must be matched to a current reality within your company. You will need to monitor your performance to this standard and provide the machinery to force compliance on your programs with your own standards. The techniques discussed in this book and elsewhere on continuous process improvement should be applied to your system engineering process to uncover weaknesses and to understand useful priorities for their correction.

You will also need a way to educate your workforce in your system engineering process. This can be done at low cost through brown-bag sessions in your plant for your motivated employees, who happen to be the very ones you most wish to retain. The SEM/SEMP can be used as the textbook for these sessions. You should also cooperate with your local chapter of the International Council on Systems Engineering (INCOSE) and a local college or university to establish or support a system engineering certificate program that qualifies for your company's tuition reimbursement program.

Given that you have a written procedure for performing system engineering, a way to educate/train your workforce in performing that procedure, a means to continuously improve upon that procedure, and a way to encourage compliance with your process, you are on the road to success in performing system engineering effectively. You will also have little to fear from the standards any customer may impose on you.

12.9 Rapid identity documentation

It is possible to maintain a current enterprise identity within an environment with a high rate of change through the use of computers. It is time to defend that claim. The reader familiar with database concepts will have already observed, through the previous material in this chapter, the image of a set of databases that could be used for this purpose — yes, databases rather than word processed documents. The suggested approach is to apply the same tools used for requirements capture to planning document capture. If you are still using word processors for specification preparation, you will be one step behind the suggested process and a few words of database encouragement are in order.

Several computer applications on the market as requirements tools at the time this book was written were adequate for planning information capture, and several of these tool companies were hard at work adapting to enterprise modeling tool sets. Table 12.6 lists several of these tools. These companies have, in the past, been very generous with their time and talent, providing classes the author has taught at universities with tool demonstrations and requirements talks. Identifying them here as potential planning tool sources is a small repayment of that generosity.

Table 12.6 Potential Planning Data Capture Tool List

Product name	Supplier/American agent
RDD-100	Ascent Logic Corp.
RTM	Marconi
Doors	QSS
Core	Vitec
Cassets	Rockwell
Spec Writer	PRC Division of Black and Decker
Document Director	Compliance Automation, Inc.

The advantages of using a database to capture planning data is precisely the same as the advantages of using a database to capture requirements. Modern tools for this purpose will not only permit the storage of and access to requirements or planning data just as documents do (in paper or word processor form), but they will also permit the coordinated capture of other related information that cannot be retained efficiently and economically in paper documents or even in the associated word processor or spreadsheet form. The principal kind of supporting data of interest is called traceability data, but there are some other subtle and powerful benefits that are not so obvious.

You have seen that the statement of work content should be traceable to the system specification, proving that we are doing no work that is not focused on the product and that is not doing what it is required to accomplish. Well, how do we prove to ourselves in a fixed-price world and show our customer that we have accomplished that goal? The discipline of traceability will go a long way toward satisfying this need. In a modern database of the kinds listed in Table 12.6, traceability fields exist that permit us to link up a paragraph in one document with a paragraph of another document with the meaning that one of these pieces of information was derived from the other or is related to the other. In the process of creating and maintaining traceability data in our database, we will find that certain content is not traceable. This should be a signal that we may have included extraneous material in one document or omitted something important in the other. So this discipline will allow you to follow the program logic built into your planning documentation subsequent to preparing it, and it will also encourage the preparation of better planning documentation.

As important as it is, simply providing traceability is not good enough for our purposes. Doing the traceability work will actually slow down the planning process, because it is added work. So how can we accomplish more work faster, allowing us to gain the benefit from traceability while completing the planning work more rapidly? If we have captured our generic planning data in a database and our tool set is properly configured, we should be able to generate the program planning data by mapping the generic work to the program work and generating the program plans and estimates from the resulting map. Unfortunately, this capability was not yet fully implemented in the commercially available tools the author was familiar with at the time this book was written.

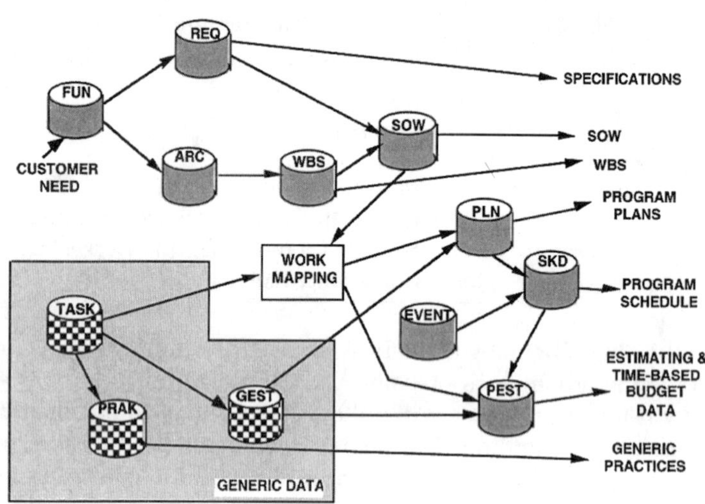

Figure 12.14   Special planning database tool set.

The author has experimented with this approach in his database toolbox, called Rosetta Stone, with encouraging results. Figure 12.14 illustrates a fragment of this toolbox for discussion purposes. The drums represent database structures. Those with a solid body are program oriented, and those with a checkerboard body are generic data systems. The arrows show the flow of work from the customer need through the functional analysis (FUN) tool, the allocation of functionality jointly to the architecture (ARC) and system requirements (REQ), the establishment of the work breakdown structure (WBS) as an overlay of the architecture and statement of work (SOW) content from the WBS, with traceability of work elements to system requirements as a cross-check on validity. Each drum is a database and traceability is retained in related databases as paragraph number pairs from the corresponding documents (in database format). Commercial tools will commonly permit parsing of paragraphs into subsets for traceability purposes to account for poorly written documents with complex paragraphs.

Generic tasks (TASK) are mapped to practices content (PRAK) and generic estimating data (GEST), and this data does not change over time except for the effects of continuous process improvement. The generic tasks are mapped to the very general SOW tasks to form the detailed planning steps (PLN). Generic estimating data is also mapped to the program estimate (PEST) based on the planned tasks. Planned program tasks (PLN) are hooked to program events (EVENT) and linked to the program schedule (SKD).

The program planning data can be created through this process quite rapidly with contributions from many sources into a common database. The planning work can be accomplished by a team formed from functional department personnel assigned to a program planning team, each mapping charter department tasks to the program SOW tasks. If the integrated management system is applied, the planning strings can be formed in the process of this mapping activity and provide the organizing codes, including significant accomplishments and accomplishment criteria drawing explanatory data from the generic data. The SEM/SEMP content would be contained in the PRAK database in this system.

The generic estimating system should be loaded with mean cost data for accomplishing corresponding tasks. As these tasks are mapped to program SOW elements, this cost data should be acted upon by schedule considerations and adjusted, if necessary, for program differences.

It is true that the data in the SOW database, for example, may be generated into a very standard printed SOW document. At any time, anyone may also directly access the networked information in the database encouraged by Figure 12.14. The generation of paper from a database is not time consuming, but the approval, reproduction, and distribution of the paper, originally and for change purposes, can be. With paper documents, you also run the risk that the file cabinets full of these plans contain out-of-date information. The only thing that saves you from errors is that no one reads the information after it is printed.

Direct use of planning database content can speed up the planning and the information access processes tremendously if your staff can become familiar with the use of raw database content rather than paper documentation. It also assures that everyone is using current information. Many of us find it difficult to use database content directly because a printed document provides an information integration function as well as simply making the information available. We find it easy to refer from one part of a document to another in paper form. More and more people, however, are becoming comfortable with direct database use and in time, this will be as easy for many people as you and I might find document use today.

If it is important to have the documents in complete document form rather than using direct database access, the listed tools can generate complete documents that may be loaded into a library hosted on a computer server. These documents will then be available to all via the network. If any one simply cannot use this information efficiently on the screen, he/she may print one out.

Most of the tools listed in Table 12.6 offer the direct database user a form with one paragraph or fragment exposed at a time. The Document Director tool listed in Table 12.6 offers the unique feature of appearing to the user like a document even though the content is in database records. Any of these tools, however, can be used to generate a screen or paper output of a document format from the database content for viewing/reading purposes.

So it is possible to document our current process, change the documentation in accordance with an effective continuous process improvement program, and quickly generate program planning data based on that generic planning data, while incorporating traceability information that ensures completeness and avoids omissions and redundancies.

chapter thirteen

Program work definition and integration using the JOG System Engineering method

13.1 Goals again

In Chapter 11 we stated some important goals to achieve in our planning efforts. These goals were spread across the programs and the functional structures. They were directed at aggregate simplicity because simplicity encourages better management. This is not intended as disrespect for the intelligence of those attracted to management, rather a statement of reality. Simplicity of process encourages more time to focus on real problems where our intellect can be applied to best effect. The mundane will generally take care of itself. Those goals were:

a. Clearly state the responsibilities of the functional departments in terms of providing coordinated quality resources to programs (people, tools, and practices) that are continuously improved in a coordinated way.

b. Clearly state the work necessary on a program to encourage understanding of the work at all levels and minimize the management difficulty, with the result that management intensity can be placed on smoothing current discontinuities and identification and mitigation of potential risks.

c. Minimize the program planning transform complexity so everyone can contribute from a position of knowledge in pouring generic planning data into the program framework, creating the minimum-cost, most effective program plan for best customer value and enterprise profit.

13.2 *Functional structures*

The functional organization is responsible for the generic capability of the enterprise. It is from this resource that all of the enterprise's programs are spawned. As we have said, this involves people, best practices, and tools. The traditional way of organizing a functional organization, adopted in this book, is by the knowledge sets generally also respected by universities. In Chapter 3 we assigned department numbers to these functional groups. Figure B.1 illustrates the structure, and Table B.1 lists the charters for these departments. These department numbers are one of the fundamental organizing markers for the enterprise. The managers of these departments depend on programs to fund the employment of their people, so they need information about program demand for these people. Therefore, programs must organize their work in such a way that it is possible to establish traceability between budgets and the functional departments as noted in Chapter 11.

These functional departments came into being in our hypothetical enterprise by mapping functionality to them from our generic process diagram (illustrated in Appendix A). These departments are responsible for supplying work corresponding to these allocations to programs that implement these process steps or tasks. When a program maps work to a functional department as a result of selecting generic tasks from the process diagram, they bring in people with specific skills and knowledge, tools, and best practices created by the principal responsible department and supporting departments.

These two entities, department numbers and generic task identifications, should be included in the program work planning structures in some way. So the program work identification for functional analysis within a particular IPT should include F41114-D2162. This covers someone from the System Requirements department (D216-2) performing this function (System and Item Requirements Analysis) on a program.

13.3 *Program structures*

An enterprise will generally have more than one program in house at any one time, so we should have some way of differentiating work for one program from another. Over any reasonable period of time there would not be more than 99 in the house so we might use the program identifier PXX, where XX is the specific program, such as P05. By extending this numbering system to the base 60 (all of the numbers, all of the capital letters less O, and all of the lower case except l), this system could identify 3600 programs.

In Chapter 11 we encouraged joining the WBS notion with architecture identification and identification of product-only entities within this structure. Under these ground rules, it should be possible to identify all work in association with the product architecture, an example of which might be A12, corresponding to an on-board avionics system which we assign to IPT

Table 13.1   Program Events

Code	Event name
E01	System Requirements Review
E02	System Design Review
E03	Preliminary Design Review
E04	Critical Design Review
E05	Functional Configuration Audit

2. The functional analysis work done by this team for this item by department D216-2 can be uniquely identified by the string P05-A12-F4111441-D2162. Note that this string ID would be unique to IPT 2 because IPT 2 has been assigned responsibility for item A12, which appears in the string. Team 3 might be assigned responsibility for item A13, for which the same kind of work would be required, but its corresponding string would be P05-A12-F411141-D2162.

Finally, a particular piece of work, especially at the higher levels, may extend over a long enough period of time to warrant partitioning into smaller and more manageable activities. Let us apply the event numbers covered in Chapter 11 to associate work with particular time frames intervening between two events. If we set up the events listed below in Table 13.1, we might designate a particular piece of work as the planning string P05-A12-E03-F4111441-D2162. This is functional analysis work performed by someone from D216-2 as a member of IPT 2 on program 05 leading up to PDR for architecture item A12.

A complete program, P05, may require thousands of these planning strings to define all work, but in so doing we will have achieved the same ends as the USAF method covered in Chapter 12, without the need to cross map so many things to the planning strings. We have used some very fundamental business entities in identifying the work directly. These are identification codes that everyone on the program can be very familiar with and not have to think about relationships between one set for program work identification and other sets used within the enterprise.

While this approach is different from that developed by the USAF, it accomplishes the same purposes, providing unique identification of all work. It does not provide within the codes themselves the significant achievements or completion criteria of the USAF model, but each piece of work identified by a function–department combination should have associated with it a clear statement of achievement expected with completion and a criteria by which one can tell when it is complete. Those two components are not properly part of the work identification string but are attributes of it in the author's view. Another way of saying this is that the parameters captured in the work identification string are all key attributes in the sense of a database that might be used to capture this planning data, and the significant achievements and completion criteria are nonkey attributes.

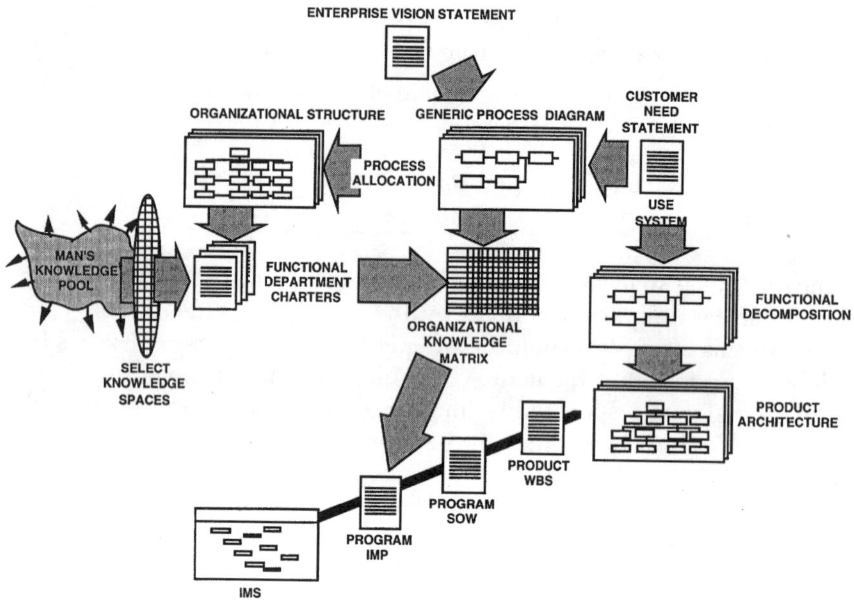

Figure 13.1   Overall work identification flow.

13.4 *Mapping generic identity to program structures*

Figure 13.1 illustrates the flow of work using this identification string concept. The enterprise vision is decomposed to identify needed processes or functionality allocated to functional departments forming charters for those departments. When a customer expresses a need, we decompose that need and allocate functionality to things of which the system shall consist. The major items in the product system become the fundamental planning centers for the program. The program manager determines at what level in the architecture it will manage the program, selects architecture elements at that level to form IPT around, and builds the program finance system around them as well. The finance people are free to call these entities by the term WBS, but they are simply architecture IDs of particular interest from a planning perspective. When we add the program and event identifiers, we have a series of planning entities such as:

        P05-A12-E01   Item A12 Planning Hooked to Event 01
        P05-A12-E02   Item A12 Planning Hooked to Event 02
        P05-A12-E03   Item A12 Planning Hooked to Event 03
        P05-A12-E04   Item A12 Planning Hooked to Event 04

These WBS entities are each expanded into more detailed planning entities by selecting the generic process tasks that relate to the event in question.

THREE LEVEL	FOUR LEVEL	FIVE LEVEL
F411 Grand System Requirements Analysis F412 Preliminary Design F413 Detailed Design	F4111 Define Product System Requirements and Concepts F4112 Define Process System Requirements and Concepts F4113 Manage Requirements at PIT Level	F41111 Product System Concept Development F41112 Product System Operational Environment Definition F41113 Perform Applicable Documents Analysis F41114 System & Item Requirements Analysis F41115 Item Concept Development and Review F41116 Verification Requirements Analysis F41117 Change Released Specifications

Figure 13.2 Task planning levels.

The kind of work that would properly occur between Event 02 (System Design Review) and Event 03 (Preliminary Design Review) for item A12 accomplished by IPT 2 could be characterized as follows:

P05-A12-E03-F4115	Form, Charter, and Train IPT
P05-A12-E03-F41114	Item Requirements Analysis
P05-A12-E03-F41131	Review and Release Specifications
P05-A12-E03-F412	Item Preliminary Design
P05-A12-E03-F4124	Item PDR
P05-A12-E03-F41133	Validation of Product Requirements
P05-A12-E03-F41124	Plan Qualification Verification Process

13.5 *Functional department mapping*

The program planning work covered above can be accomplished from the top down by a small program planning team. The remainder could also, but it is suggested that the functional departments be brought into the process to map their charter elements into the planning entities listed above. Note that it would be possible to plan the program at any level of detail desired. First, we can decide on the architecture depth, selecting either a very low level leading to great complexity in the planning data or plan the program at a high level of architecture. We might choose to plan the work at the end item (A1 air vehicle, for example), subsystem (A12 avionics, for example), or component (A125 on-board computer, for example) level.

Second, we can also choose the generic process intensity used as the basis for planning. We could choose to use a fourth-level task identification level for task F4111, as in the first column of Figure 13.2, a fourth level for task F10 as in the second column, or the fifth level of task identification as in the third column for task F102. For simpler programs, we would want to plan at a fairly high level. A more complex program may require more detail in the planning. Generally, we must fight the urge to plan a program in more detail than is necessary.

The final step, no matter the level at which we choose to do the planning, is to map functional departments to the planning strings crafted above. By way of example, P05-A12-E03-F41114-D2162 represents a complete planning string. All of this planning work would most often be accomplished during the proposal effort. The strings formed up to this point could be crafted from the top down, and people from the different departments which will have to perform work on the program could be brought in to accomplish the detailed estimating and planning. This is accomplished by each functional department representative mapping work to the process steps called out in the preliminary planning strings. These relationships should already have been created in the generic planning so the only real work to do is to estimate how much time will be required to accomplish the work on the program. In Chapter 14 we will take up methods to do this based on tailoring generic estimates for program specific situations.

13.6   *The apparent disappearance of the SOW*

In the USAF numbering system the SOW paragraphs appeared as an intermediate work organizational entity. This was a good idea and has not entirely been dispensed with in the numbering system covered in this chapter. The SOW identifies top-level work for items in the system architecture, and this is identified in the JOGSE structure by calling out high-level functions linked to top-level departments. For example, requirements analysis work is going to have to be done for item A1, which is an immediately subordinate item to system A. The string P05-A1-E01-F111-D200 identifies all of the requirements analysis and concept development work that will be done by the technical departments (D200) between program start and event E01 (requirements review). This string can be introduced into the early planning work as a lure for the program planners to expand by decomposing it to lower-tier function and department identifiers to complete the IMP content.

It may appear that we have dispensed with the WBS as well, but this is not the case. Item A1 is the WBS as well as architecture ID. So we have a way of identifying very top-level work for the new program and a way of expanding the detailed work definition. This capability should be linked to the proposal organizational structure by requiring the PIT to identify the top-level planning strings and personnel from the different functional departments responsible for detailed planning to contribute to the expansion of those strings.

13.7   *Integration of inputs into the IMP*

All of the inputs to this planning process will not represent the real needs of the program. Some department representatives will fail to identify needed work, while others will erroneously or purposely identify work that is not required. All of the inputs have to be reviewed and assembled into aggregate planning string sets and voids and excesses identified and corrected. The total of all of the planning strings arranged in alphanumeric order forms an integrated master plan (IMP) identifying all of the work required on the program. Obviously, the integration work must be accomplished by people with a system perspective from the PIT.

13.8   Integrated master schedule (IMS)

Every element in the IMP must appear on program schedules at some level. The scheduling representative on the PIT should craft the top-level IMS, showing all IPT work and lower-tier work in the context of the high-level planning strings noted above, but not necessarily detailing that work. The responsibility for expanding the schedule data can either be retained in the PIT or delegated to each team, which must craft a team schedule subject to PIT review and approval.

chapter fourteen

Task cost and schedule estimating

14.1 The three principal program planning parameters

Managers focus on managing cost, schedule, and performance on programs. Satisfying all three of these parameters is evidence of great skill on the part of the program manager. There is another parameter of success that we might add that makes it even harder to be classified as a great program manager. Dr. Marty Wartenberg is a former Vice President in Interstate Electronics Corporation and an employee of the University of California, Irvine, at the time of this writing. Marty also teaches program management classes for UC Irvine, and in those classes he likes to ask if any of the managers attending has ever satisfied cost, schedule, and performance requirements on at least one program. He gets a few hands. He then asks if any of these managers succeeded without damaging the company or the people working on those programs. He seldom gets any hands to the second question.

So you may wish to add this to the three traditional indicators of management excellence. It is postulated here that if you follow the suggestions in this book you have a good chance of succeeding even in this more demanding criteria because the people working on the program will witness and be a part of a well-planned program that proceeds in an uninterrupted pathway to success rather than bouncing between many unforeseen, constraining conditions, each one draining away some respect for the program manager. Each of these conditions also often damages the company and some people as well. So "plan the work and work the plan" is still good advice for us all.

14.1.1 Margins of various kinds

The performance of the product is defined in the specifications, and the program should have ways of predicting and measuring the evolving design for compliance with these requirements as well as a thorough verification program for proving it. Technical performance measurement (TPM) offers the manager an important tool for tracking product performance for a limited number of key requirements. The requirements that are important enough to manage through a TPM process often will be selected for assignment of margins to encourage successful management within the context of considerable uncertainty. Margins can be assigned to cost and schedule as well as technical requirements. Cost margins are often collected into a management reserve, while schedule margin is often referred to as float.

All margin approaches work essentially the same way. You determine the required value for something and then split it into two components — a target value plus a margin. You then require people to respond to the target value and withhold the margin as a management reserve to use when problems develop in the program. The overall effect of margins is to place more stress on the designers and other workers. In many cases, they will be able to handle the stress, performing the task in less time or budget than planned or designing an item better than needed. In the inevitable case where some part of the program gets into trouble, the manager can use these margins to overcome the problem by doling them out carefully to encourage a solution to the problem.

In that all margins involve taking something away from innocent people in the beginning and giving it away to guilty persons unable to satisfy their requirements, some jaded system engineers claim that margins follow a typical management technique of identifying the guilty and punishing the innocent. Well, that it may be, but it is a very effective technique and made to work by the tremendous skill and creativity of the majority of people currently working in industry in the development of grand systems.

14.1.2   Cost/schedule control systems (C/SCS)

Cost schedule control systems are often used to coordinate budget and schedule needs and resources on programs. These systems have to be computerized to be effective on large programs because there is so much data involved and because it must be changed from time to time to adjust to changing conditions. These computer systems should capture the planning strings we developed in Chapters 11 through 13 and associate with them man-hour or cost figures corresponding to available budget and time figures telling when the tasks must begin and end to fit in with other work required on the program.

As we begin a particular task on time with the planned budget, our progress is tracked as a result of everyone working on the task turning in their time against the tasks they are working. These inputs are used, of course, to compute the payroll problem for the workers, but also these inputs are charged against the budget in the C/SCS. Over time, these charges are accumulated and reports made to show progress on the task. Figure 14.1 illustrates the machinery of these systems in the form of an accumulative track of man-hours over a planned period of time. Milestone 1 was intended to be achieved at a planned number of man-hours and on a particular date. In this case, the milestone was actually achieved later than planned and at more cost than planned. When the data collected reveals this kind of situation, a

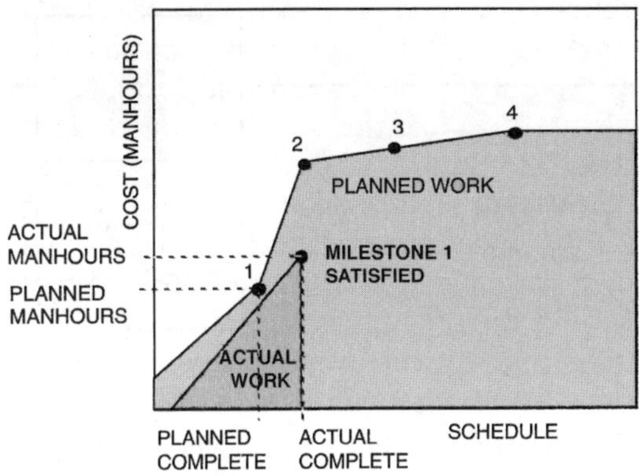

Figure 14.1   C/SCS application.

C/SCS will generally generate a variance report requiring the person responsible for this task to answer some penetrating questions about why this task is overrun in both cost and schedule. The person responsible in this case might claim that he/she started late because of difficulty obtaining staff and the staff was not prepared for the difficulty of the task. In the next cycle leading to milestone 2 this task might be more carefully watched by program management in an effort to cause it to achieve its goals within planned cost and schedule.

The milestones indicated in Figure 14.1 are mini events, of which there may be many between two major program events used in the planning work discussed in prior chapters. In each case, we should have recorded in the planning data a criteria whereby it can be determined when the milestones have been satisfied and they can be claimed.

The cost schedule control system should also include program evaluation and review technique (PERT) or critical path method (CPM) capabilities, permitting identification of the critical path of work through the program. The diagrams that these computer systems create are often called *program networks*. These networks involve tasks and events strung together by directed line segments. Both of these systems started out as activity-on-line depictions. All of the tasks we have identified in program planning become lines on the network with vertexes at related events. Activity-on-node systems have also been developed. While these tools have many powerful features, none is more impressive than their ability to display the critical path between any two nodes. Given two or more parallel work paths between two nodes, the critical path is that string of work that requires the most time to accomplish and is characterized by a zero schedule float or margin.

Ideally, we should be able to manipulate the C/SCS to search for sound management policies for situations that arise. Some people call this playing what-if games with the data. Such changes as we might make in the process should not become a part of the formal record. Only when we have deduced

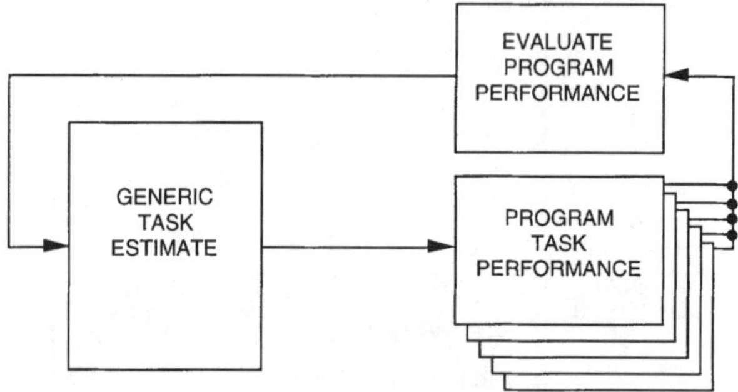

Figure 14.2   Generic estimate improvement.

an advantageous change and it has been approved, should we modify the formal plan.

14.2   Task data sheets

The generic planning data should include information about nominal cost and nominal task time span, and this data should be updated based on actual program performance. Figure 12.2 offers an example of a task definition form that might be used to capture this information. In this case, the task is a reliability allocation activity performed by someone from department D216-4 performing work on function F41114471. The program work identification string for this in the IMP might be P05-A12-E03-A12- F41114471-D2164, using the method explained in Chapter 13. The form includes an estimate of the number of heads and man-hours required to accomplish the task on a nominal program. If, when we planned this particular program, we had concluded it was a nominal program, we would have estimated the job in accordance with the figures on this sheet. If we concluded that the task was less or more demanding, we would have adjusted the estimate accordingly and documented the rationale.

14.3   Estimate improvement

Estimates appearing in generic planning sheets must, at some point, all be estimates analytically formed, but over time the accuracy of these estimates should improve because you readjust them based on actual performance on programs as suggested in Figure 14.2. At some level, the program estimates for tasks should be compared with actuals, and assessments should be made of what any difference between these numbers implies. This is a matter of a search for truth. If, on a particular program, the task is badly overrun, does it mean that the task is much harder than estimated, or was it because it was accomplished by people unskilled in the task, or was it badly managed?

Alternatively, we could simply collect the actuals by task and deal with them without knowledge of the causes. The mean of the actuals is probably close to where the estimated value should be. Obviously, this could all be done by a computer into which has been entered the task estimates and task actuals by program. The next program estimate could be adjusted for all means of all prior task implementations. Where estimate increments are trending up, we should get a warning signal to search for the cause because our performance should get better each time as a function of repetition, with some variance for program difficulty.

We should be seeking predictability in our program planning work. What we estimate during the planning process we should see occurring in actual program performance. Here is where we start to see the big value in generic planning strings. Every program is planned based on the same generic process and planning data, and we can easily capture statistics that have a lasting meaning. If an enterprise plans each program from a different process perspective, it will be very difficult to correlate the results with a planning basis. It will be very difficult to establish a servo loop effect which we can use to null out the differences between planned and actual performance.

As important as it is, it is not enough, of course, just to be able to estimate accurately what a task will require. We should also be working to reduce the cost and schedule demands for tasks, while improving the quality of the work performed. An analogy with rifle target shooting applies here. Two rifle shooters fire 10 rounds at, say, 500 yards. One of the shooters observes that his shot pattern is all over the target. The other shooter sees all 10 rounds within a 10-inch radius on the left side of the target. The second shooter will be able to adjust for wind and move the center of the shot group to the bullseye. The first shooter cannot make any adjustments to improve performance on the second 10 rounds because there is no clear pattern.

So given that we can measure performance in association with a plan and that our performance shows compliance with estimates, we should inquire if there is some way to purposely change the plan and performance so that the customer receives more value. Functional management and the EIT should be constantly reviewing program performance and reaching conclusions about process steps that could be improved. The EIT should be very careful in selecting improvement actions and do so from a systems perspective. Some improvements may have little effect in reducing program cycle time because the associated tasks do not generally appear on a program-critical path. The EIT should manage this process and encourage coordinated improvements implemented by the functional department managers in the resources they provide programs.

chapter fifteen

How-to knowledge access

15.1 Knowledge depth differences

The normal content of company system engineering documents is limited to defining what must be done, the sequence in which it must be done, and possibly why it is done. Few of these manuals include detailed content about how to do this work, at least not to the degree of detail that a textbook on the subject might cover. The system engineer, however, needs access to the how-to information that fills in the gaps in the practices documentation. If the enterprise makes no effort to define the how-to knowledge, this valuable resource is by default locked into the corporate memory of the current staff. As the staff with experience moves on to retirement or departs for other reasons, this corporate memory drifts out of the company.

It is important that the enterprise identify what must be done in the system engineering field on its programs, but it is not enough. You must move on to the next level of detail, provide access to the how-to information, and find some way to make it available to your employees. This book focuses on the system engineering field, but a company can, and should, apply this same model to its how-to knowledge.

15.2 Sources

The how-to knowledge can reside in several different places and not necessarily the same place for all of this information needed by a company. These places include: external standards, textbooks, the company's system engineering manual, training courses, and human memory.

15.2.1 External standards

There are several standards available in the system engineering field: IEEE P1220, EIA 632, and MIL-STD-499 to name three. The first two are current and maintained, while 499 has been discontinued by DoD. All three of these documents contain useful information about the systems approach, but they do not provide in-depth coverage. Rather, they provide a list of things that should be done.

Defense System Management College (DSMC) has published a volume called *System Management* that is used in a course taught by DSMC for program managers. It provides sufficient detail in some areas to qualify as how-to knowledge. This document has been released in several different versions, and some say the original one, developed from the Lockheed Space and Missiles Division system engineering, under the direction of Mr. Bernard Morais then the System Engineering Director for the company and now the President of Synergistic Applications, was the best of the lot.

The U.S. Army published a field manual, FM 770-78, in 1979 titled *System Engineering* that went out of use some time ago. It defined a sound systems approach and could be used as the basis for how-to knowledge.

In general, available, currently maintained system engineering standards are acceptable as a basis for a practices manual telling what must be done, but they contain insufficient detail to act directly as a source of how-to knowledge.

15.2.2   Textbooks

Several fairly recent books have been published that provide good coverage of the overall systems approach adequate for working system engineers to use as a reference source to find out how to do particular jobs. Some of those are:

> *Systems Engineering and Analysis*, B. Blanchard and W. Fabrycky, Prentice Hall, 1990
> *Systems Engineering*, A. Sage, Wiley, 1992
> *System Engineering Management*, 2nd edition, B. Blanchard, Wiley, 1997
> *Systems Engineering Management*, J. Lacy, McGraw-Hill, 1992
> *Systems Engineering: An Approach To Information-Based Design*, George A. Hazelrigg, Prentice Hall, 1996
> *Systems Engineering Guidebook*, James N. Martin, CRC Press, 1997

Other recent books have been published that provide how-to knowledge about subsets of the overall process. Some of those are:

> *System Requirements Analysis*, J. Grady, McGraw-Hill, 1993
> *System Integration*, J. Grady, CRC Press, 1994
> *System Engineering Planning and Enterprise Identity*, J. Grady, CRC Press, 1995
> *System Validation and Verification*, J. Grady, CRC Press, 1997
> *The Art of System Architecting*, E. Rechtin and M. Maier, CRC Press, 1997

15.2.3   Company system engineering manual

Earlier in this chapter, we mentioned that few system engineering manuals in industry contain sufficient detailed information to act as a how-to source, but a company could write one that did. The author has copies of old manuals from General Electric, Martin Marietta, and General Dynamics that did provide detailed how-to information, but companies preparing manuals today tend to model their practice from IEEE 1220 or EIA 632, which do not provide the detailed coverage. In fact, until a few years ago, the author would have encouraged a company to write a manual with considerable depth, probably because the author has always been able to write easily about our profession and failed to consider how much difficulty many others have in doing so. The author has found a better, easier-to-implement, more flexible way of capturing the how-to knowledge, and that is offered in this chapter. If your company has a budding author on the payroll, however, you may be able to satisfy the how-to knowledge need directly in your practices.

15.2.4   Training courses

All of the previous methods of capturing the how-to knowledge require the individual to read and understand the content. Another approach is to use these sources either directly or through suppliers to build one or more training courses that organize the material for more effective learning. These training courses could come from several sources such as:

a. In-house courses prepared and taught by company personnel.
b. In-house courses purchased from a supplier of system engineering courses traceable to the company process description and presented by company personnel.
c. Traveling short courses such as those provided by JOG System Engineering (the author's company), Synergistic Applications, University Consortium For Continuing Education (UCCE), Center For Systems Management, and SMi. These courses are commonly offered in various cities with an aerospace population in close proximity and/or on-site at companies by contract or purchase order.
d. Local university extension courses. This may include a complete system engineering certificate program as it does at University of California, San Diego, Irvine, and Riverside, or isolated courses that may satisfy some subset of your needs. Most university extension departments will provide their program on-site at companies, and UC Irvine has a lot of experience in doing so. By the time this book is published, they will have completed at least six on-site certificate programs in Southern California and as far away as Indiana. A few universities offer a B.S. or M.S. in system engineering, and if your facility happens to be close to such a campus, that university can, over a sufficient period of time, provide an excellent education for your system engineers.

15.2.5   Registered experts

Your company may have people on the payroll who are respected as experts in one or more fields of system engineering who could act as mentors for those who are not. To take full advantage of this resource you need to make a list of these people keyed to the areas of expertise. When someone needs to improve their knowledge in one of these areas, they contact the designated expert. In order for this to work in a large company, a funding method must be defined, such as a work order for the expert to charge when mentoring. Where both parties are working on the same program at the same time, this cost can very likely be charged to normal work on the program because it is not intertwined with normal work. Over time you will lose these experts and have to consciously refresh your source of experts.

Companies that have an active mentoring program, for college new hires especially, have this arrangement in place. Why wouldn't every engineering company have such a program? Managers in the system engineering field commonly are very hostile to hiring recent college graduates, opting instead for people with some experience in an engineering field in industry. The author, while an engineering manager in industry, was forced by company policy to hire people right out of college into a system engineering department, and while unhappy about the policy initially, he had to admit that these people turned out as well as experienced people hired from other companies. Their success was in part due to selecting good talent (avoiding the people who couldn't wait to design circuit boards, for example) and to the way we partitioned the complete system development process into tasks small enough that an intelligent person could master them but large enough to be a whole job and represent a challenge.

15.3 How-to knowledge directory

There are several ways we can get access to how-to knowledge. Now the problem is to make an inventory of the knowledge that is required and link to the source for that knowledge such that all of our system engineers can gain access to it as needed or as directed. The needed knowledge is defined by the functional department charters listed in some detail for our sample company in Table B.1 derived from the functional allocations in Table A.1.

15.3.1 Directory format

Table 15.1 lists a sample of the functional tasks from Table A.1, links them to the responsible department (L for lead in the RESPonsibility column) and contributing departments (C in the RESP column), and identifies the preferred how-to knowledge source. There may be seven acceptable sources of how to do this work, but you have done an adequate service if you identify at least one for each task. The contributing departments have been left out in the interest of space, except for one example case for task F115J.

Your directory could be improved over that shown in Table 15.1 by including separate columns (fields) for the type of source (book, standard, course, expert), and detailed identification of the source within that type that could include a chapter/section reference in the case of a book or standard.

The reader is probably convinced that it will be impossible to keep track of all these tables and the information in them. This probably would be true if they were all separate tables maintained by typewriter or word processor. All of this information should be in a relational database, and the records and fields of the tables should be set up for maintenance by appropriate assigned personnel. Alternatively, the PIT could be the only party allowed to change any content based on an agreement previously reached. Department numbers come from Table B.1.

15.3.2   Directory development responsibility

The knowledge directory must be assembled by someone. The question is who? The functional departments should be the knowledge owners for the enterprise, so the managers of these departments or persons they designate are the correct source of this information. When allocating functionality to departments, we noted that some tasks properly flow to multiple departments at some level of task indenture. In these cases, we have an obligation to identify which of these departments is the principal department, and that department should have the responsibility to prepare the first description of the task, identify any tools to be used by all parties to the task, and determine the how-to reference.

The PIT should act as the owner of the directory as they should for the whole set of information defining the enterprise identity supported by the functional departments as a function of the task responsibilities allocated to them.

15.3.3   How-to knowledge access

The final issue to resolve is to provide access to the knowledge we have identified in our knowledge directory. If we have a company or engineering department library, this is probably the right place for any textbooks or standards we have identified. There should be a clear policy defined and available to everyone on how to gain access to training course resources identified in the directory. Names and phone numbers or email addresses might be adequate for formally identified experts.

Most engineers spend a lot of time at their computer, so this seems like a good place to deliver the how-to knowledge if possible. The SEM can, of course, be placed on-line and hyperlinked to more extensive coverage in the form of references to books, more extensive coverage owned by the enterprise, audio and/or video presentation on the topic, or reference to a course in a local university or other source.

Table 15.1   How-To Knowledge Directory

Task ID	Name	Department number	Resp	How-to reference
F41114	Item Requirements Analysis	216-2	L	Book, SRA, Grady
F411141	Study System/Item Requirements	216-2	L	Book, SRA, Grady
F411142	Select Specification Template	216-2	L	Book, SRA, Grady
F411143	Item Applicable Documents Analysis	216-2	L	Book, SRA, Grady
F411144	Traditional Structured Analysis	216-2	L	Book, SRA, Grady
F4111441	Functional Analysis	216-2	L	Book, SRA, Grady
F4111442	Performance Requirements Analysis	216-2	L	Book, SRA, Grady
F4111443	Timeline Analysis	216-2	L	Book, SRA, Grady
F4111444	Architecture Synthesis	216-2	L	Book, SRA, Grady
F4111445	Expand Interface Analysis	216-2	L	Book, SRA, Grady
F4111446	Process Analysis	216-2	L	Book, SRA, Grady
F4111447	Specialty Engineering Requirements Analysis	216-2	L	Book, SRA, Grady
F41114471	Create Specialty Engineering Requirements Analysis Demand List	216-2	L	Book, SRA, Grady
F41114472	Reliability Requirements Analysis	216-4	L	Book, Reliability, ARINC
F41114473	Availability Requirements Analysis	216-4	L	Book, Sys. Eng. & Analysis, Blanchard & Fabrycky
F41114474	Maintainability Requirements Analysis	216-4	L	Book, Maintainability, Blanchard et al.
F41114475	Material and Processes Requirements Analysis	216-7	L	Expert, R. Burnes
F41114476	Item Cost Requirements Analysis	216-8	L	Book, Eng. Economy
F4111448	Service Use Profile Analysis	216-2	L	Book, SRA, Grady
F4111449	End Item Zoning Analysis	216-2	L	Book, SRA, Grady
F411144A	Interface Requirements Analysis	216-2	L	Book, SRA, Grady
F411144B	Write Requirements Paragraphs	216-2	L	Standard, MIL-STD-961D
F411144C	Lead System/Item Requirements Analysis	216-2	L	
F411144D	Edit/Update Current Content	216-2	L	Book, SRA, Grady
F411144E	Integrate Requirements	216-2	L	Book, SRA, Grady

Table 15.1   How-To Knowledge Directory

Task ID	Name	Department number	Resp	How-to reference
F411145	Item Concept Development and Review	210	L	
F115246	Manage Specialty Engineering Requirements Analysis	216-2	L	Book, SRA, Grady
F411148	Object Oriented Analysis	215	L	Book, OO Modeling & Design, Rumbaugh et al.
F41114A	Publish Specification	216-2	L	
F41121	Plan Quality Process	241	L	Book, Quality Eng., Juran
F41122	Plan Material and Procurement Process	160	L	Expert, D. Canton
F41124	Plan Qualification Verification Process	242	L	Book, V&V, Grady
F41123	Manufacturing Requirements Analysis	223	L	Expert, M. Duggins
F41125	Plan Logistics Process	231	L	Book, Logistics Engineering & Mgmt., Blanchard
F41126	Item Schedule Requirements Analysis	190	L	
F41127	Item Deployment and Operations Planning	231	L	Expert, A. Adams
F41134	Establish & Maintain Product Traceability	216-2	L	Book, SRA, Grady

chapter sixteen

System engineering training program development

16.1 Training program overview

Training programs provide people with how-to knowledge in an organized framework. We discuss ways to prioritize the parts of the process most in need of improvement and focus energy on providing training for those parts using the low-cost lunch-time seminar as an initial means to satisfy the need. Other alternatives, discussed in the previous chapter, are considered, including existing college courses in the area and arrangements that could be made with local universities to create programs that would qualify for company tuition reimbursement. Training solutions of whatever kind should be linked to your how-to knowledge base traceable to the tasks in your generic process, as discussed in Chapter 15.

16.2 Process evaluation and training prioritization

Given that your company is in need of system engineering education, and you have a written system engineering practices document that at least tells what must be done on programs from this perspective, you need to decide where to spend your scarce training funds. No matter how the training will be arranged, it will cost something, and industry in general is very frugal with training money (with a few enlightened exceptions, such as Motorola which allocates training funds to every individual and expects them to be spent each year). To proceed, one should evaluate their process for education priorities. In so doing, you should talk to program chief engineers (by whatever name, the person responsible for program technical matters) and program managers and find out what system engineering work they think could be done better with the most payoff in terms of program success.

Make a prioritized list of system engineering tasks and start trying to match the items on the top of that list with sources of corresponding knowledge in your organization and in books and standards. The granularity of the beginning task list is important. You should start at the top and outline the work definition down to the level that it would take no more than an hour to cover the topic in a classroom situation. It is at this lower tier that you should select tasks for inclusion on your prioritized list.

16.3   The low-cost lunch-time seminar

If you cannot obtain the funds you need to train your complete staff in system engineering, you have to find an alternative that will pay some dividends. At the other extreme in cost from a complete certificate program, one can create a very effective lunch-time seminar (or brown bag seminar where people can bring their lunch) series of lectures. The topics are from the top of your prioritized list. The next step is to find someone who is qualified who will volunteer to teach the topic. The experts we listed in the last chapter is probably the same list we need here.

The mechanics of the program are very simple. You need to reserve a meeting room of adequate size for each lunch time you plan to present a course. You should try to get the best room possible to give the program the illusion of management support whether or not this is true. You might begin by offering one class a week always on the same day at lunch. If your company only has a 45-minute lunch time, you may have to reduce the time from one hour to 45 minutes, but 50 minutes is probably a good compromise if the company lunch period is between 45 and 60 minutes. Those attending could be encouraged to make up any time they lose from work by attending the seminar, especially when they have to drive from one facility to another.

The fact that the lunch-time seminars will not cost your company any money (volunteer instructors from your staff, company facilities used when they would normally be vacant, and employees voluntarily attending on their own time) has advantages, of course. But, there is also a down side. It is probably not legal for you to require employees to attend these training sessions in that they are offered on the employee's own time, so you can only make a training hit with those who are willing to participate voluntarily. You also probably cannot rank your employees for pay and promotion purposes according to whether or not they attend these classes. If the training is of value, however, one should expect that those who attend will increase their value to the company, and their performance improvement on programs should be positively affected as a basis for pay raises and promotions. You should avoid a direct link between attendance in this kind of program and employee ranking.

You will also encounter hostile signals from people in the management hierarchy. In that attendance is voluntary, it is conceivable that someone from another department, not in system engineering, may choose to attend and that the ideas expressed are in conflict with those of the attendee's management chain. There are also control freaks in organizations who might try to interfere only because they were not consulted. The Engineering Administration Director approached the author when he set up such a program at

General Dynamics Space Systems Division and inquired how he was paying for the program. The answer was that it was costing the company nothing. Whereupon the director asked what the employees were charging their time to, and the author responded that they were on their own time and did not need to account for their time any more than if they sat at their desk and ate lunch. The director offered that there was something wrong with the arrangement that he had not yet understood. Well, the author simply continued and expanded the program because he was right. If we wait for approval on activities such as this, we will always find others with reasons that inhibit action.

The author's program included several categories besides system engineering how-to knowledge. In addition there was a product line series consisting of one-hour classes on the systems in the Atlas Centaur launch vehicle. These were often taught by recent college hires who were thus forced to learn how these systems worked and locate experts in related fields they might not otherwise have met so early in their career. We also offered a gray-beard series where very senior or retired engineers were asked to come in and give of their experience. One of these conducted by a former chief engineer from the early Atlas program was attended by several of the engineering department directors. Finally, we offered a series on what the different departments did on programs. The lecturers for these programs were functional department engineering chiefs responsible for these departments. The highlight of this series was when one structural analysis department chief was lecturing and another department chief attending broke into the conversation with the comment that his department did that work on programs. All of these categories provided valuable learning opportunities to the college new hires in the author's system development department.

16.4   Local university cooperation

There are some advantages in acquiring system engineering training from local universities and chief among them is the shift of this cost from department training funds, always in short supply, to the company's tuition reimbursement funds, which a company generally tries to sustain at whatever demand dictates. If you are successful in making this transfer work, it may cause top management to reevaluate your company's tuition reimbursement policy, but it is unlikely that the demand that is placed on the program from this source is not enough to stimulate a protective response.

16.4.1   University readiness for system engineering training

Few educational institutions are prepared to independently champion the cause of improving system engineering practice in industry by better preparing college students for this work. Few educational institutions today offer students an effective course of instruction in system engineering as practiced in industry. Fewer still recognize the existence of a problem in an unrestrained commitment to specialization. Textbook and reference work publishing houses are organized into very specialized components as well, and this tends to encourage the *status quo*.

The gestation period today for a system engineer is several years following graduation with a B.S. degree from an engineering program. Universities are preparing specialized engineers in a number of engineering fields. Graduates from these programs obtain employment in specialized engineering departments in industry and gain experience in their profession. Over a period of years, these engineers acquire maturity and some demonstrate an interest, aptitude, and ability in generalism that transcends their specialized beginnings. Some of these engineers gravitate to a system engineering organization or responsibility with varying degrees of knowledge and experience.

Engineers trained on the job as system engineers are supplemented in industry by ex-military personnel who gained hands-on experience with complex military systems and thus developed an appreciation for and skill in applying thought processes useful in system engineering work.

In some companies, years of progressive engineering specialization (encouraged until fairly recently by specialized government customer agencies and reinforced by specialization in academia) without corresponding improvements in generalism, have yielded islands of specialized, serially and autonomously employed talent. These islands are not easily connected into effective concurrent engineering teams capable of simultaneously creating an integrated product solution consistent with a corresponding production and logistics support capability, all optimized for the desired measures of effectiveness (e.g., performance, cost, operability).

Companies that have been able to build and sustain an effective system engineering capability within this environment are indeed fortunate. If a company does not happen to have an effective system engineering capability, it can be a very difficult task to acquire it. Also, companies with a sound capability are faced with the problem of sustaining it. There are few localities where an infrastructure external to a company exists to support generation and maintenance of a sound system engineering capability. Companies have been forced to become self-sufficient in igniting and sustaining their own system engineering capability. As a result, knowledge of the process currently resides unevenly in industry and in some government agencies.

16.4.2 *Impediments and overcoming them*

Given that the premise is accepted that American industry needs an improved process for educating system engineering talent and improving system engineering capability as a counterbalance to specialization, how do we go about responding to this need? There are many impediments to changing the current condition, but the consequences of continued movement in the direction of unbalanced specialization are so severe that we should make a serious effort to understand how we might restore a condition of balance between specialization and generalism. Let us list five impediments and then attempt to remove them.

First, companies that have an effective system engineering capability may be reluctant to share their knowledge for purely competitive reasons. Since process knowledge resides largely in industry, it appears necessary for industry to take a lead in changing the current condition of equilibrium through sharing their knowledge. Second, educators now staffing colleges and universities are, for the most part, specialists well qualified to perpetuate the specialized stereotypes and not well qualified to encourage generalism. Third, college curricula in specialized engineering programs are fully loaded now with needed subjects required for accreditation, and it is very difficult to add one or more courses encouraging an appreciation for generalism without lengthening the program. Fourth, there are few good books on system engineering, publishers having succumbed to the same imbalance observed in academia and industry. Finally, the whole educational system is in a state of equilibrium that cannot easily be nudged off the specialization fixation.

Industry must be the principal source of process knowledge in any effort to improve the educational experience for system engineers. This process knowledge is unevenly held in industry. Those who have it have to be convinced it is in their enlightened self-interest to share their knowledge.

It is true that the management of a company that now has an effective system engineering capability in a geographical area that also includes competitors who do not, may feel protective about their advantage. Those who do not have an effective process and those who share a geographical area with none of their most serious rivals will probably not be concerned about loss of competitive advantage through cooperative development of a local system engineering education capability.

Even in an area where there is one company with an effective process that refuses to join forces for the reason of maintaining competitive advantage, that company will eventually join in once it sees that the others are benefiting and eroding the difference between their capabilities. While it is helpful for at least one member of a college system engineering support group to have a sound system engineering process, it is not essential so long as all of those who are willing to help share a desire to acquire it.

Despite any initial reservations about competitive position, it is in the interest of all the industries in a given locale to have local resources for system engineering education available for all. Given that system engineering capability is a valuable commodity, availability of a means of attaining it is obviously in the interest of those who do not have it. However, it is also in the interest of those who now possess it to help give it to others.

There is no company so blessed with system engineering effectiveness that it could not benefit from trying to help others attain it. Any teacher will tell you that the best way to understand something is to try to teach it. It is also true that a fundamentally sound way to improve a process is to expose it to critical comment from as many people as possible. Local support for a system engineering education program can be an effective part of a company's continuous process improvement regimen because it will find no one more critical of its best efforts than outsiders with different experiences. You can be certain that at least some of these outside ideas will be better than your own.

Every company in a given locality, including those that have a sound system engineering process, will benefit from improving the qualifications of the population pool from which they all draw the majority of their employees. By sharing, we will all become better than we are now. Those with a more effective process at the beginning will retain that advantage if they continue to work at it.

Working system engineers and managers in most companies recognize the benefits from shared access to system engineering education capability and knowledge. Most see a greater benefit from sharing for their own company than from withholding knowledge to protect competitive advantage.

This is the basis for a lot of sharing that takes place now through national organizations such as the Institute of Electrical and Electronic Engineers (IEEE) and American Institute for Aeronautics and Astronautics (AIAA). The Defense System Management College (DSMC) in Fort Belvoir, Virginia, offers an excellent system engineering management course open to industry.

These existing efforts are all very commendable, but we need a firestorm of cooperative programs between industry and educational institutions across the country to reinvigorate industry with effective system engineering practices and break down the growing autonomy in engineering organizations fed by galloping specialization at the very time concurrent engineering initiatives offer a hope of rebirth for the system engineering process.

Many practicing system engineers, educators, and government persons familiar with the acquisition process and the intended role of system engineering banded together in 1990 stimulated by exactly these problems into what has come to be known as the International Council on System Engineering (INCOSE). This organization is a support group for improving the practice of system engineering in industry, academia, and government agencies and improving the conduits through which talented engineers may pass to become effective system engineers. The charter of INCOSE is:

1. Foster the definition, understanding, and practice of world-class systems engineering in industry, academia, and government.
2. Provide a focal point for dissemination of system engineering knowledge.
3. Promote collaboration in systems engineering education and research.
4. Assure the existence of professional standards for integrity in the practice of systems engineering.

INCOSE encourages companies to establish in-house system engineering training programs, changes in college engineering education programs to better prepare graduates for effective contribution in the evolving work atmosphere in industry, and college extension programs to prepare engineers now in the workplace for improved system engineering skills.

It is industry and government customer agencies that have the process knowledge and have sensed the need for improvements in the preparation of engineers for system engineering work, and it falls on these sources to provide the ignition energy for those improvements. The fuel to sustain the resultant fires of improvement must be supplied jointly by government agencies, industry, and educational institutions.

16.4.3   UCSD Extension case study

The author's purpose here is to expose a pattern by which a local improvement in system engineering education can be realized based on a case study that took place in San Diego, California, in 1990 involving the University of California San Diego (UCSD) Extension Program and local industry. In September 1990, very soon after the meeting in Seattle, Washington, that formed NCOSE (later changed to INCOSE), the author called on Ms. Judy Cottrell, UCSD Director of Corporate Relations in the Industrial Liaison Office of the Division of Engineering, with a proposal to introduce system engineering courses, such as requirements analysis, into the UCSD engineering curriculum.

After some discussion we agreed that the UCSD Extension program would be the best place to try out these courses. UCSD actually offered a master of science degree in system engineering that was research oriented and focused heavily on control theory at the time.

Ms. Cottrell and Professor Dave Sworder, of the AMES Department, had in August 1989 hosted a workshop, stimulated by General Dynamics Space Systems (GDSS) Division Systems Engineering department, on improving preparation of graduates for participation in system engineering work in industry. This meeting adjourned with an action item for Professor Brian Mar from University of Washington to host a second meeting in Seattle in 1990, which became the beginning of NCOSE.

With Ms. Cottrell's encouragement, the GDSS Systems Engineering Director, Mr. Dave Clemons, and the author, both founding members of INCOSE, contacted the Director of Business and Engineering at UCSD Extension, Dr. John Peak, with an offer to help establish an extension program in system engineering subjects. Dr. Peak accepted the offer and astounded his guests by expressing an interest in moving to a certificate program quickly, thereby demonstrating that one of two needed enabling functions were in effect in the San Diego area.

These two enabling functions for creating a local system engineering education capability are: (1) a local educational institution with the spirit of adventure and (2) one or more local industries with an effective system engineering capability or interest in building one that will share their knowledge. If these two enabling functions are present, the several impediments listed earlier can be dealt with effectively within a cooperative environment.

The initial meeting between representatives from industry and the educational institution cannot take place without some preparation on the part of industry representatives. The industry representative in this case, GDSS, was formed in 1985 by separating the space and energy functions from Convair Division of General Dynamics to reform what had been earlier named the Astronautics Division, which had produced the Atlas ICBM. Systems engineering management of the new division determined that a priority had to be placed on improving its requirements analysis capability and began a program of research into effective methods and tools useful in a distributed, concurrent engineering environment.

A flexible approach was developed featuring three strategies from which a program could select with emphasis on structured development. A progressive requirements writing style was developed to match the degree of expected formality with program phase. Commercially available computer tools were studied for requirements capture capability with the conclusion that there were none available at the time that satisfied the perceived needs. This resulted in an internal effort to develop a tool in a prototyping environment simultaneous with efforts to define a more effective process. The synergism between these two efforts accelerated both, rather than retarding the process as might be expected.

It was immediately obvious that there would be no way to fund the reeducation needs of the Systems Development department from available indirect sources once a more effective process was defined, so the genius of the GD Corporate System Engineering Seminar, which had been offered for several years at that time, was replicated in the form of a Division System Engineering Institute offering lunch-time and after-hours classes to avoid an indirect cost burden. Procedures prepared to define the system engineering process became the textbooks for courses, since there were few good textbooks on system engineering subjects and insufficient budget to purchase them anyway.

As a result of this prior work, when the meeting took place with Dr. Peak at UCSD, GDSS Systems Engineering had been teaching requirements analysis and other system engineering topics to its engineers in-house in its institute for several years and had evolved, at the expense and hopefully to the betterment of its own population, an effective system engineering teaching capability.

Shortly after the meeting, Dr. Peak formed a System Engineering Advisory Committee including representatives from General Dynamics (Convair, Space Systems, and Electronics divisions), SAIC, TRW, General Atomics, and Teledyne Ryan Aeronautical. The combined population of system engineers in these several companies provided a sound student base for the planned courses. The committee met several times to define a certificate program and selected the author to teach the first course, in requirements analysis, in the spring of 1991. That course has been presented seven times in the program at UCSD since the program began, and all of the courses in the program continue to be well attended as this book comes to life in 1999.

There are other examples of this process at work in other parts of the country. In each case, there seems to have been a special relationship between local industry and an educational institution. The relationships between Texas Tech at Lubbock, Texas, with Texas Instruments as well as Boeing in Seattle, Washington, with the University of Washington are two other examples. Loral Aeronutronic inquired of University of California Irvine about system engineering education opportunities and, while UCI did not at the time have a program in readiness, Mario Vidalon, Extension Director for Engineering and Information, arranged with the author to present a series of courses at Loral for the University as he began to build the program on campus. That program has delivered six complete six-course on-site system engineering certificate programs at industry sites and begun on-campus courses as this book comes through the word processor. The University of California Riverside got into the system engineering certificate program business in response to a visit from a delegation of local INCOSE chapter representatives led by Mr. Howard Korman of TRW. This is the perfect cooperative relationship — an INCOSE chapter that took on a project to stimulate a local university to build a system engineering education capability for the area served by the chapter and the university.

Universities, always in need of funds, can, of course, always apply money contributions provided by local industry to encourage particular programs of interest to industry, but the commodity of greater value is access to people and their knowledge of the process within which engineering graduates must function. This does not have to cost the company a significant amount of money. In the case of GDSS and UCSD, the total cost to GDSS was no more than 10 man-hours for certificate program meetings. The rest of the time was a matter of individual effort away from the job.

Clearly, system engineering management in industry has a responsibility to provide a means to educate their personnel in system engineering topics, and local educational institutions provide a cost-effective resource that can help managers meet that responsibility. These institutions cannot create effective system engineering education programs out of whole cloth, however, because they do not have the know-how. Industry must provide the ignition source with its knowledge of process and desire for continuous improvement.

It is conceivable that the initial approach could be initiated by an educational institution or by local government representatives responsible for acquisition of large systems and concerned about the state of system engineering education in their geographical area. This initial appeal might be focused on one or more companies that could be encouraged to demonstrate their commitment to civic involvement. However, the process is ignited, all three parties are needed to sustain the improvement efforts.

American industry needs to improve its system engineering capability. Your locality probably includes at least one educational institution that would like to participate in helping to satisfy local industry's need, but they have to have industry's help to structure the program and provide knowledgeable instructors. You should consider taking the initiative soon to talk to someone in a local educational institution about your needs and how you can help that institution help you to satisfy those needs. The author's motivation in joining up with UCSD in the beginning was that he could not get access to department training funds, but there was an almost unlimited availability of college tuition reimbursement funds.

The UCSD program probably succeeded from the university's economic perspective through the work of Mr. Charles Zumba, a vice president at Scientific Applications International Corporation (SAIC). He encouraged management to offer SAIC graduates of the certificate program a gift of company stock. As a result, over the years the program has been in existence, nearly half of the class members in courses the author has lectured in have been SAIC employees.

16.5 *Purchase of a tailored commercial certificate program*

One solution to the enterprise system engineering training program is to simply purchase it from a training provider. In so doing, the company should insist on the content of the courses being mapped to their process definition in the form of a list of your process steps (blocks on your process diagram) in one column and the course content reference in another column of a matrix. In order for this to work, you must first have a process definition, of course.

In some companies where the author has discussed their training needs and provided training, the company representatives have expressed concern that the course material support their written process description only to find that their process description contains only the what-to-do information in such general terms that anything could conceivably map to it. The company may be generally applying a specific set of techniques below the level of process documentation, but it is not captured in writing and exists only in corporate memory. A good compromise in process specificity is to capture the what-to-do knowledge and link it to one or more specific techniques by reference to a book or standard. This makes it unnecessary to essentially write a textbook while also clearly indicating the preferred technique where several alternatives exist.

For example, the author's company provides a complete system engineering certificate program of six three-day courses (Foundation of System Engineering, System Requirements Analysis, Design and Integration, Validation and Verification, Specialty and Concurrent Engineering, and System Engineering Deployment) that involves 18 days of training that can be spread over any period of time. It can be tuned to any product line by adapting the techniques covered to the company's needs. Each course involves team projects that can be linked to the company product line rather than to the specific examples that are used when courses in the program are offered at a university as part of an on-campus certificate program or as a public short course where it is necessary to appeal to a broader product experience base. The courses also cover several alternative ways of accomplishing some tasks permitting the one preferred by the company to be emphasized. Each course comes with a 300-page handout, including text material and a copy of the presentation materials. The program can be contracted directly (and continuing education units awarded to a permanent record if desired) or through a university extension office. This program is also available in six one-day formats as a refresher or overview program.

16.6 A system engineering help window

Training accomplished closest to the need best serves that need. Most systems engineering training opportunities, where they do occur, are too far removed in time from the need for that training. Commonly, good external training opportunities cost more than we can afford to spend and are not scheduled for our convenience but rather for broader market demands. The combination of these factors makes it very difficult to improve one's skills and knowledge rapidly in this field. This section offers a system engineering organization a way to parlay a prior investment in networking and systems engineering process definition into an affordable solution to this problem. The solution also permits the use of a simple generic enterprise system engineering process manual by providing the detailed how-to information in the form of just-in-time training materials linked to the process manual. The author believes this to be the ultimate in-house system engineering training solution for the current state of technology. This section is based on a paper delivered at the 1997 INCOSE symposium.

16.6.1 Current realities

Over the past decade, the desktop computer has become as common as the pencil and paper of years past as the preferred instrument for recording the results of system engineering work. A system engineer may very well spend half his/her time working at such a machine, recording the results of the work directly in its final media. More and more, the product of our work resides in computer media commonly riding on a network server such that it is, or can be, more or less universally accessible within some domain.

While at General Dynamics Space Systems Division in the early 1990s, the author championed an on-line specification library that provided desktop access to all program specifications whether the person used a Macintosh or an IBM machine. McDonnell Douglas on the FA-18E/F Program has gone one better by providing company and customer personnel a view of the requirements in an RDD-100 database expressed on their screen through an Interleaf application accessed via an intranet web site. These and other applications are a wonderful advance compared to paper copies of specifications in document cribs and out-of-date copies in piles on all four corners of the engineer's desktop, but they are only offering people the results of work well done. The content had to first be placed into those documents or databases by people who knew what they were doing.

What if you have just been assigned the responsibility for a task that you have never done before? Would you reasonably expect in your current situation that there would be an available course or detailed description of how to do this job so you could start the job from a position of knowledge rather than learning while doing? Perhaps your company has a written system engineering practices document, but when you open it you find only a brief paragraph on this subject. Your practice, however, does include a reference to a book that goes into detail on this task. You take the time to go to the company engineering library only to find that all of the books were recently given to a local university as a cost-cutting move.

Well, all is not lost. You have a supervisor who knows everything and peers who have done this kind of work before. Unfortunately, your supervisor is working on a program at another facility to reduce overhead expenses, and your peers all complain that they are working too many hours now just doing their own job because the company is still running lean after several rounds of downsizing. There is a local university system engineering certificate program, but the course covering the task you have to do requires one evening a week for three months, and it does not run for another six months. There is a three-day course running next week put on by a short-course company in a town close enough to commute to every day, but there is no department training budget.

You have no choice but to start doing the work and learning as you go. The work product will likely cost more to create, take longer to create, and be of less value than it would have been if you could have started the job from a position of knowledge. There must be a better way to deliver just-in-time training to our system engineers.

16.6.2   Enter the help window

Since we spend so much time working at a computer, it is perfectly natural to use that computer to deliver the training experience we are looking for. Your company has probably already made the investment in networking and is committed to maintaining that capability. Perhaps your company created a system engineering practice document by editing EIA 632 or IEEE P1220. As a beginning, we could put the company practice on-line, but that probably will not fill the need for details. What we need is a system engineering process help window on our desktop computer to expand upon the terse content of our practices document.

This window could be opened by an employee in exactly the situation described a moment ago. Our system engineer, John, must start on this new job tomorrow. This afternoon he opens the system engineering help window and receives a 30-minute pitch on technical performance measurement. He finds out that he must identify a short list of key parameters and gets some help in identifying those most beneficial to the program. John discovers that each parameter must be assigned to a principal engineer and that each principal engineer should have two charts on his/her parameter. First, he/she needs a tracking chart that shows the parameter required and demonstrated values in time coordinated with related events. Second, he/she needs an action plan telling how to achieve coincidence between these two values and when and why we should believe that will occur. This information will enable John to organize an effective TPM plan that otherwise would require lost time and a less effective implementation.

16.6.3 Route to the initial window

We are assuming that your company has a current systems engineering practices manual or other process document. If you do not, then there is some prerequisite work that must be accomplished before you can take advantage of the suggestions in this paper. The lack of a computer network is another inhibitor.

The first step we should take is to place our systems engineering practice on-line so anyone can call it up and check a particular point on screen or print out a copy. Many companies have placed a note on the cover of their on-line documents to the effect that a paper copy may not be up to date and to check with a person at a phone number given on the paper copy for the current version. This document should be located on a server generally accessible from all programs in your business unit. For our purposes, we will call this server storage space allocation SEPLANS.

This is a simple move, but we should not trivialize it. The fact that your process definition is universally available in the current version is no small achievement however easily it was attained. The remaining problem is that our process description is not sufficiently comprehensive to transform an engineer reading it into an instant power-house of skill. So we have two choices: (1) change our process manual to include the necessary depth to support help window needs or (2) supplement the manual with additional information keyed to the manual content. The second alternative is encouraged, and that supplementary information forms the system engineering help window.

Your preferred system engineering process description should, of course, be coordinated with the way your company chooses to do the system engineering work on programs. It is not adequate to have a process description and ignore it. The reason for having a process description is to guide people in performing work on programs and to encourage repetition in applying that process so people's performance will improve over time. We should insist on continuous incremental improvement of this process rather than periodic revolutionary changes, because we wish to maximize the common portion of the process between any two applications.

Our manual is on-line. It is maintained consistent with the way we wish to perform systems engineering work. Our programs are applying our standard process described in our generic systems engineering process manual. Given this baseline, there is hope for us in implementing the help window.

It is possible, of course, that we could have a terrible mess as a system engineering process description. One way to ensure that we are close to quality in our process description is to map what we think is our customer's preferred process description to ours. If we have three primary customers and each has a different view of system engineering perfection, we may have to make three maps and, to the extent that they do not all describe the same process, we may have to tailor one or more of them to a common process that we describe in our internal process description. Out of this experience will come a conclusion that you have accounted for all of your customers' needs in the area of system engineering, while providing your people with a single common process description that can be used on all programs.

To pick up the story of the window, we must decompose the process described in our manual into components or modules about which we can prepare real-time instructional material. These components should be no larger than can be described in a half hour using an application such as Microsoft Power Point, and they must map cleanly to your process description. Ideally, this will involve simple references to a single paragraph, rather than bits and pieces spread throughout.

We should make a list of these modules and prioritize them based on some criteria: (a) which ones are most in need of improvement, (b) which ones are the easiest to implement, giving us early successes, or (c) some combination of these factors. Somewhere in our organization we should be able to find people who know enough about these topics that they could, within a month or two, even with all of the other demands on their time, come up with a set of self-teaching charts covering an assigned topic.

It remains to manage this whole activity to closure, resulting in a series of computer files, each providing the how-to instructions on a part of our generic system engineering process description. The training materials should be reviewed and approved for inclusion in the growing library of system engineering knowledge available to your engineers.

These files are then placed in our SEPLANS directory/folder using file names that are self evident. For example, we could call the file on technical performance measurement simply TPM. Other files may present a more difficult naming chore so we may wish to simply number them in accordance with the paragraphing structure of our system engineering process document, thus gaining good traceability, something we should have regardless of how we name the files.

The understanding is, at this point, that we only have some charts containing graphics and text that collectively convey to an intelligent engineer how to perform the related task. Clearly, this is not the most powerful training environment in that it does not appeal to the full range of senses of the person being trained. But it is a beginning.

16.6.4   *The external and mixed alternatives*

If your company does not have the talent to develop these training materials, you can bring in an outside agent to develop them. A more likely alternative is that your company can develop some of the modules, while others would be developed more cost effectively by an outside agent.

The materials developed by an external source should be identified with respect to their source. Some materials will come from your own company. Others will come from the external source and may be copyrighted, resulting in the need to pay a license fee for a finite period of time. Other materials may be created by the external agent while employed to do so, and these should be your intellectual property.

16.6.5   *Preplanned or incremental window improvements*

We could begin our attack on systems engineering ignorance without a lot of serious thought about the future. As is generally the case, a grand plan would come to a better result, however. The only case where we might want to proceed without a plan is where we perceive a planning stalemate, either for political reasons or due to lack of ideas about long-range needs. In this case, movement is preferable to stagnation. This one exception aside, we should recognize our help window initial operating capability as a beginning with a planned or incremental growth path.

Once the foundation is in place we will find many ways to improve upon it. The way we have described the window so far, it is assumed that the user would have little trouble navigating through the material to find the part of immediate interest. If our window content becomes complex, we may need to employ a hypermedia concept to encourage easy navigation. In this arrangement, the user clicks on something of interest, possibly even a paragraph in our practices manual, expanding the view on that topic. This is an alternative to the hierarchical structure of the data implied in the earlier discussion.

While graphical window data can be very effective as a just-in-time training source, it can be made more powerful by including voice clips briefly explaining particular points. Microsoft Power Point and other presentation packages have this capability. Your delivery workstations must be equipped for sound in order to deliver the complete message in this event.

Video clips can also be added to the charts in many presentation packages (including Microsoft Power Point). Some points can be communicated much more powerfully via video than by a single graphic or even using the build function of a graphics package where you can link several charts together in a sequence, giving the illusion of motion or change in time.

The ultimate application of the system engineering help window would be implemented through a database application. Even inexpensive databases such as Microsoft Access and Borland's dBase V For Windows are compatible with graphical, audio, and even video clip field data.

The records would correspond to the charts already developed. Each chart becomes a graphic in the database supplemented with administrative and configuration management data as well as text, voice, and video clips

Table 16.1   Suggested Data Organization

ID	Topic
01	System Engineering
01-01	Requirements Analysis
01-01-18	Technical Performance Measurement (TPM)
01-01-18-05	TPM Action Plan

as appropriate. These records should be organized into modules of closely related information identified by key fields defining the broad subject area (systems engineering), perhaps a secondary organizational layer like requirements analysis, and a module such as TPM.

Table 16.1 offers a way to structure the data. We have assigned 01 for the system engineering subject area, possibly 02 for design, 03 for manufacturing, and so forth. Topic 01 within the systems area is assigned to requirements analysis, and 18 within requirements analysis has been identified for TPM. Finally, we now get to database records with real content. The record identified by the four ID field values 01-01-18-05 would contain a chart showing an example of a TPM action plan and a voice clip saying, "The parameter principal engineer should maintain an action plan telling what actions are being and will be accomplished in the future to cause the design to attain the required value. In this case, we see that Bill, the principal engineer, intends to perform an analysis of new data derived from a development evaluation test to be completed in two weeks to see what the new demonstrated value is." A video clip could then come up in a window showing Bill talking to a test engineer with some apparatus in the background.

The necessary code can be written to link these records into a stream of images of value in preparing an engineer for performing this work in the near term. You will also need the code to control the whole help system, making it possible to locate and call the desired material. This can be done through a hierarchical menu system or through a hypermedia application as desired.

16.6.6   Work tool linkage

So far, we have been discussing the systems engineering help window as an isolated utility. The optimum way to deliver this service would be through the very tools the engineer uses to do work. True, many of these tools have a help function giving support for tool-related considerations. The suggestion is that the tool makers consider providing links to a company's process help system. Alternatively, this may be a fertile area of expansion for tool makers, providing a process-oriented how-to help window.

16.6.7   *Help window conclusions*

The computer will continue to change our lives at an ever-increasing pace. Sometimes those changes will be compatible with a more enjoyable as well as effective work environment. The system engineering help window is one of those changes. Ultimately, we should cause these machines to support our needs so we can almost effortlessly apply our skill and knowledge to the problems we are most effective in solving, the difficult ones requiring the application of judgment and interpersonal interaction.

The more of the mundane burden that we can remove from management and each other, the more time and energy will be available with which to ply our profession and serve our customers.

It will be a long road from the rudimentary applications of the kind discussed in this chapter to a really effective capability useful to new-hires thrown into the fight, but it is a road that will come to a useful end.

chapter seventeen

System engineering assessment and improvement

17.1 Is a static identity adequate?

If we are successful in completing our process description, can we then rest and apply it for all time thereafter? Some companies have taken this route in the past, especially when they are flushed with contracts and have the indirect budget to spare to engage in procedure writing. The business cycle eventually returns to lean times, and these practices fall on hard times along with the rest of the firm. The procedures are not maintained. Another reason for a departure between written and practiced methods is the replacement of the manager who stimulated the written practices by one who has a different value system. It may take another managerial generation to return to the maintenance of these documents. In one company the author visited, their system engineering practice was not bad, but it was identified by the company that had owned the division some time previously, a signal that some other things may have changed as well. One would find very few enterprises with a current set of practices.

The enterprise identity described in the author's book *System Engineering Planning and Enterprise Identity*, and assumed in this book as discussed in Chapter 1, is not intended to be a fixed entity but a continuously changing one. Yes, we will have to maintain our process definition for it to have any lasting value. Change is an essential part of the life of a person or a corporation, and we have to accept it and adapt to it. This is the reason for mounting and sustaining a continuous process improvement feature in our process. We have to be careful, however, to avoid revolutionary change in a continuous stream. There has to be enough sound content to begin with so we can incrementally change our practices so as to preserve the vast amount of the current content at any one change, thereby preserving the learning benefits from process repetition. This may be very hard to do. We wish to have a documented process and yet be constantly changing that process description.

17.2   Procedures media

In what form shall we publish and retain our process definition? A word processor leaps to mind immediately, since the computer is perfectly adapted to continuous improvement. It is very easy to edit these documents stored in electronic files. Alternatively, some computer tool makers, like Ascent Logic, at the time this book was being written, were becoming very interested in applying their tools to enterprise modeling as well as the product system engineering process. It is possible to enter all of your procedures into a database from which practices can be printed. Some of these systems even allow you to simulate operation of your enterprise in accordance with the way you have described it in an effort to find better ways to define your process.

The past standard approach to this would involve printing approved procedures and distributing them to manual holders. There is a lot of unnecessary labor involved in this process including reproduction of paper, distribution of paper, manual update, and periodic audits of manuals to ensure they are current.

You are encouraged to use an on-line, networked medium for publication, distribution, and access. In this approach, your approved procedures are uploaded from the workstation where they are maintained to a network server library of procedures. When users need to access one of these procedures, they only have to call it up on their workstations. Everyone has access to the same configuration of the current procedures.

The combination of computerized procedures, a central electronic library, and universal read-only access through a network will enable an economical approach to maintaining your procedures to take advantage of incremental improvements. If one needs a copy of a complete practice or a few pages, they can simply print them out, use them as needed, and throw the paper away. In order to signal the reader that a paper version may be out of date, some companies place a warning on the cover or in a header that this version may not be current and to check the on-line library for the current version.

17.3   Lessons learned and continuous improvement

17.3.1   Continuous improvement process

Figure 17.1 illustrates a suggested process for incrementally improving our process continuously. Lessons learned from programs are studied (function F21) for ways to improve the current process description maintained in function F26. In function F23, we study external process descriptions and compare them with our internal procedures. This may lead to insights into improvements or knowledge about our process maturity relative to our competitors.

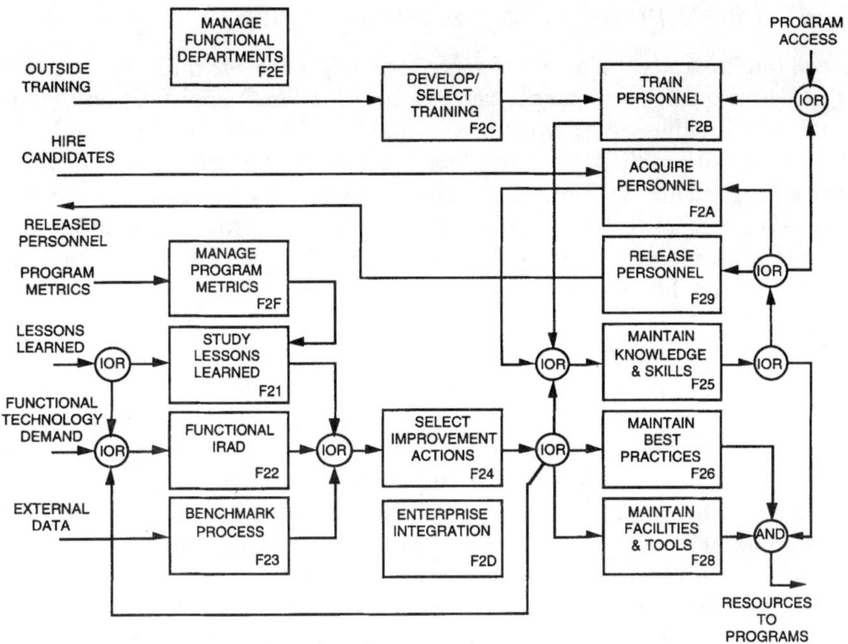

Figure 17.1   Overall continuous improvement process.

The results from these two steps are opportunities for improvement actions. In function F24, we determine which of several improvement approaches we should implement. This may involve personnel training, tools improvements, or procedural improvements. The selected action or combination of actions is accomplished in functions F25, F26, and/or F28. All of this work should be the responsibility of the functional portion of the matrix, providing an infrastructure within which programs can apply a common process, possibly adjusted for program peculiarities.

17.3.2   The servo loop analogy

Control theory, as applied by engineers in the development of systems or networks to control processes and machines, offers us a mechanism of vital importance in the continuous improvement of our processes. We should seek out closed loop processes with degenerative feedback that tend to follow-up to a standard. In these kinds of control loops, an error signal is created through comparison of the commanded value and the actual value achieved. The result is an error signal that should drive the actual value closer to the commanded value. Our practices are the command value, and our current practice is the actual. The error signal is the delta between these. In Figure 17.1 that error signal is created in F21, F22, and F23 by comparison of some measurement with our internal process.

17.4 Performance measurement systems

In order to take advantage of the servo loop approach, we must have a means of measuring our performance relative to our process specification. A common term for these measurements is metrics. There are several kinds. We should identify metrics for our generic processes that tell how well we are performing them on programs. It should be possible to make comparisons within a program to track performance and across program boundaries to see differences in performance that may lead to insights into how to improve practices. They have to be numerical in nature in order for these benefits to be realized.

Two measurement systems are commonly applied on programs to identify how well the program is responding to its contractual requirements. These systems are *cost/schedule control systems* (C/SCS) and *technical performance measurement* (TPM). The former catalogs all program budgets in accordance with the work breakdown structure, integrated master plan (IMP), and integrated master schedule (IMS) discussed in Chapter 12. The latter tracks the values assigned to key requirements in specifications.

An enterprise also needs one more system providing program feedback to the functional organizations. This system should be driven by the content of the written practices. For each task at some level, one or more metrics should be defined that can be tracked on programs with relative ease.

17.4.1 Cost/schedule control systems

As work is accomplished on a program, the personnel fill out time cards noting what work order they are working on, and this is entered in a computer that tracks program cost and schedule performance. All program tasks are also identified with completion criteria for each milestone or event and when that criteria is satisfied, the corresponding milestone is claimed. These systems will track performance in terms of man-hours and time consumed relative to milestone achievement. If you claim a milestone before the allocated budget and schedule are exhausted, you are in a favorable underrun condition with respect to both. The worst case is that you have overrun both cost and schedule allocations when you achieve the related milestone.

Figure 17.2 illustrates a typical cumulative track of cost and schedule on a program where the milestone is claimed after having consumed more schedule and cost than planned. If a task being tracked is in an overrun condition beyond the gray threshold area on the figure in either budget or schedule (threshold selected by program management), the C/SCS will issue a variance report that requires the person responsible for the task to report why the overrun condition exists and requires submission of an action plan that tells what will be done to correct the overrun condition. These variance reports should be reviewed by management and either accepted or further inquiries made to determine the veracity of the report, its assignment of cause, and chances of success in achieving the action plan goals. Where these

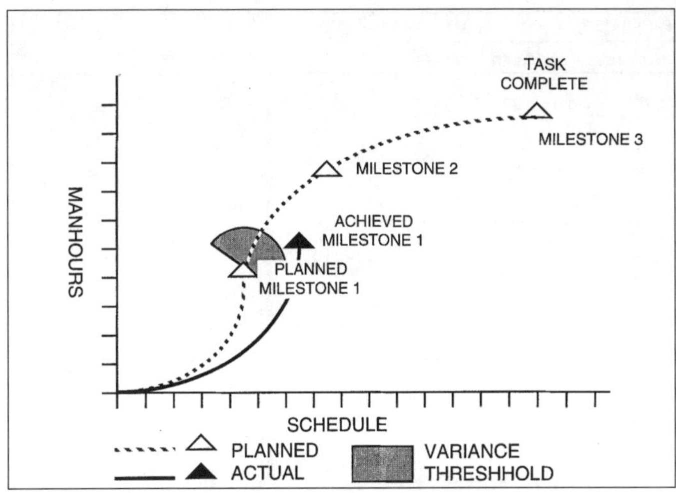

Figure 17.2   C/SCS reporting.

are accurate and valid showing a cost and/or schedule problem that cannot be closed within the current resources, management may consider turning loose some budget margin or schedule float to boost the task resources to a realistic level. Or they may simply require corrective action within existing resource constraints. Where the claims are invalid or fraudulent, the program may have to consider replacing the person responsible for the task.

17.4.2   Technical performance measurement (TPM)

TPM complements C/SCS on programs by providing a set of metrics keyed to risky system and item requirements in program specifications. We first select a small number of requirements that satisfy a criteria defining our concerns. An example of such a criteria might be:

a. The requirement is in the system or an end item specification.
b. There is some risk that the requirement cannot be satisfied within the current program constraints.
c. The enterprise has not been fully successful in satisfying this requirement on past programs.
d. The requirement is closely linked to program success or failure or has a great deal of leverage in determining program success.

On a small program, we may select as few as 10 system requirements for this purpose. On a large program with multiple teams at two or more levels, we might identify as many as 100 parameters with each team responsible for several.

For each TPM parameter we should assign a responsible engineer who must create and maintain two charts: (1) a TPM tracking chart and (2) an

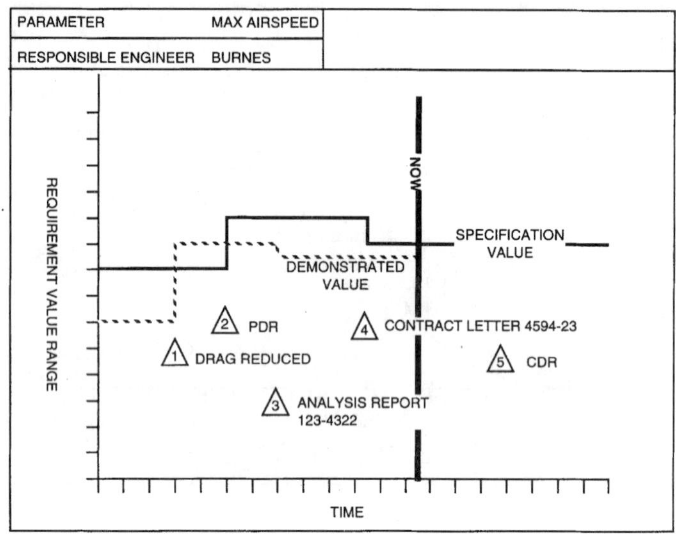

Figure 17.3   Typical TPM chart.

action plan chart. The first one should show a graph with the requirement value range and trace the history of the requirement value and the achieved value in time identified on the horizontal axis, as illustrated in Figure 17.3. Notes offer keys to significant events in the change of the requirement and demonstrated values. The second chart, only necessary where the demonstrated value has not achieved the required value, should offer a specific plan for closing the gap between demonstrated and required values, as well as a time table for achieving closure.

17.4.3   Generic process metrics

Each generic task selected for program application should have one or more metrics identified for it and a simple means whereby the data can be collected without excessive burden for the program. These metrics should be collected on each program and made available to the EIT for formal capture and reporting.

17.5   Formal post mortem

It is common practice for a program to hold some kind of review in association with major milestones or program events. In these reviews, the program compares performance against program goals for that program task, event, or phase, reaches conclusions about that performance, and makes decisions about subsequent planned activity. This same approach should be applied to program performance relative to the quality of the process applied.

The U.S. Army has developed a technique for formal evaluation of performance subsequent to a training or real military action called an *after action review* (AAR), which we might do well to understand and apply to programs as a means to gain insight into our performance on programs leading up to task milestone, program event, and program phase terminals. If we have planned these activities in accordance with the content of this book, each task at all levels and for all milestones and events have associated with them clear definition of the work to be done and clear criteria for the conditions under which the work can be said to be complete. It remains for us to capture our actual performance statistics and that can be done through our cost/schedule control system (C/SCS) and other metrics. Given that we have the plan and the actuals, we have the detailed information needed for a thoughtful and rapid evaluation of our performance.

The Army's AAR focuses on three important factors: (1) What was the plan? (2) What did we do? and (3) What was the outcome? The answers to these questions can form a useful narrative coordinated with the metrics collected for that activity under review. This review must be held close up to the conclusion of the related work in order to access the memory of the participants about related activities. Also, some or all of the people involved may be departing the program upon completion of the activity.

The AAR, by whatever name, would have to be planned as a part of an activity so that there is some time and budget to accomplish it. One way to do this is to simply make it a part of every formal review, the final element in a review. For example, at the end of the preliminary design review, the participants should answer these questions candidly and report the results. This could be done with the customer in attendance or subsequent to his departure as one chooses. If the customer has been a real team member during the work performance, he should be included in the review. This may be more than a little painful in that the review will expose things that went wrong (in some cases, because of flawed judgment of those participating).

This is one of the reasons that some persons in a position of leadership do not encourage this kind of review. Mistakes will be made. Many important mistakes are made by persons in positions of responsibility, and some people cannot stand the ego impact of having these mistakes exposed. This is something that an organization and its members must overcome if they are serious about improving process and performance. Interestingly, the Army allows criticism of superiors by junior officers and enlisted personnel in these meetings. It required a lot of adjustment in the Army to tolerate this, as you can imagine. Somehow the Army's discipline system survived this change and so can yours. Army personnel apparently came to understand that they needed the truth about how well they had done, and this truth would include exposure of problems caused by its own people as well as the enemy. The basis of change should be reality not fantasy.

There are other expressions of this approach in the management literature as well. There is a technique where the participants quickly list the positive and negative things that happened in a meeting and discuss them. The AAR simply extends that notion to the preceding activity about which the meeting or review was called.

In addition to finding out the truth about our performance and having the truth as a basis for corrective actions, we are also likely to find that the policy of honestly reviewing our performance at the end of a task will lead to more thoughtful performance of the work *during* the task.

Programs would do well to enlist the help of functional department managers to participate in these AARs as well as the program reviews they should follow. This permits the functional managers to remain current on how their people, tools, and processes are doing on programs and provides program work to unburden a probable overhead limitation. The AAR could be a quality inspection function to check how well the process specification was implemented on the program tasks covered by the review.

The results of this review along with the metric captured on the related tasks provides the EIT with the information they need to assess the quality of performance of the program and determine what, if any, process changes should be made based on the resulting lessons learned.

17.6 Two-dimensional process audit

We have gone to a lot of trouble to insist on a common enterprise process. Our rule is that all programs must implement this standard process. How do we ensure that this actually happens? First, proposals should be screened to ensure that they reflect this rule. Second, we must audit programs in accordance with some predetermined sampling rule to verify that they are applying the standard process. A two-dimensional audit process is suggested. The first dimension verifies that programs are properly implementing the generic process. The second dimension determines how well our practices compare with an accepted standard, thus benchmarking process maturity.

Figure 17.4 shows how this process can operate. We select specific practices to audit from our generic documentation, such as a particular SEM/SEMP process, that maps to particular program process steps. An auditor studies how the program process is performed, using a question list prepared for this purpose. Several alternatives are possible from this audit. The auditor may conclude the generic process is good and that it is being implemented well, resulting in no need for corrective action. Detected practice problems should result in practice changes. Unfaithful implementation on the program should result in changes in implementation. Figure 17.5 shows a process diagram that accomplishes the task implementation faithfulness assessment.

It is possible that bad practices are being faithfully implemented, and we need another process to detect this event. This same practice discussed above should also be compared with an accepted standard, resulting in a conclusion either that our practice compares favorably or unfavorably with

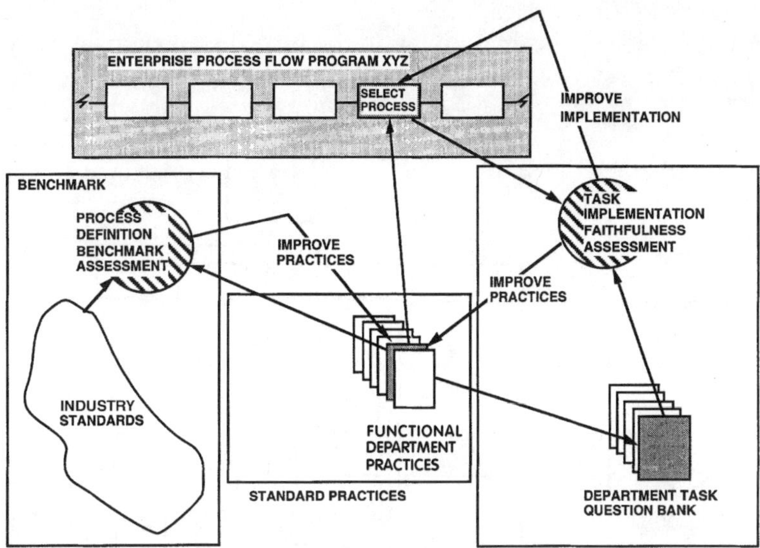

Figure 17.4   Dual track assessment.

the standard. If unfavorable, it should lead to an improvement in our practice.

Our audit process should collect numerical data that can be used to track performance over time. Figure 17.6 suggests a means of converting audit results into metrics that can be tracked over time. The answers to implementation assessment questions are converted into a score between zero and 100, just like an exam grade is computed. The process benchmarking grade must be more subjectively determined in the same range, though it would be possible to define ranges for particular classes of comparisons. Figure 17.6 illustrates several extreme but possible results during audits. We would prefer to receive scores of 100,100 on all audits. Where we fail to achieve a dual high score, a need for changes in our practices, implementation, or both are suggested.

The implementation quality assessment checklists are relatively simple to construct. You begin with the standard practice that covers the activity in question and form a list of questions, much like building a quiz that tests class members' knowledge of the most important points covered in a lecture. The next question is, "Who creates these questions and when?" When you begin this process, you will have no audit questions at all. You can either form a team to generate the questions for selected activities as an audit process implementation step, delaying the beginning of audits until finished with the questions, or proceed with the audit and build the questions as you go. The latter course is encouraged, because the process can then be improved incrementally as you go.

Let us say that you choose to audit one process per month per program per major functional department. Each functional department must select

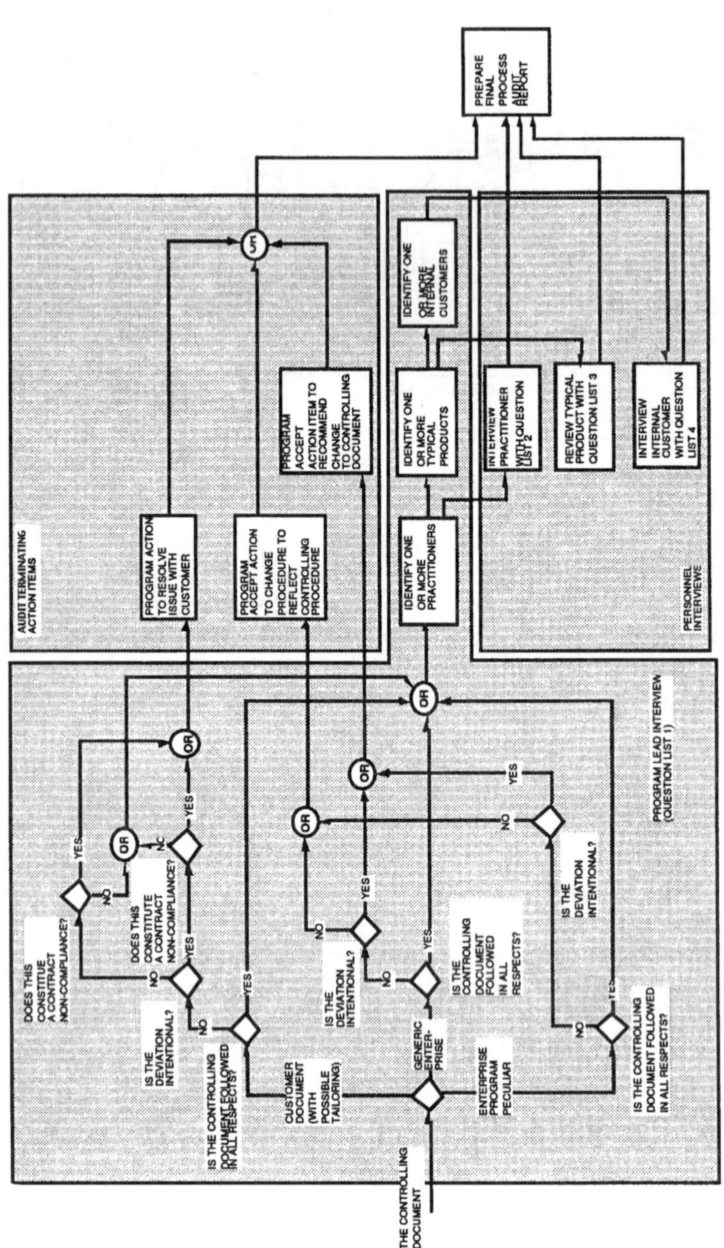

Figure 17.5   Detailed assessment decision network.

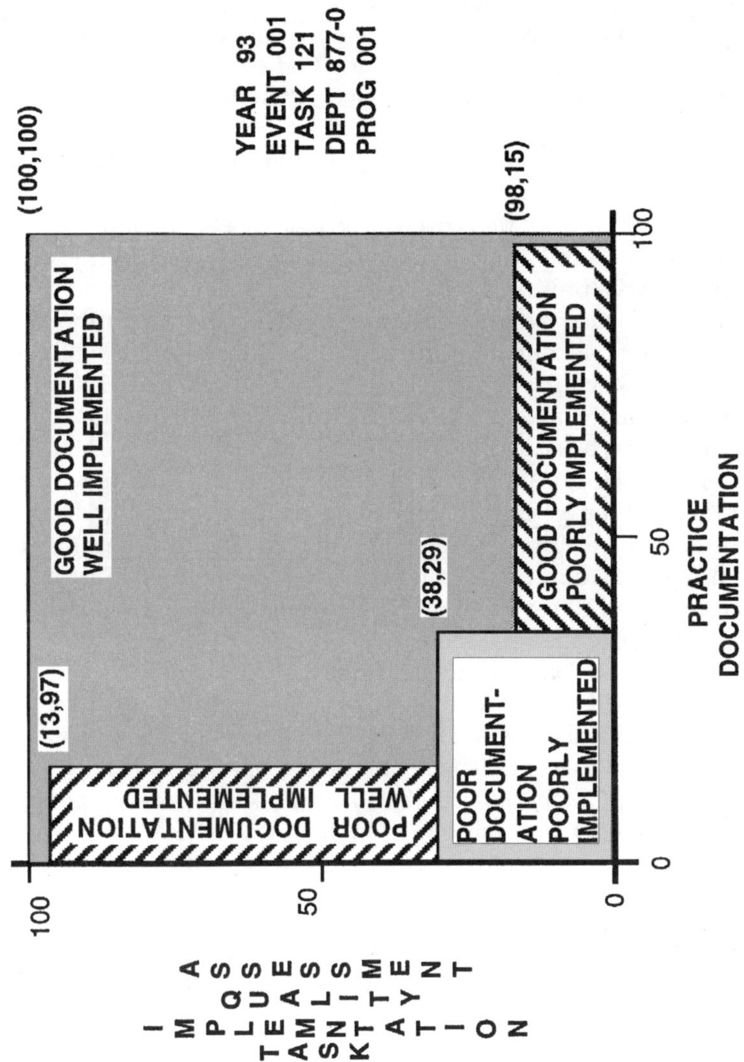

Figure 17.6 Two-dimensional assessment scoring system.

the persons to conduct the audits and assign responsibility and a schedule for these audits. The person responsible for a particular audit must first determine if a set of questions exists for the audit. Initially, a set of questions will not exist. If there are none, the auditor must first make the list, as discussed above, and gain functional management's approval. This list becomes the baseline for subsequent audits, with adjustments over time.

The suggested list of questions has four parts, as noted in Figure 17.5. The first part is defined generically in Figure 17.5, and the questions are answered by the program lead person for that function. If the problems encountered involve contract noncompliance, then the audit should be curtailed and the compliance issue resolved. If no contract compliance issues are detected, the auditor identifies a practitioner on the program for that activity and applies question list 2, focused on task performance in accordance with the controlling procedure. The practitioner identifies a typical product generated on the program in accordance with the practice, and the auditor evaluates the product against question list 3. Finally, the practitioner identifies an internal or external customer for the activity, and the auditor applies question list 4, containing perhaps only three questions: "Does the product conform to your requirements?," "Are you satisfied with product quality and accuracy?," and "Was it made available to you in a timely way?" Other questions can be added.

Scoring for these questions can be easily done. The questions may be weighted the same or differently depending on their content and management's attitudes toward their relative importance. The total score is scaled for 100. For any one question during the audit, the auditor must define a grade between 0 and 100 or 0 and 10. Scores are accumulated and influenced by scaling to produce the final score.

The most difficult part of the benchmark quality audit is to select the standard against which you will measure your capability for a particular functional discipline. The evaluation of the internal practice against this standard can only be accomplished subjectively, barring an extensive analysis and creation of a checklist that will very likely be more difficult to maintain than the practices themselves. One simple way to put this in numerical terms is to count the number of paragraphs where you comply and form a ratio of that number to the total number of paragraphs in the benchmark document. The question can be stated relatively simply, "Does our practice conform to this paragraph of the selected standard." If it does in all cases, you should award a score of 100. If it does not, you must assign some numerical grade less than 100. You could do this based on a simple paragraph count as noted above, or you could adjust the ratio as a function of the difficulty in closing the gap through changes in the in-house practice. This factor would not only involve the written practice but possible associated social issues in our organization. The score should be accompanied by an explanation.

Simple computer database systems can be built to capture the results of these audits, provide administrative support, and perform the arithmetic to generate final scores. At the time this book was written, the author knew of no readily available products on the market that support such audits.

Given that we have accumulated information about our performance through this audit process, we must have the machinery in place to do something about the results derived from it. That machinery must include methods to change our practices where needed and enforce work performance on programs in accordance with our standards where deviations are not in our best interest.

17.7   External maturity models

Several organizations have developed models of excellence against which you can compare the features and capabilities of your own enterprise and derive an indication of how your organization is performing relative to a well-recognized standard. Some of these models focus tightly on system engineering, while others are software oriented or broadly tuned to the whole enterprise. In all cases, questions have to be answered, and the answers or products made available in response to questions are evaluated by someone to determine organizational qualifications for a particular rating.

17.7.1   Malcolm Baldridge National Quality Award

A gentleman named Malcolm Baldridge was the Secretary of Commerce during the Reagan presidency. On August 20, 1987, the 100th Congress passed the Malcolm Baldridge National Quality Improvement Act of 1987, establishing the machinery to identify and reward American firms characterized by quality. This action was motivated by the results of congressional studies indicating that the leadership of the United States in product and process quality had been challenged strongly by foreign competition, with the potential for very serious consequences in our economic viability in the long run.

The act established a Board of Examiners to evaluate performance of candidates for the award and a Board of Overseers to define the standards for the award, to review the evidence of company qualifications, and make recommendations each year to the Director of the National Bureau of Standards within the Commerce Department. The criteria for selection in the original act were:

a. The company must apply for the award in writing.
b. The company permits a rigorous evaluation of the way in which its business and other operations have contributed to improvements in the quality of goods and services.
c. The company must meet the requirements and specifications identified by the Board of Overseers and the Director of the National Bureau of Standards.

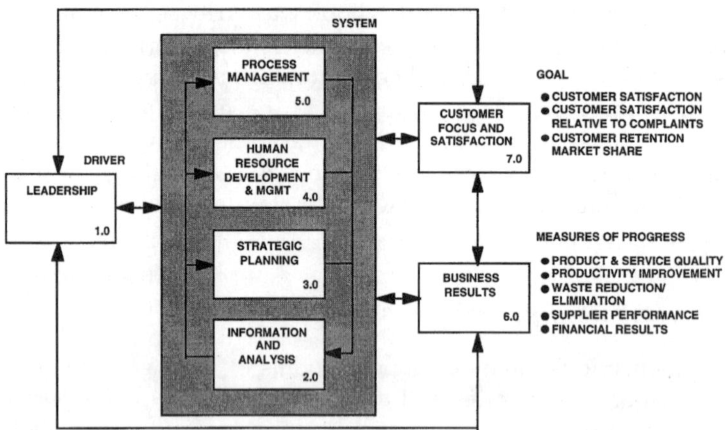

Figure 17.7   Malcolm Baldridge Award evaluation pattern.

Figure 17.7 shows the evaluation process applied by the examiners. The categories, eight of which are indicated on the figure, can be changed by the Board of Overseers as can the details supporting each category. The examiners are given free hand to evaluate materials supplied by the firm and to visit the firm for more detailed evaluation of those materials if desired. There is no cost for the evaluation, but the company attempting to win this award most often ends up spending a tremendous amount of money to attain it through improvements it has to make to be a contender.

The management literature contains papers on the sinking spell winners sometimes experience after winning the award, linking it to the same effect that makes it difficult for Super Bowl winners to repeat the following year. The intensity, concentration, and focus that one has to sustain has a draining effect, requiring a period of psychological rest. Even if you have not been on a Super Bowl winning team or won the Malcolm Baldridge award, you can perhaps identify with this feeling because of your experiences on a recent proposal effort.

17.7.2   *ISO-9001 certification*

The International Standards Organization has crafted a standard for development of products numbered ISO-9001, titled *Quality systems — Model for quality assurance in design/development, production, installation and servicing*, within the quality standards series numbered 9000, which has been applied by many organizations. Some have gone to the expense of becoming certified in its application, while others have claimed conformance through internal audit and adjustment of their process. There is nothing in this standard to which any reader of this book would take exception. All of the content makes good sense. The authors of the standard have distilled the basis for excellent enterprise performance down to its fundamentals. In reading a standard like this, and it is only six and a half pages long, one can't help wonder why companies have such a difficult time implementing these simple and sound practices and why any company would want to do anything else.

Any company can purchase a copy of the ISO quality standards and audit its own performance against those standards — changing practices, culture, and implementation to comply. Alternatively, it is possible to hire a certification firm to accomplish the audit, report the results, and help you attain certification status. One of the fundamental steps in the ISO certification process is that one have a written process and that you actually follow it. This is quite simple, but few companies comply with this requirement, especially the part about following their own internal practices.

While helping one company with some system engineering work where the company had become ISO-9000 certified not too long before, the author asked to see their system engineering manual to make sure he was not creating anything in conflict with their written practices. It turned out that the company did not have one. Their reasoning was that since they did not have a system engineering organization, it would not be proper for them to have a manual for an organization that did not exist. The ISO certification firm apparently agreed with them. The author agrees that it is not necessary to have a functional organization in order to do system engineering work on programs because system engineering is more of a process than a professional discipline and department name, but it would help to write down what it is and how one does it whether it is required by a standards body or not.

17.7.3   *Carnegie Mellon software capability maturity model*

Carnegie Mellon University was approached by DoD several years ago to establish a Software Engineering Institute (SEI) to develop a means of determining the maturity of an enterprise software development capability. DoD had become very concerned about the difficulties they were having maintaining software for fielded systems and anticipated these problems getting worse as their software dependency grew. SEI developed an organized process for satisfying DoD concerns involving a means of awarding a numerical score in the integer range 1 through 5 inclusive, 5 being the best. This is called a *capability maturity model* (CMM).

This process has worked out well, with only a few detractors. In general, it has been well received, and most defense contractors have worked very hard to move their software development capability at least to level 3 as encouraged by DoD. Table 17.1 defines the five levels in this system. The certification process involves answering a series of questions and producing evidence of compliance. The questions used in evaluating maturity are based on a structure of maturity levels (1 through 5), supporting key process areas and key practices which specify key indicators, as illustrated in Figure 17.8 from SEI CMM literature.

At the time this book was being written, one could count on one hand the number of firms in the world rated 5 under the SEI CMM software model.

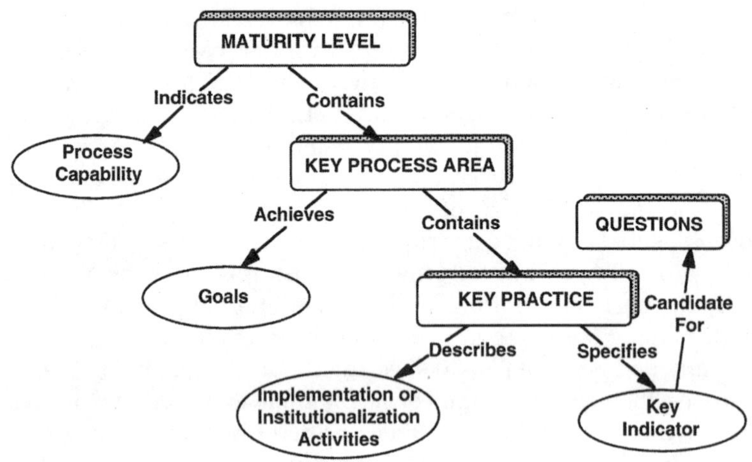

Figure 17.8   SEI software CMM.

Table 17.1   SEI Software Capability Maturity Model

Level	Name	Key words	Description
Level 5	Optimizing	Continuous Improvement	The organization has quantitative feedback systems in place to identify weaknesses and strengthen them proactively. Project teams analyze defects to determine their causes; software processes are evaluated and updated to prevent types of defects from recurring.
Level 4	Managed	Predictable	Detailed software process and quality metrics establish the quantitative evaluation foundation. Meaningful variations in process performance can be distinguished from random noise, and trends in process and product qualities can be predicted.
Level 3	Defined	Standard and Consistent	Processes for management and engineering are documented, standardized, and integrated into a standard software process for the organization. All projects use an approved, tailored version of the organization's standard software process for developing software.
Level 2	Repeatable	Intuitive	Basic project management processes are established to track cost, schedule, and functionality. Planning and managing new products is based on experience with similar projects.
Level 1	Initial	*Ad hoc* and Chaotic	Few processes are defined, and success depends more on individual heroic efforts than on following a process and using synergistic team effort.

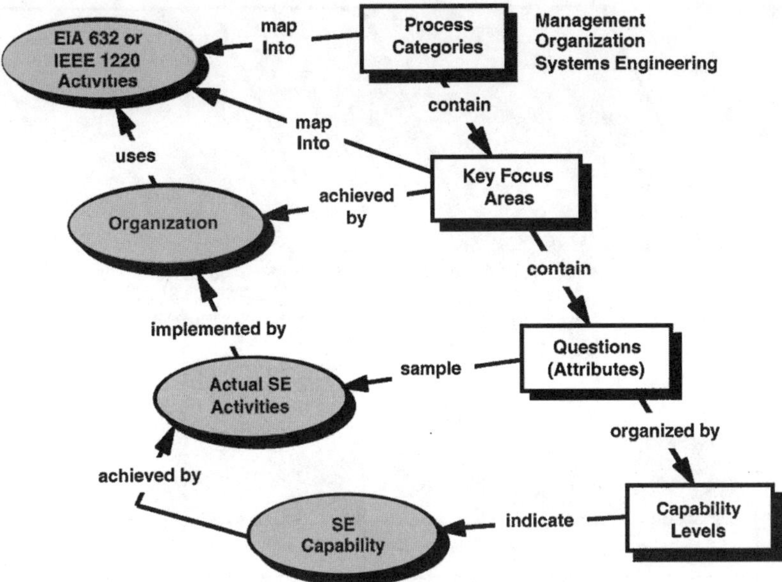

Figure 17.9   INCOSE SECAM interim model.

Of course, every firm developing software had not been evaluated by SEI, but the percentage of those evaluated level 5 was very low. Despite the simplicity of the requirements and their great good sense, it is very difficult to achieve something close to perfection in one's process and its implementation. This is true in software, in system engineering, and in industry in general in all fields. This delta between the relative ease in describing a sound system engineering process and the difficulty in implementing one was, in fact, what attracted the author's attention to the subject of this book.

17.7.4   *INCOSE system engineering capability maturity model (SECAM)*

The International Council on Systems Engineering (INCOSE) has created a capability maturity model as well. Its structure, illustrated in Figure 17.9, is composed of four model structures depicted by rectangles linked to four corresponding real-world system engineering elements depicted as ovals joined by directed line segments indicating relationships between these entities. The whole system engineering process is partitioned into the three process categories indicated in Figure 17.9, and each of these consists of several key process areas (KPA). Each KPA has associated with it a series of questions, organized by six assessment levels, that are the basis for the assessment process.

Figure 17.10 illustrates a sample set of scores for an organization based on assessment of their process in accordance with the questions in each of

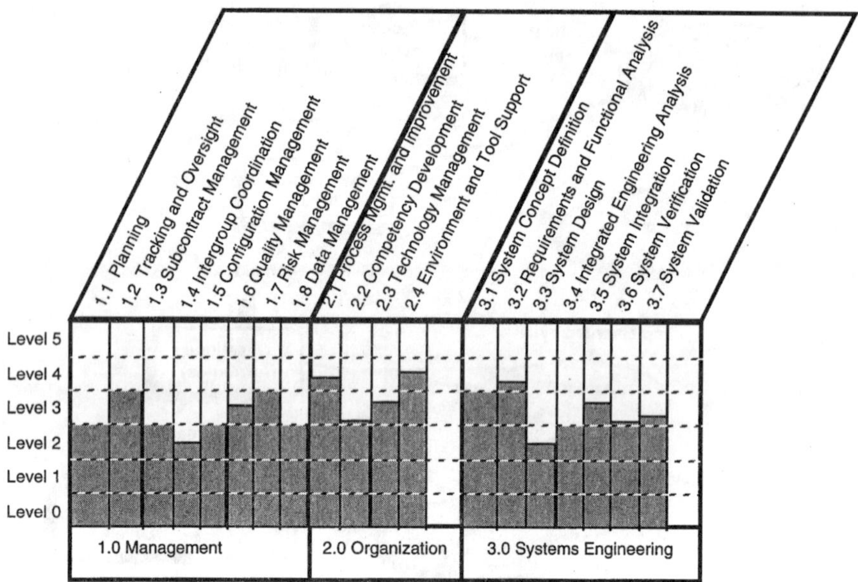

Figure 17.10   INCOSE SECAM scoring profile example.

the KPAs subordinate to each of the three process categories. Note that the INCOSE SECAM carries six levels rather than the five used in SEI CMM.

Table 17.2 identifies the six INCOSE SECAM capability levels and corresponding process attributes. The model also includes nonprocess attributes not included in Table 17.2. Level 5 is the highest or best capability.

Table 17.2 INCOSE SECAM Capability Levels

Level Number	Capability Term	Process attributes
5	Optimizing	Program process effectiveness goals are established upon business goals Continuous process improvement of program processes Continuous process improvement standards
4	Measured	Metrics derived from process data Quantitative understanding of program processes Ability to predict performance Program process-induced defects identified Program processes improved
3	Defined	Processes are defined by organizational standards Standards are tailored and used Tailoring is reviewed and approved Program processes data is collected Customer feedback is obtained
2	Managed	Policies define need for activities Activities are planned, tracked, and verified Work products are reviewed for adequacy Corrective actions are taken Work products are controlled
1	Performed	Activities done informally Non-rigorous plans and tracking Dependency on heroes Work products are in evidence General recognition of need for activity
0	Initial	General failure to perform activities No easily identifiable work products No proof something was accomplished

chapter eighteen

Motivation of the systems approach

18.1   The logic is not accepted everywhere

Throughout this book and in so many others on system engineering, the logic of the systems approach to the solution of complex problems rings true to the faithful, but there are many people who do not accept the notion that the systems approach well-applied will result in reduced cost, better product performance, and on-time delivery relative to the results obtained by an *ad hoc* approach. The available statistics quoted in many books on program management, system engineering, and system acquisition seem to support the utility of careful planning, definition of the problem as a prerequisite to solution, active integration across cross-organizational interfaces, an energetic search for potential problems (risks) followed by active reduction in their potential for occurrence and degree of damage to the program, and clear proof of requirements compliance as ways to develop better systems.

Clearly, spending money well in the early program phases to mitigate risks and prevent them from appearing later in the program has the effect of solving problems in their least expensive form when the investment is minimized. Solving problems after the engineering drawings, test plans and procedures, manufacturing and quality planning, customer training materials, and technical data are complete costs more than solving problems when a simple conversation in the hallway can preclude the problem from ever materializing. For example, a 15-minute conversation between four engineers costing the company $50/hour results in a total of (4)(1/4)($50) = $50. This same problem could cost $100,000 or more to correct in the late program phases and have a significant adverse impact on the delivery schedule. The logic of these ideas still fails to penetrate the defense systems of some of the people who work on these programs.

If it is true, as claimed by psychologists, that humans make an error in each 1000 operations and that the work involved in solving a very complex problem involves millions of operations, then we should expect that we will make thousands of errors in solving the problem. Some of these errors may be easily overcome or never even come to the surface as a problem. Others, discovered late, may force us into very costly corrective actions. We would therefore be wise to have an active mechanism to detect these errors as early as possible and avoid others. This is what the systems approach is intended to do by providing a structure and encouraging risk identification and mitigation.

It is very unfortunate that many of the people who harbor hostile feelings for the systems approach are in positions to cause deployment of the systems approach to be stillborn or otherwise make it a success or a failure in that they are program managers. One of the key reasons companies are not successful in the application of the systems approach is a failure of program managers to demand that it be applied and applied well. One other responsible position within an organization where we will often find resistance and sometimes open hostility to the systems approach is, of all places, the vice president or director of the engineering department, a position commonly held by a person who rose through the ranks of functional management in design or the analysis departments of engineering. A third string of resistance is often found down through the directors and managers of the design and analysis domain in the functional organization. Active resistance seldom comes from the working level.

18.2   The causes of poor system engineering performance on programs

System engineering is poorly done on programs for reasons that can be grouped into supply, demand, and deployment causes. Either the enterprise has a questionable system development process, there is no demand for an effective process (insufficient budget provided to do it well, for example), or the machinery is not in place to transfer a capability for doing the work well into a program which wants to apply it well.

18.2.1   Supply problems

In order to perform system development work well, the people in the organization must have a set of skills and knowledge that is fairly well defined and agreed upon in the many books and standards on system engineering. Unlike the past, there are now many books on this process. There is a growing list of universities that provide instruction in system engineering, either in the form of an elective course that can be taken in an engineering degree program or a complete system engineering certificate. Three University of California campuses (San Diego, Irvine, and Riverside) have certificate programs through their extensions. The universities of Washington and Arizona have been teaching the systems approach for years. There are many more across the country that offer the student some insight into this process. There are many consultants available in this field, stimulated by the premature departure of many managers from industry through downsizing. That is a vanishing resource but it will be available through the final decade of this millennium.

The point is that there is a way for an organization to improve the skills and knowledge of its personnel, as well as individual professional system engineers, to independently improve their understanding of the process even if their company is not so inclined. Nonetheless, there is a great need for system engineering education in industry because the process is not self renewing in most companies. There probably is not a written process description nor a consciously provided way for the people in an organization to acquire the knowledge to excel in implementing it. Seasoned system engineers depart the company for retirement with no way to transfer their knowledge to those just beginning their careers. While there are some universities that provide some form of educational opportunities in system engineering, it is nothing like the availability of mechanical, electrical, or chemical engineering educational opportunities. Also, these fields have been practiced well in industry in fairly standardized ways long enough that the transference of knowledge within the company in these fields is relatively easy.

However, the supply of effective system engineering capability is not abundant in industry despite the increasing availability of the resources necessary to make it happen. This condition is probably improving, but it has a long way to go to become a renewable resource within most companies.

The good news is that the engineers at the working level in the system engineering field seldom offer any resistance to learning and applying the systems approach to the best of their ability. The supply side problem is caused more by poor management in functional system engineering departments or the absence of departments charged with maintaining the company's system engineering knowledge than by working-level staff purposely thwarting the development of an effective systems capability. There are also a refreshing number of companies working to prepare system engineering practices manuals and find ways to effectively deploy those practices. The number is increasing, but many others have not discovered the simple secrets booming out of ISO 9001 and system engineering capability assessment models.

18.2.2 Demand problems

The author believes that the single biggest problem in system engineering is a failure of programs to demand excellence in this field. This failure can be traced to the desk of the program manager who, either through ignorance of the benefits to be derived or in spite of them, continues to insist on application of the *ad hoc* approach. Believers in the system approach shake their heads in disbelief that anyone could fail to understand that the system approach provides obvious benefits that result in programs that are easier to manage, satisfy all of the customer's requirements, and come in on cost and on schedule. There are reasons why program managers refuse to accept this logic.

Many people who seek out program management and are creative people and good *ad hoc* thinkers. Balancing on the edge of the cliff can be such a great challenge for these people that they may secretly not want to replace that opportunity with good planning that prevents these situations. Many of these people believe that high risk is a normal part of management rather than risk management (and mitigation) being a component of good management. The author knows of no way to retrain these people for service in later phases of programs. They may be very useful in earlier program phases except that they will have already buried many mistakes in the program by the time they hand off to a more stable individual to take the program through manufacturing and engineering development. The same characteristics that drive these people to become program managers are the same ones that lead to their failure to apply sound system engineering practices.

The only way the author would encourage that this kind of person be placed in a program manager role is with a system thinker as his/her deputy responsible for the day-to-day leadership role. One risky alternative that could be tried is to convince the risk-loving program manager of the amount of perverse enjoyment he/she might gain from applying the system approach for the first time.

There may still be some people in companies who lead programs from a position of ignorance about the benefits of the systems approach. The recommendation is that these people be identified and an attempt made to educate them. If that attempt is successful, and they have other qualities to encourage their continued employment as a program manager, then problem solved!

There is another class of program managers — those who know about claims for the improved management environment on programs that apply the systems approach but who have not yet experienced it. They continue to encourage the *ad hoc* approach because that is where their experience is. Despite knowing that their current methods will continue to result in problems, they do not have the self-confidence to apply the systems approach on their first program. These people need some handholding from the company system engineering manager. One of the things that encourages this attitude is that once money is spent, it cannot be restored. Thus program managers know that if they withhold money early in the program it will be available. If they spend it doing the system work early when it can be effective in reducing later risks, it will not be available later if the early work does not have the desired effect on later program phases.

One of the great truths of system engineering is that one must accept the notion of delayed gratification. Most often, the benefit derived from spending program money for system engineering work is not immediate. It is delayed for weeks, months, or years. Every parent understands how difficult it is to convince a teenage son or daughter of the wisdom of foregoing a social event in favor of doing assigned homework based on delayed gratification. Many people in earlier generations were forced to respect this discipline because the Depression and World War II interrupting their lives. With the U.S.A. at the pinnacle of world power and the good times continuing to roll, it is hard for people to relate to delaying gratification. This leads to an attitude of encouraging only those things that lead to immediate benefit, and system engineering simply does not work that way.

18.2.3   Deployment problems

This book will, it is hoped provide a resource for those who find themselves in an organization with an acceptable process defined in the functional organizations, qualified people to accomplish the work on programs which demand good performance in this field, but they do not seem to be able to deploy an effective capability into those programs. The earlier chapters of this book are laid out to support use of the book in a prescriptive way, building each layer of the deployment capability until complete. The reader is encouraged to apply it in that fashion, but the enterprise must have an effective functional system engineering capability, programs that demand it, and people ready to do the work. Generally, this is enough to encourage effective deployment but not always. It takes some experience to be able to do it well and, if that hasn't happened yet, it might be difficult to do well. As a minimum, you need to use these early programs as a source of critical feedback for ways to improve the process and your organization's ability to deploy it.

Many companies that have not had a proud and effective history of system engineering have established a system engineering council, by this or some other name, charged with bringing the company to a point where it is practicing the process well. This group writes a system engineering standard, monitors and studies program performance in this field, selects training needed to bring up staff knowledge, and researches computer tools useful in performing the work. In companies with an existing system engineering organization, it can act as this agent, of course. The author's preferred approach in this matter is the establishment of an enterprise integration team (EIT) to act as this catalyst, as discussed in earlier chapters.

18.3   Changing manager attitudes

Those in positions of responsibility for system engineering in companies should make an effort to find out the attitudes presently among managers in program and functional roles within their company in order to map out a bottom-up strategy for improving the company's systems capability. Your company will not improve in this area without the active support of these people. They will not necessarily have a positive outlook. While the logic of the systems approach will carry the day with some people, many others will be guided by their own experiences with the systems approach. Since is has seldom been applied well, that experience may not be good. This is the fundamental problem in improving the systems approach in companies. You need to build on past successes, but if you have not had any, how do you get this process moving? This is the typical boot strap problem.

Persons interested in deploying an effective systems approach can make the mistake of looking for the same motivation among all of those in opposition, and this is not the case. So we must apply a broad-band approach with different messages for different people. Two specific targets should be program managers and functional domain managers.

18.3.1   *Functional domain managers*

It sometimes happens in organizations that the functional domain managers (structural design, electrical engineering, etc.) come to feel a real dislike for system engineers and their organization. There may be many reasons for this, but let us trace just one of them. The author has talked to many managers about this since leaving captive employment and found a good deal of similarity in their responses. A common response is one of dislike for system engineers who failed to do a good job of requirements analysis (often referred to as requirements flowdown by the managers), leading to poor design solutions discovered after the fact in qualification testing, manufacturing, or subsequent customer use of the product. The situation is made all the more memorable for the domain manager when the system engineer is then rewarded for coordinating the solution to the big problem, which the domain manager believes he/she caused in the first place.

While employed as a system engineer, the author seldom went out of his way to find out what domain managers thought about system engineers or the way the process was applied in companies within which he worked. That was an opportunity lost that you should not repeat. You need to collect information on current attitudes both as a way of finding out how difficult the conversion process is going to be and to provide an opportunity to spread a positive message.

A survey involving forms to fill out in isolation is not recommended. You need to meet one-on-one with the domain managers to suggest the depth of your concern, which cannot be faked easily so don't begin this until you have managed to elevate your own motives to what is honestly best for the enterprise. Prior to these discussions, you do need to map out the questions that will provide the information you need and to prepare the supportive message you should deliver in the process of inquiring about managers' attitudes and experiences.

The person making these contacts should be from as high up in the functional system engineering tree as possible, such as the director, top staff, or department managers. Of course, if the organization reflects the author's belief in lean functional organizations, there may not be very many people who fit the bill. No matter. Even if the only person available to do this work is the director of system engineering, it should be recognized that there are few things more important to him/her than finding out how the current capability is perceived as a basis for improvements in perception or reality, as appropriate.

Ideally, the person making the survey has a good personal and professional relationship with the domain managers. If this is not the case, regardless of what else is done, improving interdepartmental management relationships should be given some effort. In this regard, the system engineering manager will find that all of his/her peers are vitally interested in what they do and will most often be delighted to speak about their job and the importance of it. So one approach to building a good relationship is to listen to these explanations. Another way to kick off an improvement is to ask these managers to participate in a no-cost lunch-time series of lectures for members of the system engineering department (and anyone else wishing to attend) on the different engineering departments and what they do. The author did this at General Dynamics Space Systems Division, and it worked so well that engineering chiefs would sometimes attend from other departments.

Many functional design department group managers have concluded that the system engineering process progressively damages their creative designers because of the brain-deadening appeal to rigor and discipline. These are characteristics of a misapplied system engineering process. The reality is that a program needs a variable environment with respect to creativity and discipline which can be thought of as being on opposite extremes of a continuum. In early program phases, one needs to appeal to the creativity of the staff to come up with good alternative approaches. In later program phases, as the investment in past decisions becomes great, the program needs to transfer into rock hard discipline, making it purposely hard to change the design. It is very hard to master the skills and techniques to be equally effective as a system engineer in both of these situations. Many system engineers came to this field through early or late program experiences over many years and have concluded that there is only a single balance point of worth between creativity and order. When such a person is first assigned to a program on the opposite end of the development spectrum, he/she will generally run into trouble because their model does not fit the need. Either they are being too open to change in later phases or too closed to alternatives in the early phases.

The reader may feel that this is not a significant problem, but as evidence one should consider what several communist governments did to their people and countries from 1920 through 1990. The practice of communism in the 20th century is not a bad model of system engineering run amok. Engineers who have experienced this environment while working as designers become very hostile to the process as managers.

The best way to cause design and analysis managers so afflicted to reconsider their attitudes toward system engineering is for the system engineering manager to convey to them the system engineering policy and belief that the system development process is based on providing the creative genius of design engineers as much solution space as possible and otherwise facilitating the communication of specialized knowledge between engineers to support that genius in the development and selection of preferred design concepts. This is difficult for some system engineers to do because they have incorrectly come to believe that all programs revolve around them. It is the creative genius of the design engineer from which the design flows. Take that away and everyone else becomes frozen in their tracks. Take away the system engineer and life goes on, not efficiently perhaps, but it goes on.

18.3.2   Program managers

We have already discussed the special problems of lack of demand for effective system engineering traceable to the program manager. If this is the only weakness in the chain of acceptance of the system approach, it may be helpful simply to draft company policy signed by the chief executive that the system approach will be applied on all proposals and programs in accordance with a released written standard. It is much harder for a program manager to ignore this kind of direction than when the policy has not been written down.

Another approach that can be helpful is to work with the estimating department, which may not have good skills and experience in developing system engineering estimates. It often happens that the estimate turned in by the system engineering organization does not survive a proposal team scrub because of lack of understanding of the utility of the services offered. It is much easier to convince the proposal manager of this when the estimating department is supportive than when they do not have the information they need to advise the manager on the real benefits of delayed gratification.

If the company has a department of program managers from which all program managers come, the person who manages this department needs to be approached first to find out whether they have a supporting, detracting, or flexible attitude about system engineering and to make an effort to change that attitude in the latter two cases. If this person is a supporter, it may be possible to establish policies encouraging the demand for sound system engineering practices and a staff that knows how to deliver it.

18.3.3   Managers within programs

In addition to the overall program managers, we will also find on programs several managers with system responsibilities that depend on how the program is organized and run. The program may be organized into functional departments (functional department subsets) responsible for technical development and may include WBS managers to coordinate the product axes of the program in terms of cost and schedule. Alternatively, and more ideally, the program may employ cross-functional teams in which the team managers have technical, cost, and schedule responsibilities. All of these managers within programs should have a positive attitude toward the systems approach, but that is often not the case.

One manager at McDonnell Douglas referred to those who were not supportive of the process as "misdirectors." This is not to point a critical finger at this company, which is doing a lot to improve its practice of system engineering, but rather to recognize a reality which was also poignantly made clear in a piece by U.S. Air Force Captain Tom Schorsh in the November 1996 (Volume 9 Number 11) issue of *Crosstalk,* the Journal of Defense Software Engineering, published by the Software Technology Support Center at Hill Air Force Base (reprinted from the August 1996 issue of *Nexus*). The piece was titled "The Capability Im-Maturity Model (CIMM)." It added a null and three negative maturity identifiers to the positive 1 through 5 identified in the Carnegie Mellon SEI software maturity model. Captain Schorsh was offering this model extension with tongue in cheek to a certain extent, but those familiar with system development programs in industry can relate to the levels he identified. Table 18.1 modifies both the Software CMM level descriptions (1 through 5) as well as Captain Schorsh's additions (0 through −3) to translate them from a software-specific to a system development generic meaning. Some of these words might be shocking, but that may be what is needed to cause some persons to reconsider their attitude toward effective development practices.

18.4 *Customer demand for improvement*

Failure to apply practices in the best interest of the organization has a lot in common with an individual's failure to change his/her habits to eliminate cigarette smoking, excessive eating, or excessive intake of alcohol. You know logically that these things have an adverse impact on your well being, and it is very simple to stop the bad practice. It is not *easy* to do it, however. It is often the case with alcohol and illegal drugs that the person must hit bottom before he/she can break out of the destructive pattern. So it is with some companies and their managers with regard to the poor practice of system engineering. One will hold onto the *ad hoc* approach almost to the company's grave and sometimes all the way to the end of the company. Perhaps we should form an organization for recovering *ad hoc* managers, complete with a 12-step process of renewal.

In the aerospace industry, it is often the government customer that provides the incentive for a company to cease and desist from its self-destructive behavior. Many companies have been told to either improve their system engineering capability or have a large contract withdrawn. Following replacement of a few clueless misdirectors, hiring of some qualified system engineers, training of the staff, establishment of a system engineering department to sustain improvement in this field, and preparation of written practices, these companies have some chance of survival.

In the past, it has been fairly easy for a company to hide the fact of its system engineering incompetence for some time in that the fruits of its failures ripen late in a program at a time when the customer's need for the

Table 18.1   System Capability Maturity Model

Level	Name	Key words	Description
Level 5	Optimizing	Continuous Improvement	The organization has quantitative feedback systems in place to identify weaknesses and strengthen them proactively. Project teams analyze defects to determine their causes, processes are evaluated and updated to prevent types of defects from recurring.
Level 4	Managed	Predictable	Detailed development processes and product quality metrics establish the quantitative evaluation foundation. Meaningful variations in process performance can be distinguished from random noise, and trends in process and product qualities can be predicted.
Level 3	Defined	Standard and Consistent	Processes for management and engineering are documented, standardized, and integrated into a standard development process for the organization. All projects use an approved, tailored version of the organization's standard process for developing systems.
Level 2	Repeatable	Intuitive	Basic project management processes are established to track cost, schedule, and functionality. Planning and managing new products is based on experience with similar projects.
Level 1	Initial	*Ad hoc* and Chaotic	Few processes are defined, and success depends more on individual heroic efforts than on following a process and using synergistic team effort.
Level 0	Negligent	Indifference	Failure to allow successful development process to succeed. All problems are perceived to be technical problems. Managerial and quality assurance activities are deemed to be overhead and superfluous to the task of system development. Reliance on silver pellets.
Level –1	Obstructive	Counterproductive	Counterproductive processes are imposed. Processes are rigidly defined, and adherence to the form is stressed. Ritualistic ceremonies abound. Collective management precludes assigning responsibility.

Table 18.1 (continued)   System Capability Maturity Model

Level	Name	Key words	Description
Level –2	Contemptuous	Arrogance	Disregard for good development engineering institutionalized. Complete schism between system development activities and system development process improvement activities. Complete lack of a training program.
Level –3	Undermining	Sabotage	Total neglect of own charter, conscious discrediting of peer organizations' system development process improvement efforts. Rewarding failure and poor performance.

product may be sufficiently desperate even with serious performance, cost, and schedule problems, to encourage them to take whatever they can get. It is relatively easy to describe an effective systems capability in a proposal because there is a lot written about it, but it is very difficult to deploy an effective capability even for a contractor with experience. The Department of Defense is rapidly becoming skilled at keeping track of past performance and sensing early signs of failure. They often require the application of the planning machinery necessary to clearly manage programs and identify true status. Cost/Schedule Control Systems linked to sound planning data defining clear task goals and exit criteria accomplished by a product-oriented team structure combined with team-oriented product technical performance measurement (TPM) parameters encourage accurate assessment of the current situation and early identification of program risks. On large contracts, DoD will also insist on having members of its program office in the contractor's teams, making it very difficult to disguise any attempt to mislead.

18.5   Moving the top person's attitude

All of the above techniques to improve the practice of system engineering are effective if the top executive in the enterprise is not hostile to the idea. In the face of high-level hostility or indifference, people in the lower tier will only be locally successful at best and then only temporarily. The organization takes its lead from the top and will generally support the leader's attitudes, especially where they are shared. Those who try to make improvements from the bottom often place their career in jeopardy because they are making waves amidst a sea of tranquility dominated by incompetence. Small victories are generally followed by great disappointments and a general background of rejection.

The best strategy for winning over potential detractors in an organization is to have a president or division general manager who is a solid supporter of the systems approach. People in this position will always have many things about which to feel passionate, but if system engineering is one of them, the enterprise is blessed. If this person also understands how to translate passion into action on the part of the people reporting to them through the application of a simple list of priorities repeated to the point of redundancy, then the organization is doubly blessed. If this is not the case, anyone who would want to make improvements in the practice of system engineering should put the first priority on improving the top manager's attitude toward the systems approach. Failing this, one might next check the job opportunities at a company already managed by a believer.

Let us say that the reader is intent on improving the systems approach at his firm, wants to stay there, and it is currently managed by a person who does not see the wisdom of the structured approach to development of product. How might this person enlighten the leader without adversely affecting their career? The first question is, Where is this person in the current chain of management? If this person reports to the top executive, there is an opportunity for the direct approach. If not, the person must work up through the chain starting with the person he/she reports to either on a program, in the functional organization, or both. At each stage of the organizational climb, the resistance may be different so one should not be so overwhelmed by an easy victory at the manager level as to fail to consider changes in the approach at the director level. The initiative has to be well thought through before it is taken to the lowest level in the chain in the interest of minimizing the overall time span. Ideally, the management hierarchy will remain stable while the initiative is being worked up the chain. A change at the director level after the previous director has been won over may cause a reset in the plan, forcing a recycle through the new director. In a volatile organization, the improvement initiative may fall by the wayside just from exhaustion of those leading the effort.

People who make it to the top levels of industry commonly think in terms of cost benefits, so your proposal for improvement should show clear measurable benefits, probably in terms of savings. This is hard to quantify where this data has not been collected in the past, and it is especially hard to allocate potential savings from the practice of an effective systems approach. The place to look for savings that could have been realized on past programs is in the context of the cost of correcting major errors. Try to find ways that the normal systems approach would have prevented the mistake and consequent costs. For example, you may find that there were 5000 engineering drawing changes subsequent to critical design review, where 95% of the originally planned drawings should have been released, signaling the near completion of the engineering design effort. This level of changes will have been driven by a faulty understanding of the requirements. It may be possible to trace at least some of these errors to omissions or mistakes in particular specifications and show that an effective structured analysis process would have prevented the mistake. Other design changes may be traceable to great problems that only came to light late in the program, and it will likely be possible to show how an effective risk management process would have encouraged the identification of the cause of the problem earlier, leading to a much lower cost of correction or mitigation.

A little humor can help. The brief story about the software engineer who says he does not have time to prepare a specification because he has so much debugging to do is one for which it is hard to avoid the point. The Dilbert cartoon strip has produced some classic statements like the one where Pointy Hair is telling Wally that he doesn't want to spend any money on defining the requirements because they are so rushed. However, Pointy Hair says he wants Wally to start designing so that people won't think they are doing nothing. The next view shows Wally, feet on the desk, reading the paper, daydreaming, "Of all of my projects, I like the Doomed ones best."

It is not enough to parade to the top executive, or any of those en route, with a demand for improvements. It is going to be necessary to offer a complete solution. This complete solution should have at least these parts:

a. Formation or continuation of a functional system engineering organization responsible for nurturing the capability, building its effectiveness, developing its process and documentation, acquiring needed tools, and training the people who will do the work. In a large enterprise with multiple programs, a matrix organizational structure is inescapable in order to provide the focus for both continuous process improvement and product-oriented leadership. Some organizations with no system engineering organization history have established a system engineering council, by whatever name, as a transition agent.

b. Building the deployable capability to do system engineering work on programs, using the resources to be provided by the functional system engineering organization.

c. The outline for a standard process in the form of a flow diagram, a brief description of each block, and a map between the functional organization and each block of the process diagram at some level such that it is clear how the department responsibilities form up into charters in the context of the process definition. Where more than one functional department maps to a particular process task, it should be made clear which of these has the principal responsibility and therefore must show leadership in process definition, tool selection, and personnel training.

d. Identification of one or two smaller near-term projects to which the new process can be applied to prove the process.

e. A list of risks in implementing the proposed changes that management should be aware of and how the likelihood of those risks coming to pass can be reduced, along with the severity of their consequences.

It should not be necessary to have the complete solution for all of these matters, but the plans should be laid out for accomplishing them, along with any cost estimates and an estimated payback time when improvements in productivity and reduction in program risk will compensate for the cost of the changes proposed. To the extent that your company has collected data on past program costs categorically, you may be able to craft some believable numbers. Otherwise, you may have to use estimates that appeal only to faith in the savings generally accepted, though this will probably be a hard sell to management.

18.6 The workers

The enterprise that would do this work well needs people who understand the process and know how to apply it. If the staff does not have the right skills at the beginning, a way to improve staff capabilities must be found. One route is through changing the people in the staff. Those without the necessary prerequisites should be moved to other company functions or separated. Those with the right background but not yet educated, need to be trained. If there are insufficient numbers of qualified people to perform system engineering work on real and anticipated programs, then a hiring or retraining process must be undertaken. If we must acquire new people, we should have in mind the characteristics of the people we want to attract. The list of characteristics below is a start:

a. Determination. Recognizing that it is sometimes very hard to accomplish system engineering work because of the numbers of people involved and their different views, system engineers must have determination to see something through to a successful conclusion.

b. Discipline. System engineers must work toward a condition where others follow a grand plan in good order, so they should not think, act, or speak in a way that detracts from this essential aspect of system development.

c. Creativity. It may be more than we should ask of one person to be both creative and disciplined in that the two are often at cross purposes, but system development work is complex and system engineers should be alert to ways to change the product or the process to enhance movement toward program goals faster, at less cost, and without compromising product performance, if possible.

d. Communication. System work crosses boundary conditions of the product and among the people working on the development. Therefore, those who would be system engineers must be good listeners, good speakers, and capable of writing clearly.

e. Logical Thinking. Given one or more related facts, one should be able to apply the rules of logic, whether or not they can be listed from memory, to achieve a valid conclusion based on the facts.

f. Technical Competence. No one can know everything, but a system engineer must have a founding engineering knowledge in some field. Good abilities in language, mathematics, and physics encourages the ability to communicate with other engineers who have skills and knowledge in a specialized field where the system engineer is weak, with full comprehension of the ongoing work and its consequences.

g. Functional Thinking. No one person can master every technology, so the broad system engineer must be able to think about product entities in terms of input–process–output without necessarily being able to master the details of the internal functionality in a particular technology.

h. Broad Interest. A system engineer cannot focus on the narrow interests of any one engineering domain, rather across several. You do not want engineers who want to design circuit boards doing system engineering work at the system level.

i. Inquisitive. The system engineer must be eager for knowledge.

j. Tenacity. One should not easily release an inquiry before it is complete. Problems have to be solved and issues closed despite distractions and purposeful attempts to deflect.

k. Mental Agility. People who would do this work should have the ability to change their minds based on the facts and adjust their approach quickly where warranted.

These kinds of lists appear to be a resumé for a god. We should not hope to find people who match our prescription precisely, but this list can be helpful in rejecting people who fail to match any of these characteristics. While an engineering manager in industry, the author interviewed many prospective employees and sometimes made mistakes in failing to hire qualified people and in hiring unqualified people. Sometimes, however, he made the right choices in excluding candidates. Two of these experiences remain vividly in mind. In one case, the recent college graduate said three times during the interview that he couldn't wait to get to the drafting board and start designing printed circuit boards. Clearly, this young man had not come to understand system engineering in school nor did he satisfy the broad interest criterion. In another case, the author could not tell the candidate where the work would be performed because it was a secret at the time that Teledyne Ryan Aeronautical unmanned reconnaissance aircraft were being flown into China and North Vietnam from Bein Hoa, South Vietnam. However, the author thought he should at least tell the candidate that some employees had developed stomach and other problems from apparent stress caused by hostile action around the base of operations (mortar and rocket attacks). The candidate replied that there was no reason to worry on that score because he had already had ulcers. Logic was not one of this candidate's strong suits. As a minimum we should at least prevent the wrong people from getting into the system engineering staff.

18.7 Need for motivation

The most we can expect from the general population at work in the system engineering field is that they will apply themselves to learning the relevant techniques and practices and do their work on programs in accordance with that teaching as well as they possibly can. One might ask, "Yes, but what is the motivation for doing so?" The author believes this is called accepting a paycheck. Those on the payroll in a system engineering function should, like any employee, do their work professionally. Those who cannot or will not do so should be first approached in an attempt to correct performance problems and, if that is not successful, these people should be separated from the system function as well as any other function.

Beyond this simple binary motivation, it is possible to expand the appeal to those not influenced by simple compensation. Pay increases and promotions should be allocated to those who do or can lead this work well. In order for this to happen there has to be someone with enterprise responsibility for system engineering, a functional department manager who can maintain continuity and individual records of performance. This is the same person who must also be responsible for building the enterprise system engineering competence in cooperation with the enterprise integration team through process definition, staffing decisions, tool selection, education, and practices development and maintenance.

Most everyone working in industry is interested in advancement. A sound career path is not always identified in companies for people doing system engineering work. Three pathways should be identified: project leadership, functional management, and technical excellence. The use of cross-functional teams provides many opportunities for team leadership opportunities, and people with system knowledge are good selections for these roles. Those who excel in team leadership should have an open road to program management. Functional management opportunities are generally fewer in number than program opportunities, because there is only one functional entity but many programs. Ideally, these leadership roles should be open to promotion from within. Every organization includes people who are not interested in leading the work of others but are motivated to continuously improve their skill and knowledge in their chosen field. There should be opportunities for advancement for these people as well, but they need to understand that this pathway will seldom be as financially rewarding as leadership roles.

Many system engineers are mission-oriented people motivated inwardly to do the best they can without external encouragement, but they are probably not in the majority. Everyone needs and probably deserves recognition for their accomplishments, preferably in public. If we accept this idea, we should also accept that those who do not do well need and deserve critical feedback, preferably in private.

All of these means of motivating people to work well are extensions of good leadership, which we all deserve. There is no greater motivating force for most of us than the knowledge that we are working for someone who knows what they are doing and is interested in causing us to do our very best in working toward known goals. Persons who offer good leadership are involved in the education, encouragement, and guidance as well as direction of those in their charge. Some feel that teams should be self-directed, but the author has not succumbed to that attitude. Self-directed teams will never be as efficient as teams with good leadership by someone skilled in giving it.

chapter nineteen

Closing

If it is true that a structured approach in the development of product systems will yield better results than an *ad hoc* approach, shouldn't we also apply these same techniques to our process system? In so doing, we will evolve a clear definition of our process, recognize a valid need for a functional organizational structure and how best to task that structure, and acquire an efficient program transform from generic program planning data to program-specific planning data.

This book has shown how to identify all of the elements of an enterprise planning infrastructure and connect these elements into strings that can be selectively pulled into program-peculiar planning, where time and manpower intensity factors are added.

In that we have encouraged the use of cross-functional teams on programs that are aligned with the product architecture and insisted that all program work and planning data be linked to this same architecture, we have simplified the planning data as much as possible. By cataloging all of this work within the context of an understandable pattern, we have found a way to communicate subsets of the whole to those responsible for accomplishing it.

The content of this book is thought to be consistent with any and all system engineering assessment models, ISO-9001, and process standards. We have encouraged that an enterprise prepare its own process specification or standard mapped to one or more selected standards known to be favored by their customer base. It is imperative that having once prepared a process standard that it be followed, so management energy must be expended to gain acceptance and willing compliance with the content of this process standard.

It is insufficient to create a one-time condition of excellence and expect that the condition will survive in a changing world without maintenance. An enterprise needs a means of assessing how well it is doing on programs that results ideally in numerical data that can be compared over time. Based on information and lessons learned from experience, we should determine changes that will be most effective in improving our program performance and implement those changes.

In an organization that does not have a history of excellent systems capability, an outsider often finds an attitude of near despair openly commented upon. Comments are expressed in gallows humor like, "I'll give you odds we'll screw this one up too." This is very self-destructive behavior in any organization, whether it be a Boy Scout troop, football team, military unit, family, or business enterprise. Where this kind of behavior is permitted to continue, it eats away at the foundations of any hope for rebirth. Workers who think these thoughts have most often been badly led, and it may be impossible for those same leaders to change worker attitudes, restoring or creating an attitude of exuberance about their work.

Yes, a lot of the work described in this book is not always fun, or even rewarding, but the development of solutions to complex problems should be a joy, an adventure, and should contain the makings of a fine, professionally fulfilling career. When management finds the right structure for work organization, manages that work well, and is at least halfway decent to the people doing the work, they perform a great service not only to those who own the business and to its customers, but to the working troops as well. The differences between the employee who works for a paycheck despite bad work experiences and the employee who can't wait to get to work do impact the workplace, but they flow far afield from the workplace gates as well.

The consequences of failure to do these things must rest at the feet of the manager, whether engineering chief, CEO, or someone in between. The direction to change and improve must come from the manager as well, because change stimulated from the bottom cannot be sustained. It burns out even the individual zealot because there is so much resistance to it. The only way to sustain improvement from below is a conspiracy that has the discipline to rotate its leader as burnout nears, reflecting the logic of flocks of birds that fly in formation. It takes more energy to fly the lead position than it does to be a wingman. Where this process must go on covertly, however, it challenges the normal patterns of leadership and management and cannot help but create an unhealthy working environment by undermining management, despite the best motives.

While each person in the management stream must share the blame for a poor development process and should do what he/she can within the management space, the enterprise is a system and isolated positive actions will seldom cause that system to move monotonically toward the mountain top. If a system is performing badly, one can defend it by seeking out all of the systemic causes or reengineering the whole. This book focuses on the latter in order to communicate the whole pattern of behavior, but there is nothing to prevent the practitioner from applying structured analysis to the organization as you would to a major modification of your product. In this approach, you are mindful of the existing reality and seek the most cost-effective way to change system behavior to achieve new functionality.

In any case, management must take the lead in recognizing the failures of an organization and either act smartly to begin to correct them or have the good grace to step out of the way. If you can see beyond the next management buzz word to a long-term, never-ending improvement process following a sound systems approach to process improvement, you are probably well suited to lead the charge, and good luck to you.

Appendix A

Enterprise process definition

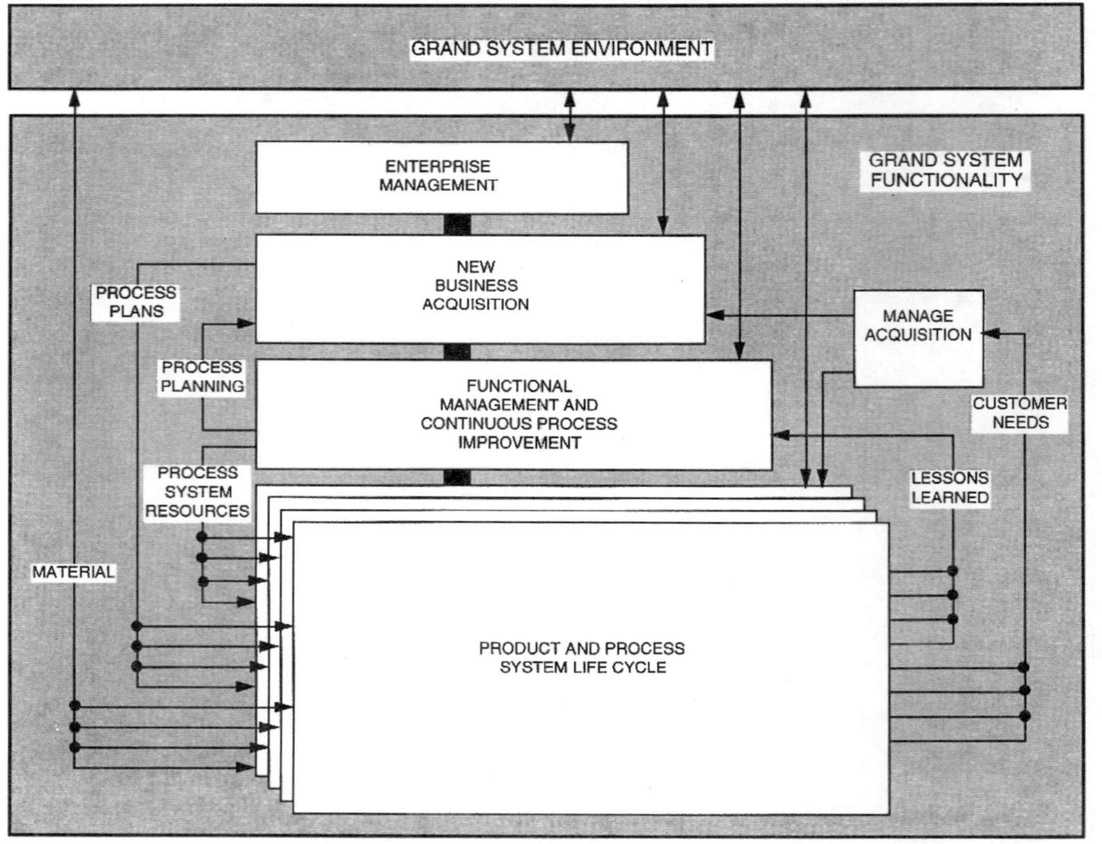

Figure A.1 Grand system functionality.

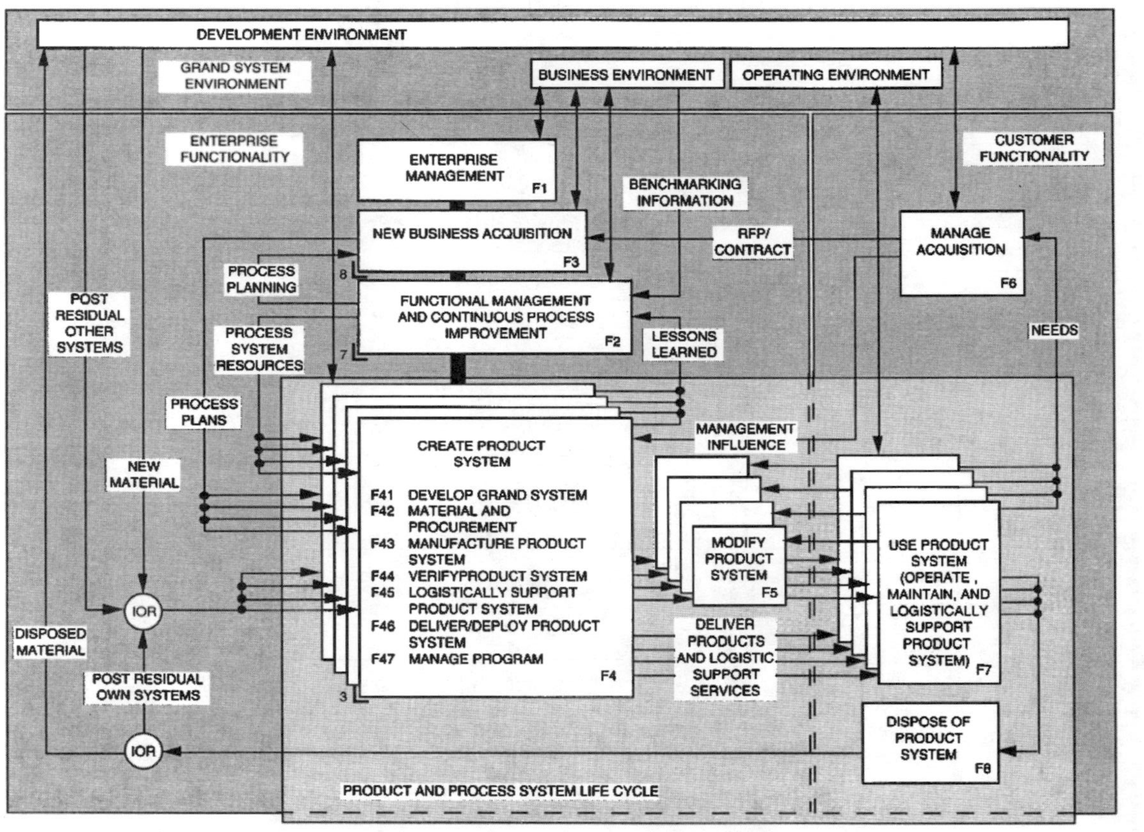

Figure A.2   Life cycle master flow diagram.

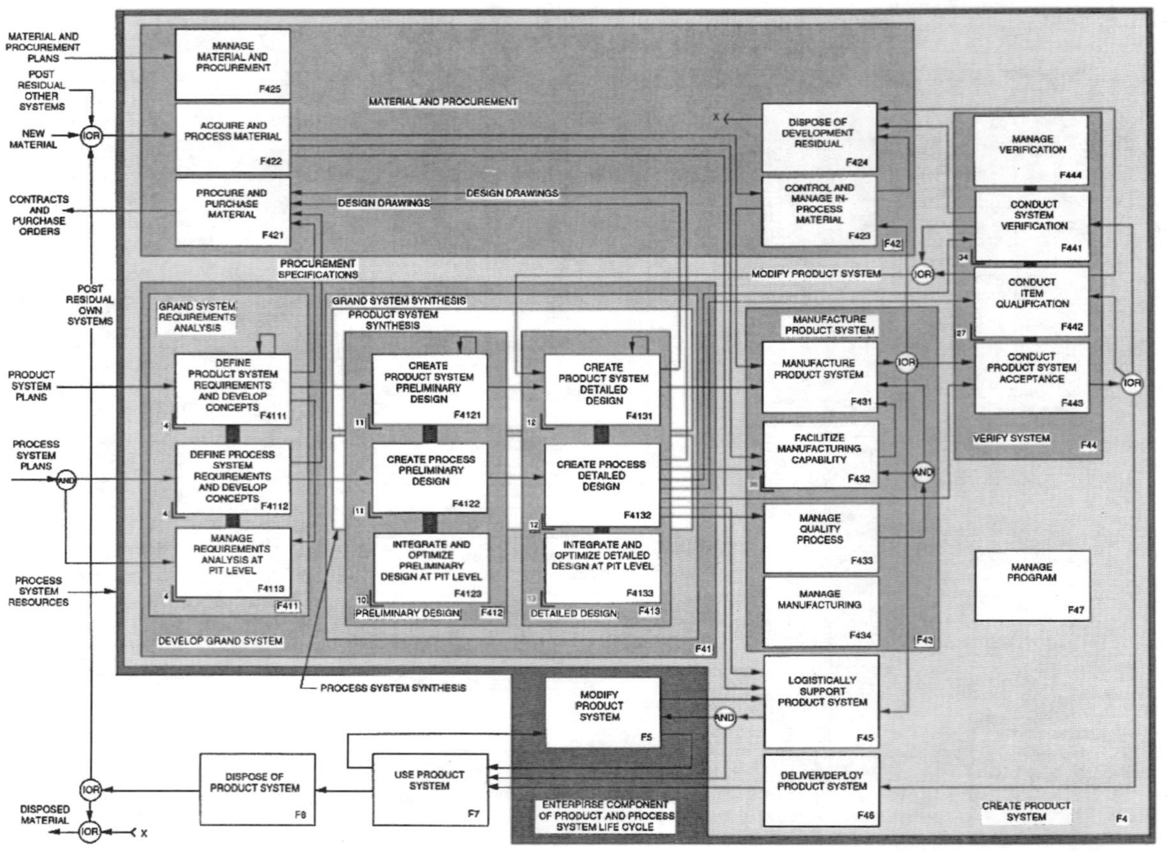

Figure A.3 Create product system, F4.

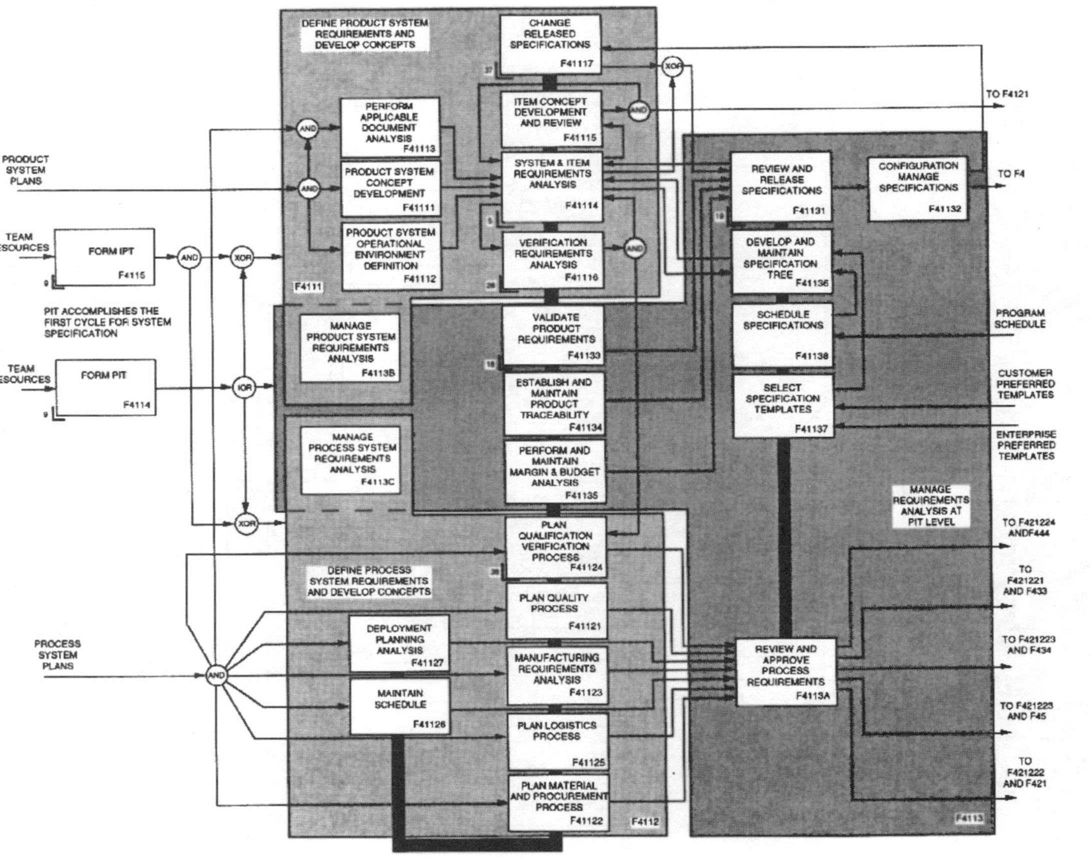

Figure A.4   Grand system requirements analysis, F411.

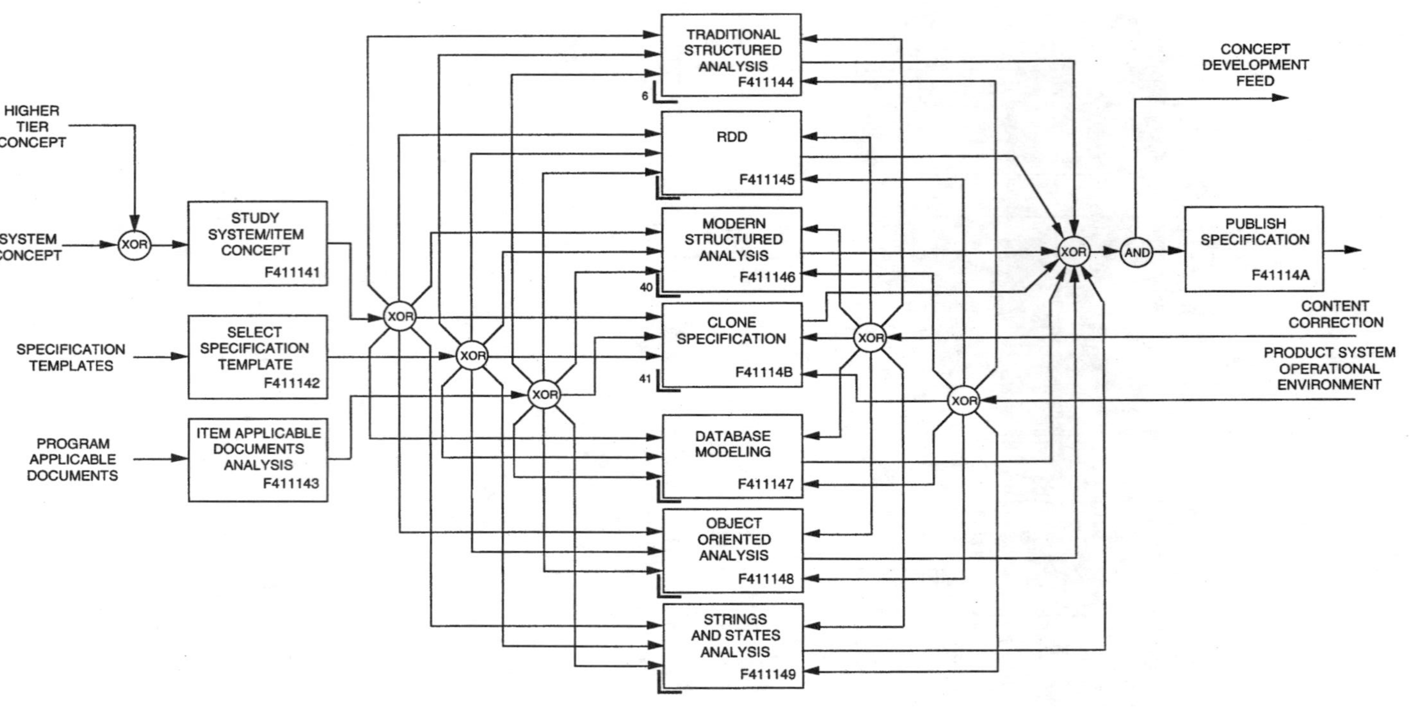

Figure A.5 System and item requirements analysis, F41114.

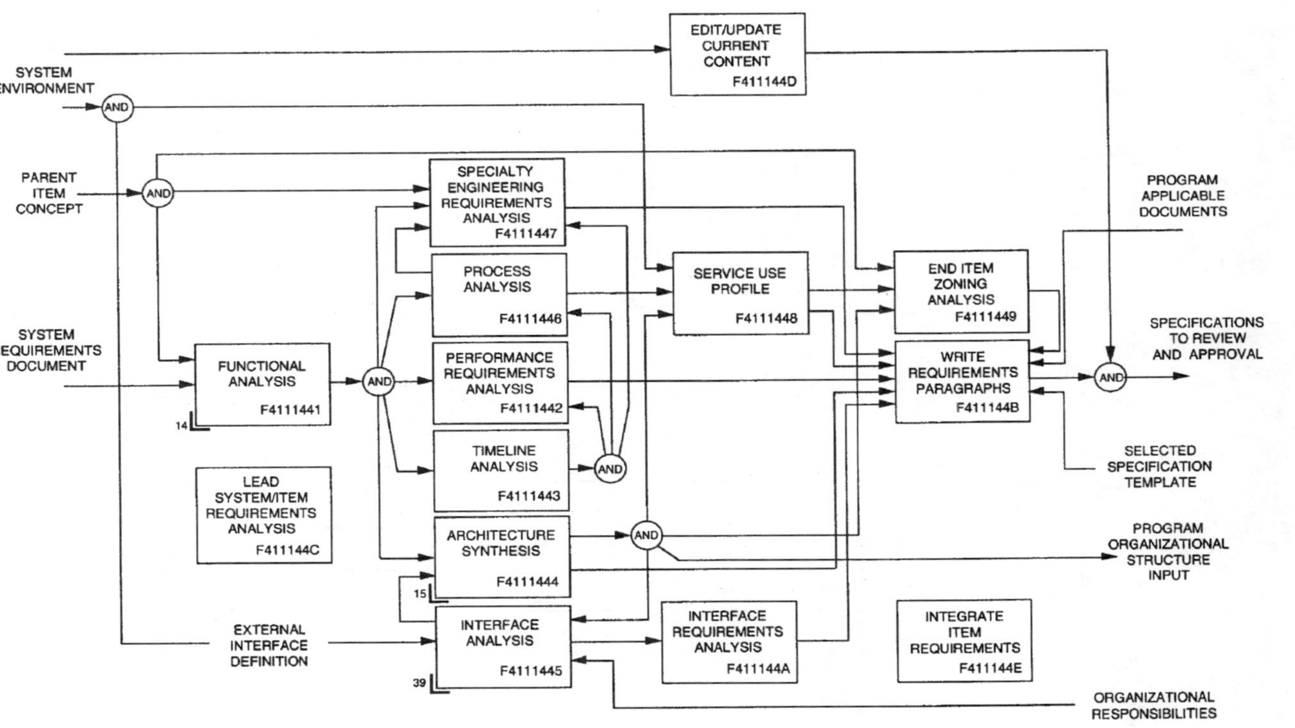

Figure A.6   Traditional structured analysis.

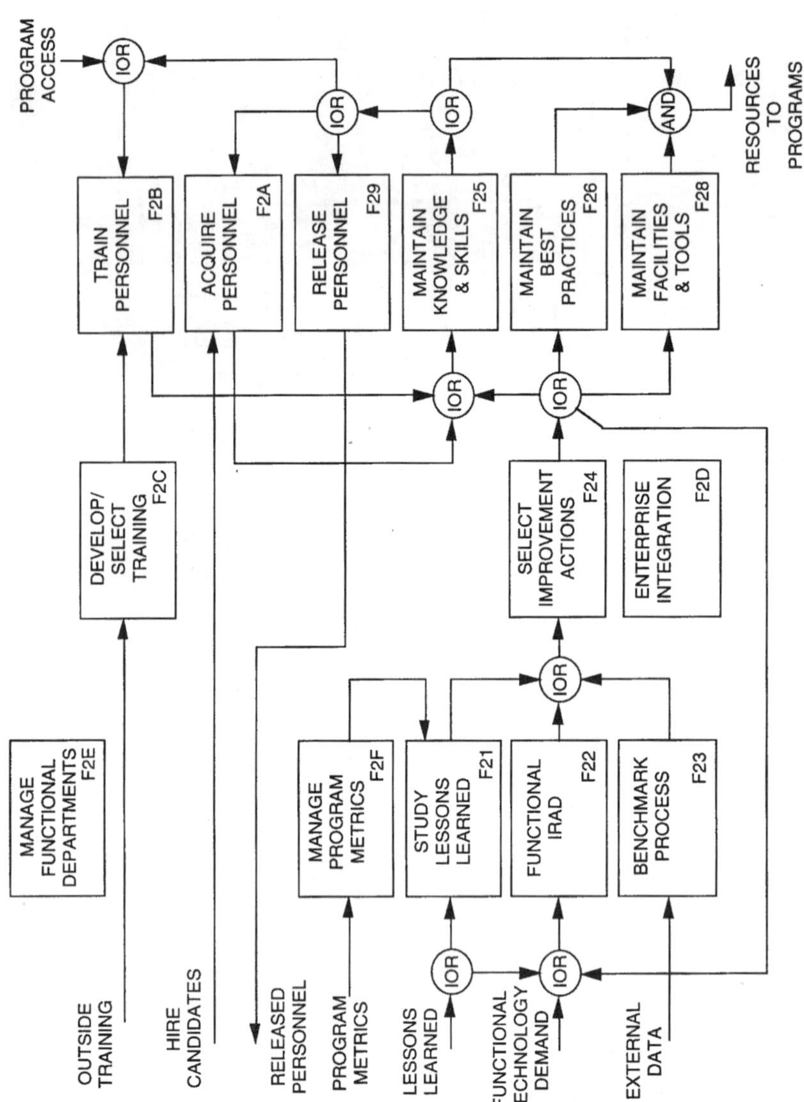

Figure A.7 Functional management and continuous process improvement, F2.

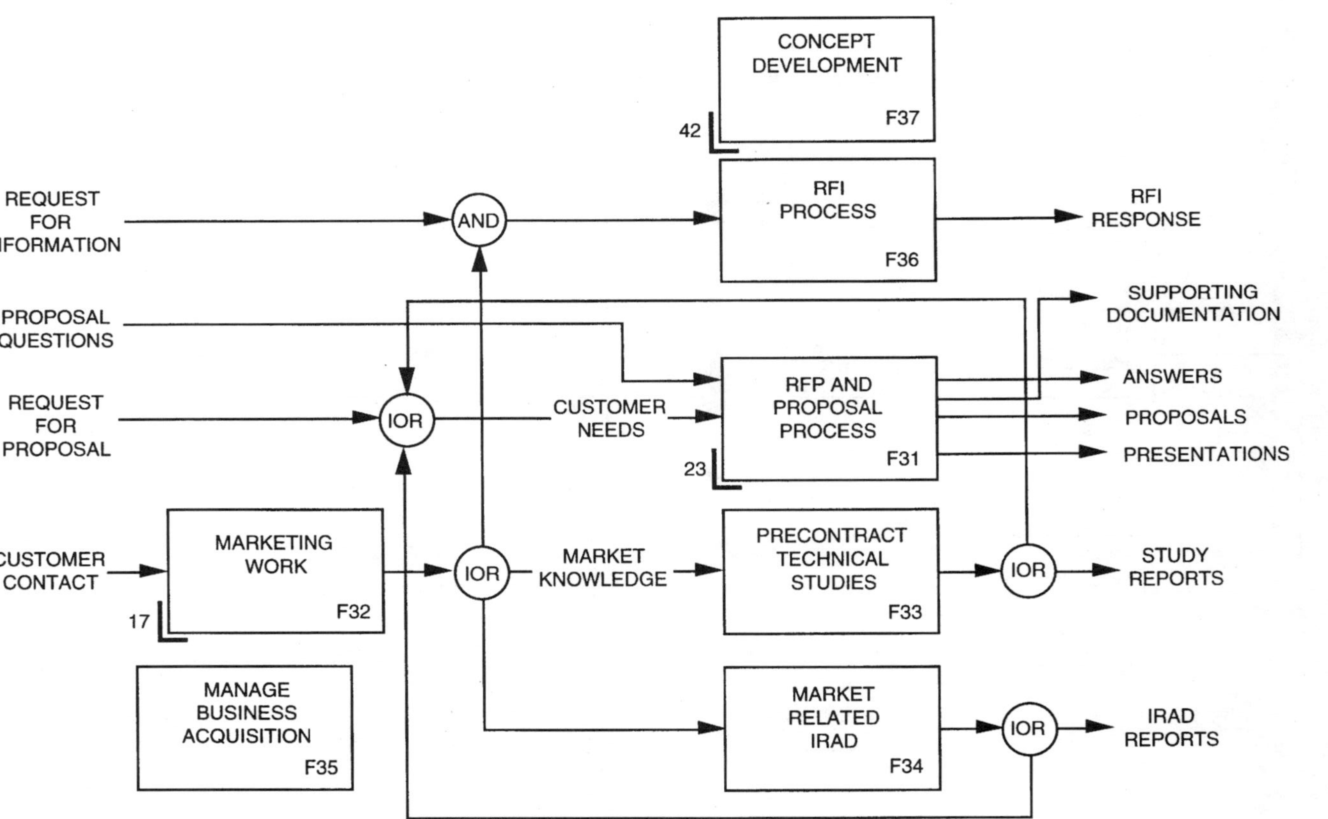

Figure A.8   New business acquisition, F3.

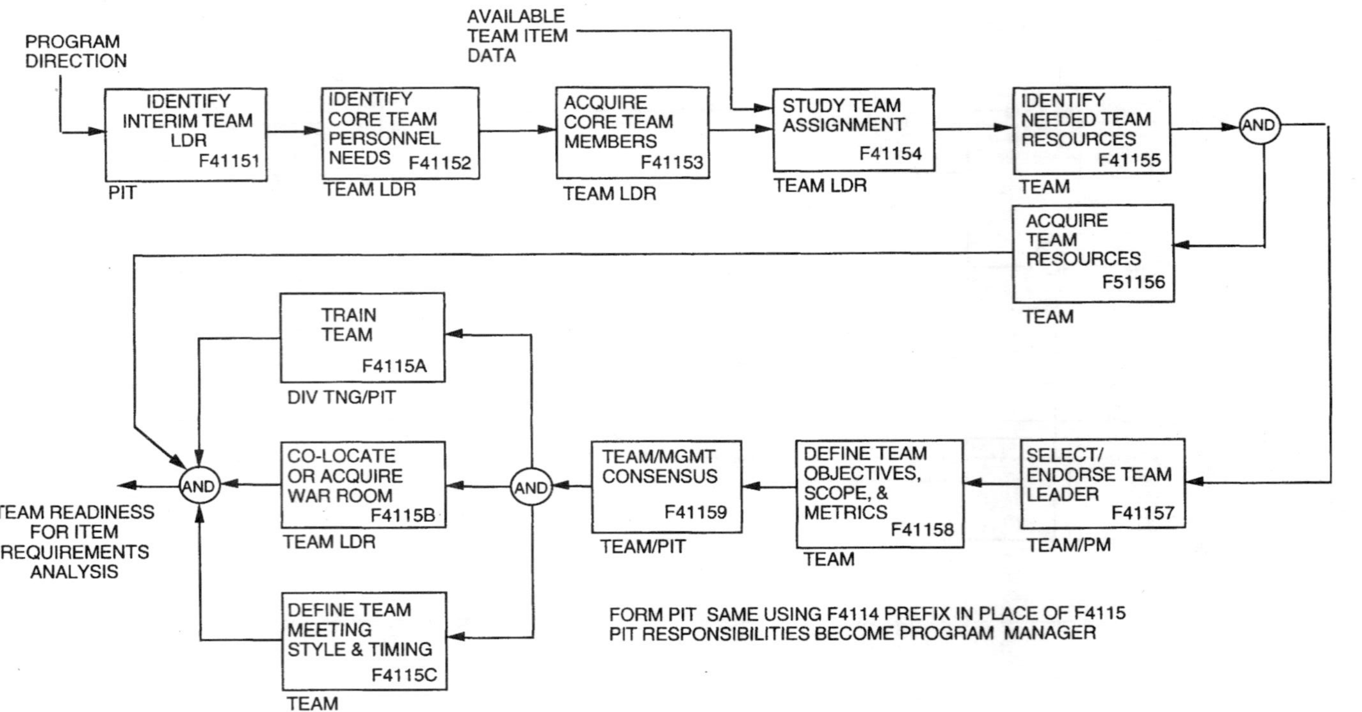

Figure A.9   Form PIT, F4114 and form IPT, F4115.

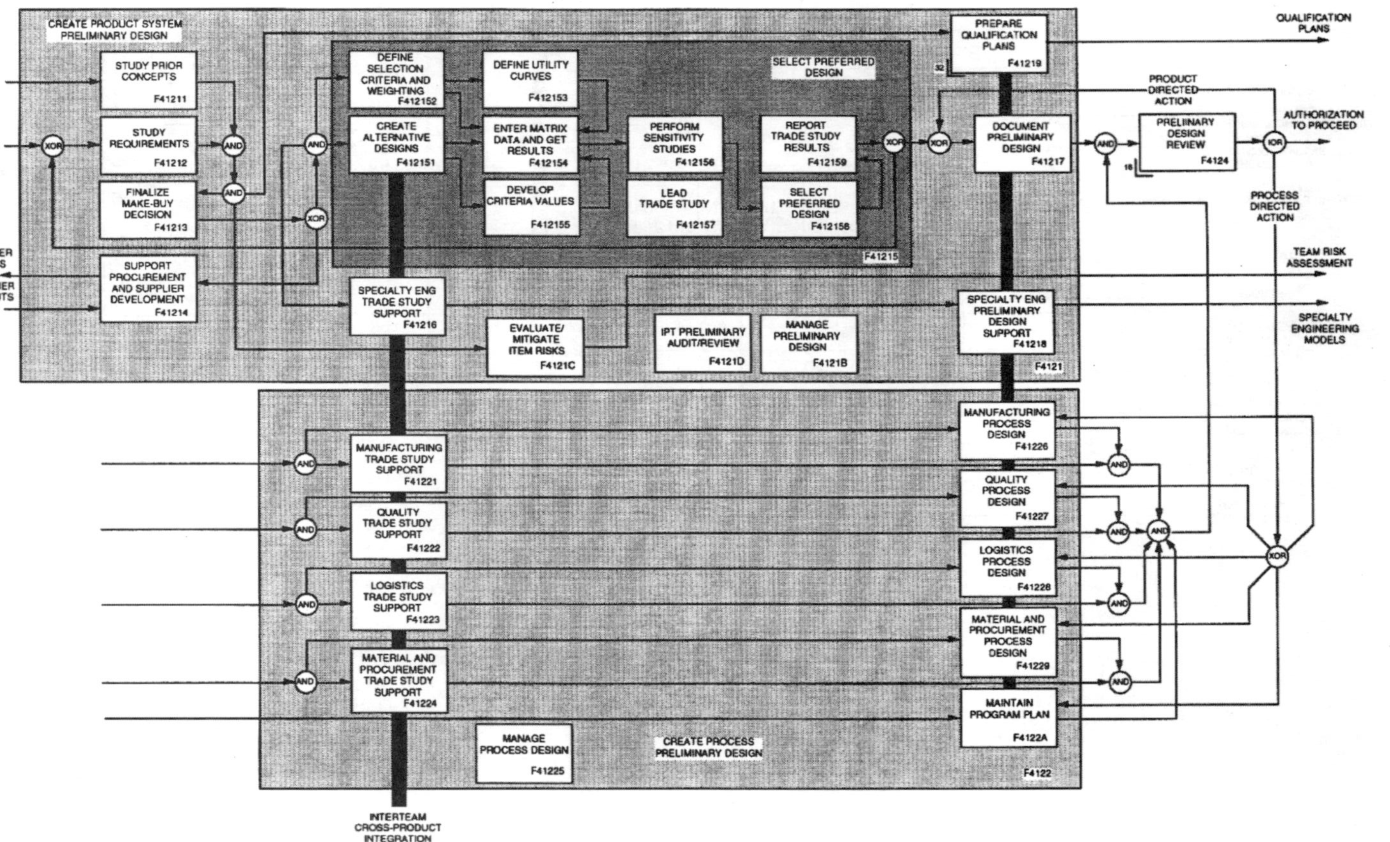

Figure A.10 Preliminary design, F412.

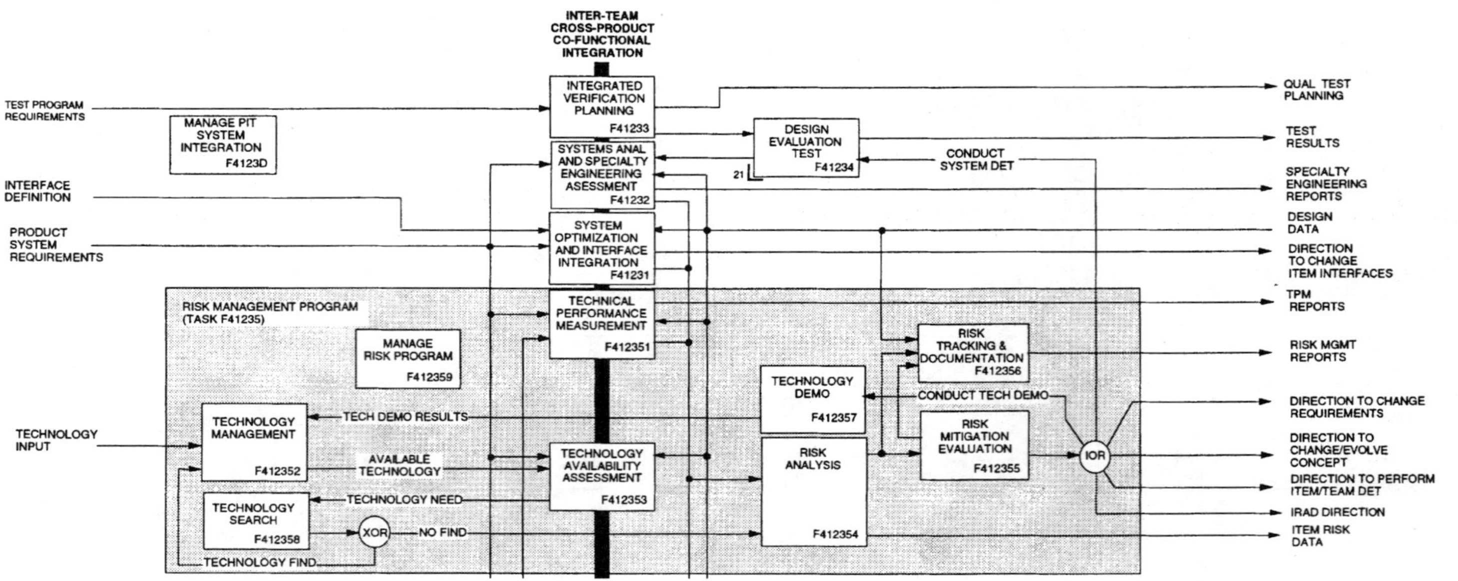

TEST PROGRAM REQUIREMENTS
INTERFACE DEFINITION
PRODUCT SYSTEM REQUIREMENTS
TECHNOLOGY INPUT
MANAGE PIT SYSTEM INTEGRATION
F4123D
INTER-TEAM CROSS-PRODUCT CO-FUNCTIONAL INTEGRATION
INTEGRATED VERIFICATION PLANNING
F41233
SYSTEMS ANAL AND SPECIALTY ENGINEERING ASESSMENT
F41232
SYSTEM OPTIMIZATION AND INTERFACE INTEGRATION
F41231
TECHNICAL PERFORMANCE MEASUREMENT
F412351
DESIGN EVALUATION TEST
F41234
21
CONDUCT SYSTEM DET
RISK MANAGEMENT PROGRAM (TASK F41235)
MANAGE RISK PROGRAM
F412359
TECHNOLOGY MANAGEMENT
F412352
TECHNOLOGY SEARCH
F412358
TECH DEMO RESULTS
AVAILABLE TECHNOLOGY
TECHNOLOGY NEED
XOR
NO FIND
TECHNOLOGY FIND
TECHNOLOGY AVAILABILITY ASSESSMENT
F412353
TECHNOLOGY DEMO
F412357
RISK ANALYSIS
F412354
CONDUCT TECH DEMO
RISK TRACKING & DOCUMENTATION
F412356
RISK MITIGATION EVALUATION
F412355
IOR
QUAL TEST PLANNING
TEST RESULTS
SPECIALTY ENGINEERING REPORTS
DESIGN DATA
DIRECTION TO CHANGE ITEM INTERFACES
TPM REPORTS
RISK MGMT REPORTS
DIRECTION TO CHANGE REQUIREMENTS
DIRECTION TO CHANGE/EVOLVE CONCEPT
DIRECTION TO PERFORM ITEM/TEAM DET
IRAD DIRECTION
ITEM RISK DATA

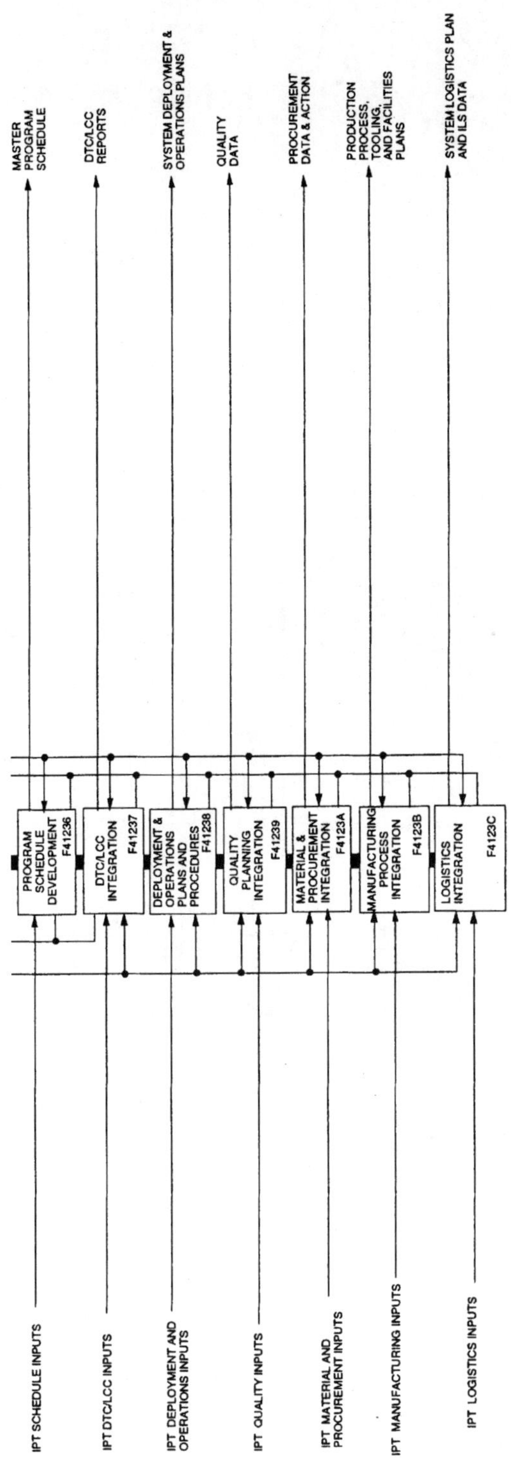

Figure A.11 Integrate and optimize preliminary design at PIT level, F4123.

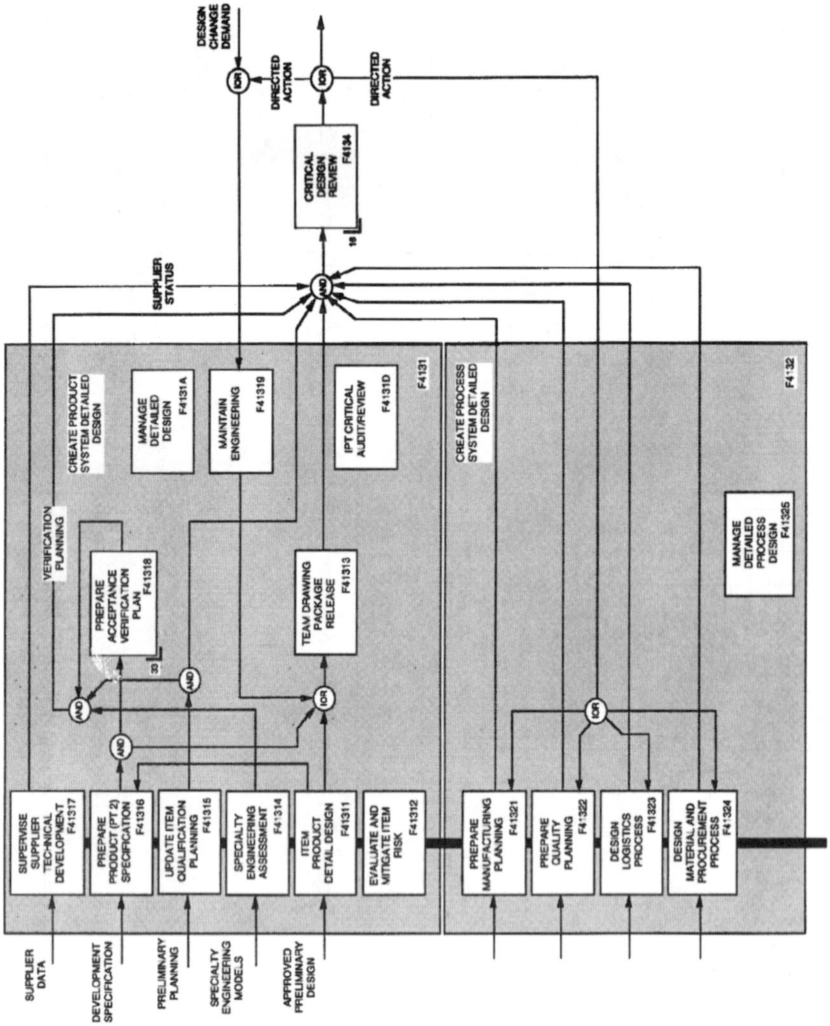

Figure A.12 Detailed design, F413.

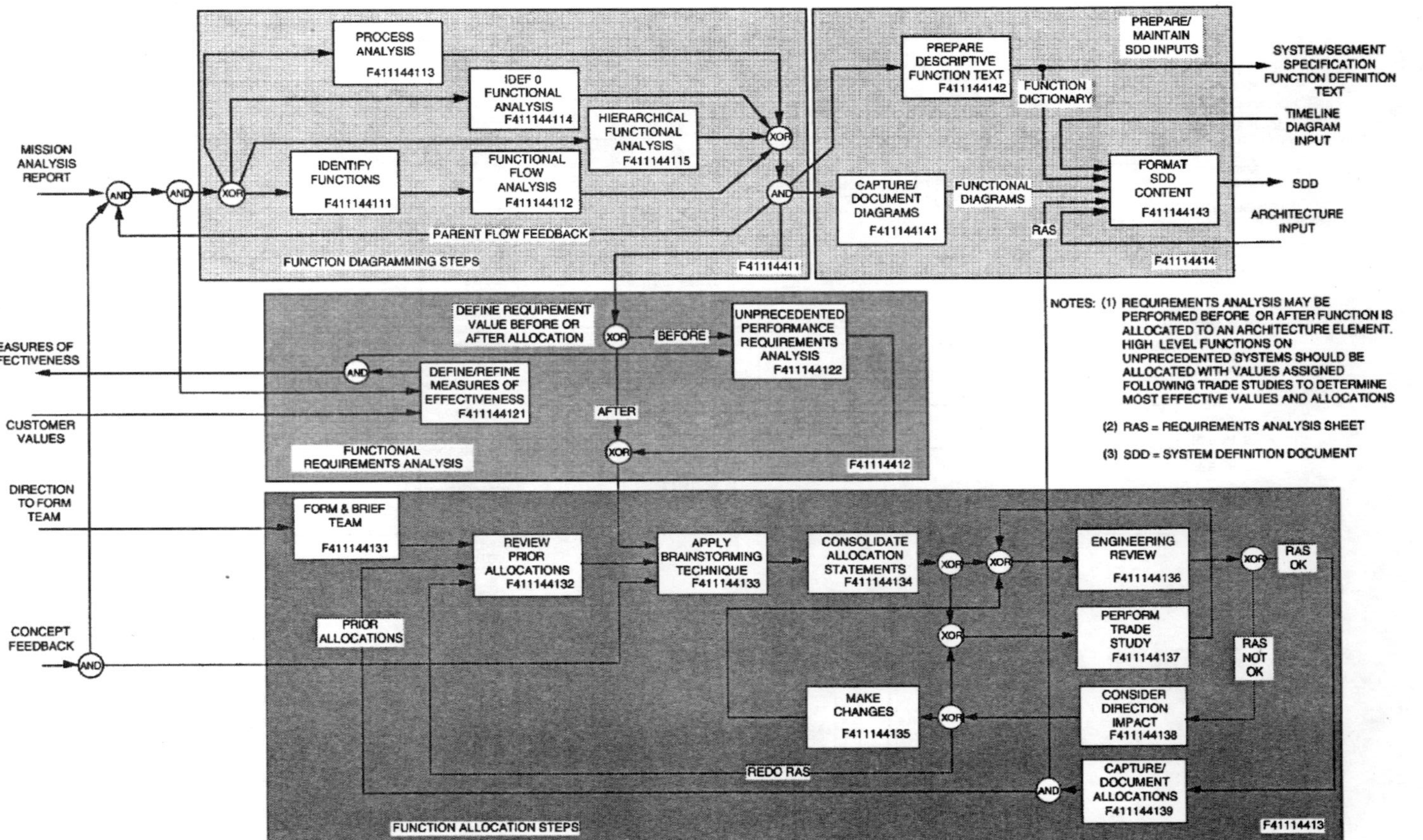

Figure A.13   Functional analysis, F4111441.

303

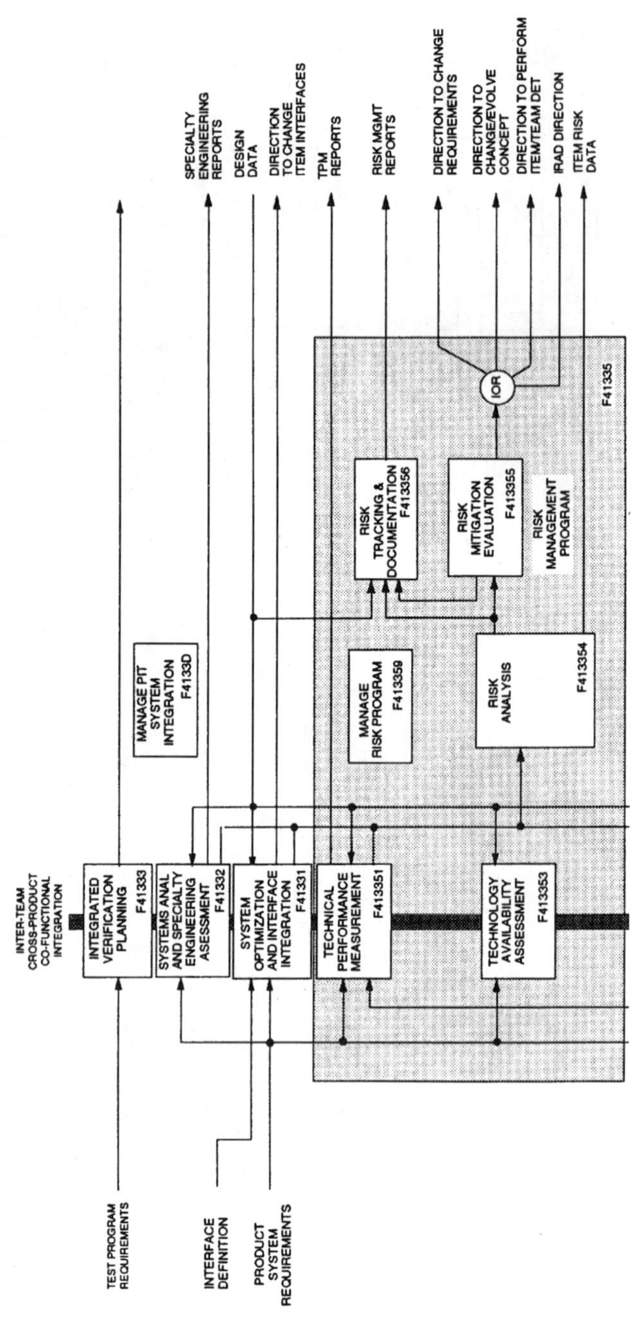

TEST PROGRAM REQUIREMENTS
INTERFACE DEFINITION
PRODUCT SYSTEM REQUIREMENTS
INTER-TEAM CROSS-PRODUCT CO-FUNCTIONAL INTEGRATION
INTEGRATED VERIFICATION PLANNING
F41333
SYSTEMS ANAL AND SPECIALTY ENGINEERING ASSESSMENT
F41332
SYSTEM OPTIMIZATION AND INTERFACE INTEGRATION
F41331
TECHNICAL PERFORMANCE MEASUREMENT
F413351
TECHNOLOGY AVAILABILITY ASSESSMENT
F413353
MANAGE PIT SYSTEM INTEGRATION
F4133D
MANAGE RISK PROGRAM
F413359
RISK ANALYSIS
F413354
RISK TRACKING & DOCUMENTATION
F413356
RISK MITIGATION EVALUATION
F413355
RISK MANAGEMENT PROGRAM
IOR
F41335
SPECIALTY ENGINEERING REPORTS
DESIGN DATA
DIRECTION TO CHANGE ITEM INTERFACES
TPM REPORTS
RISK MGMT REPORTS
DIRECTION TO CHANGE REQUIREMENTS
DIRECTION TO CHANGE/EVOLVE CONCEPT
DIRECTION TO PERFORM ITEM/TEAM DET
IRAD DIRECTION
ITEM RISK DATA

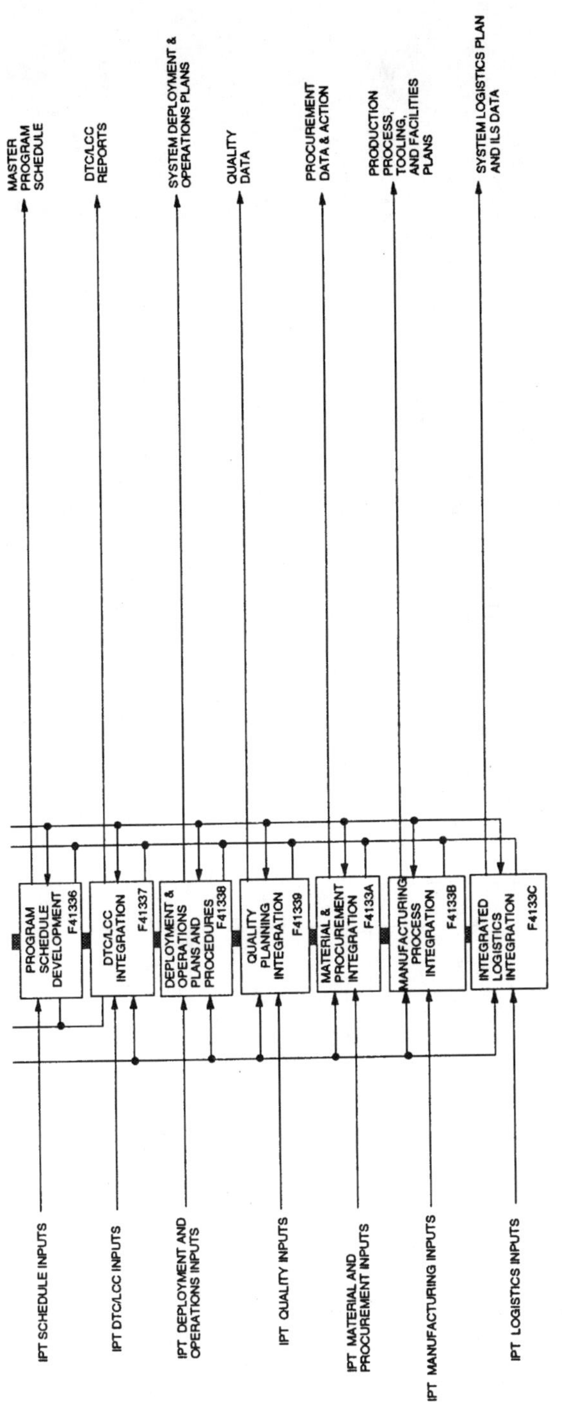

Figure A.14 Integrate and optimize detailed design at PIT level, F4133.

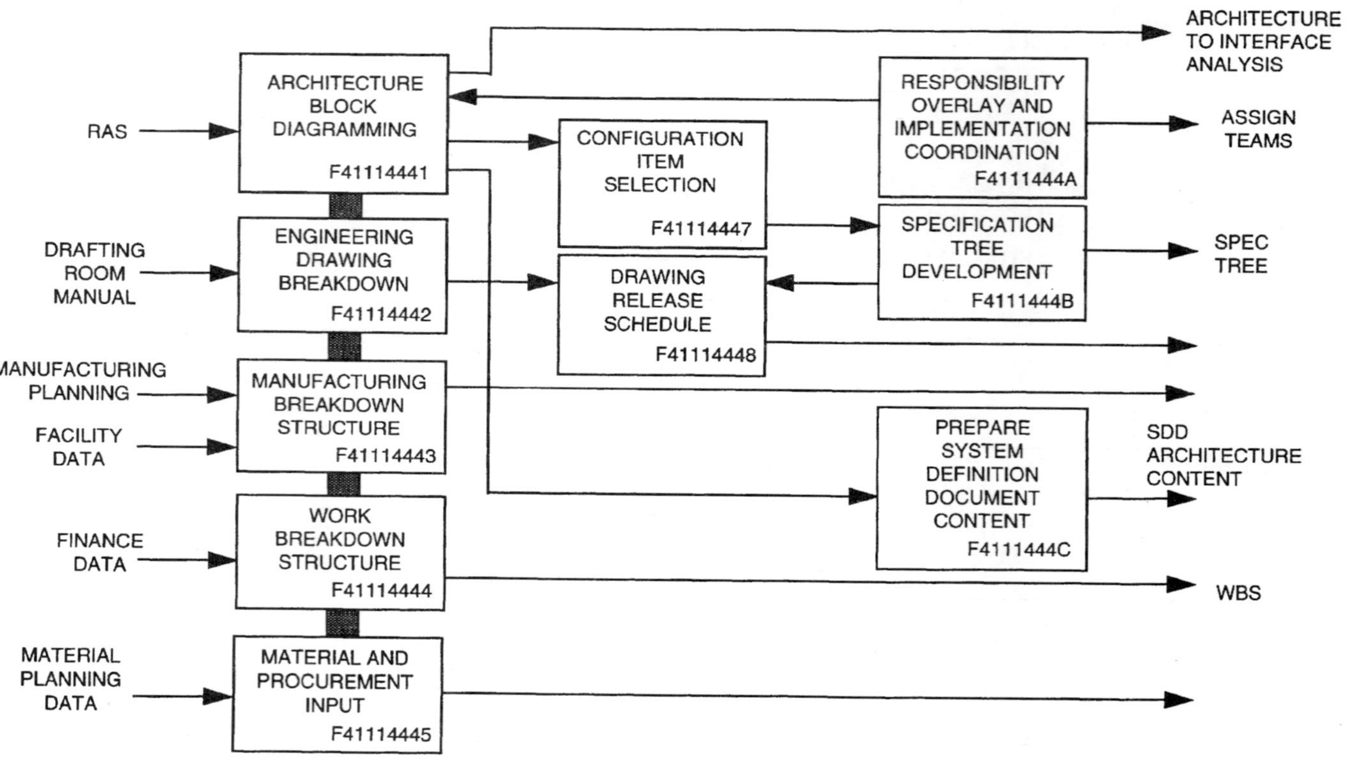

Figure A.15 Architecture synthesis, F4111444.

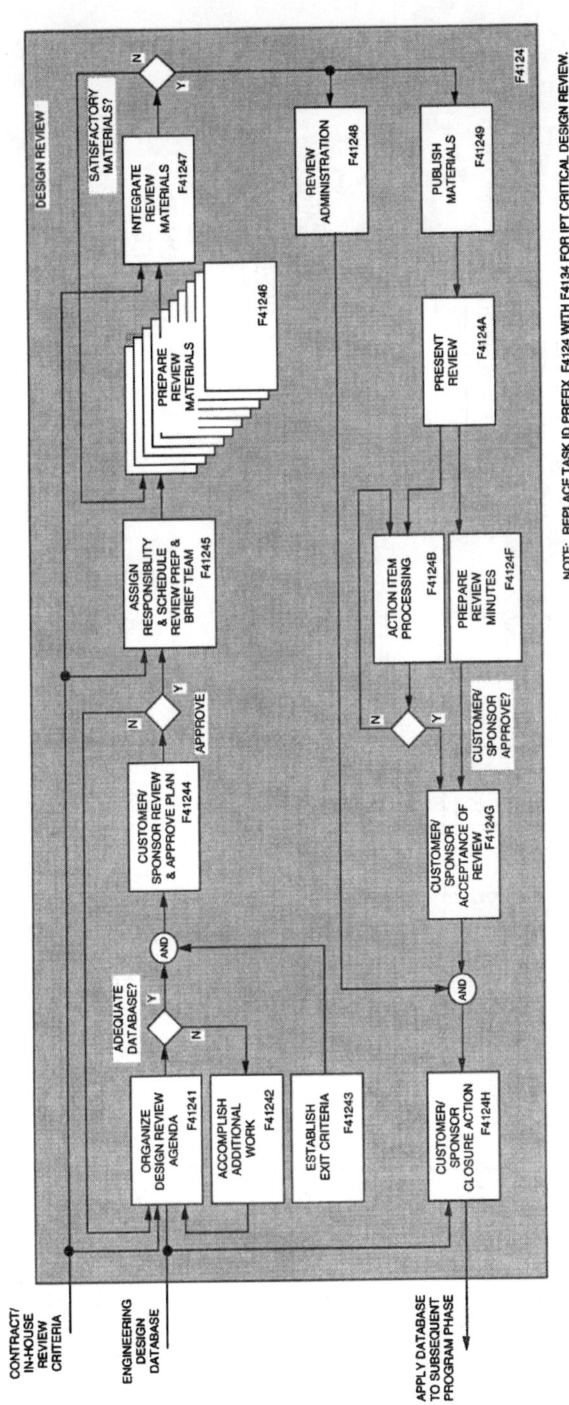

Figure A.16 Design reviews, F4124 and F4134.

System engineering deployment

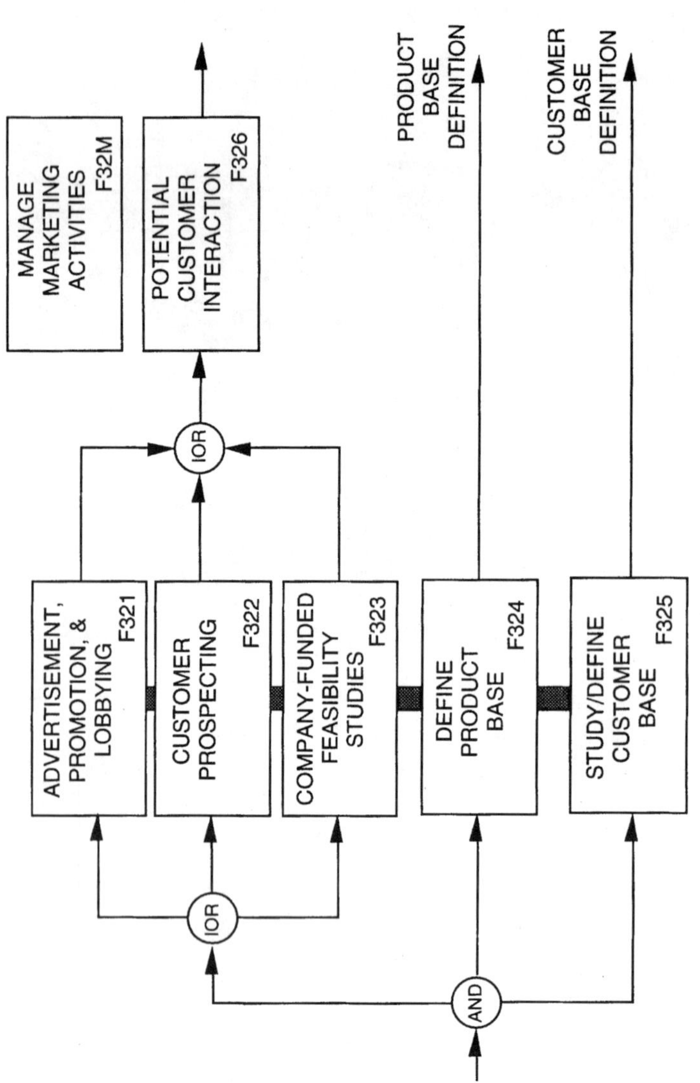

Figure A.17 Marketing work, F32.

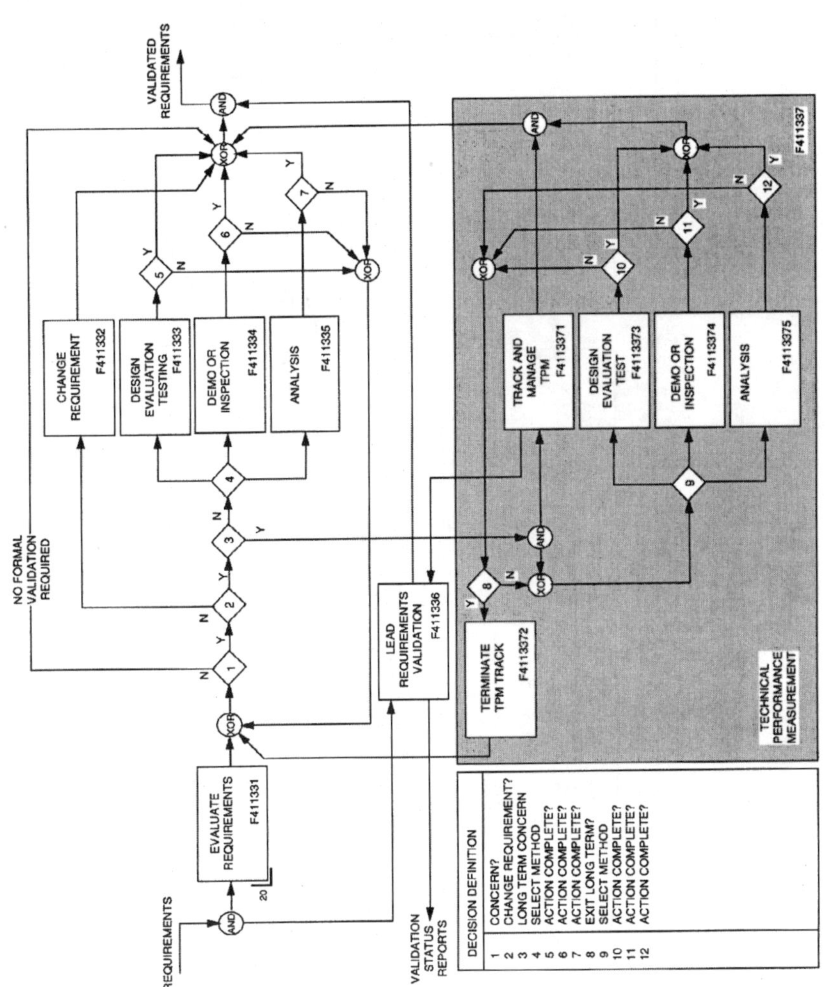

Figure A.18   Validate product requirements, F41133.

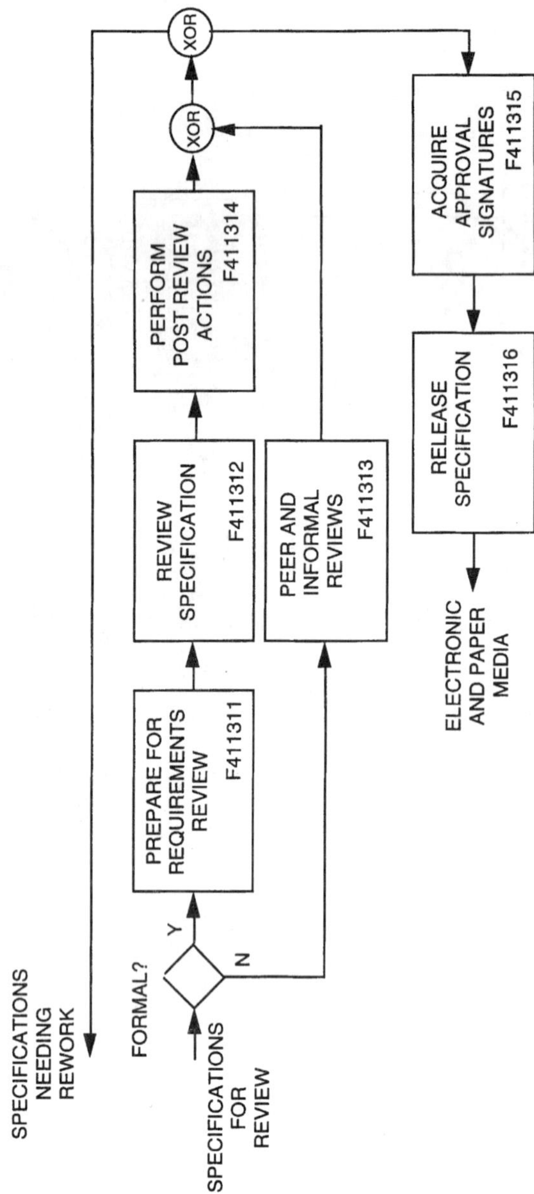

Figure A.19 Review and release specifications, F41131.

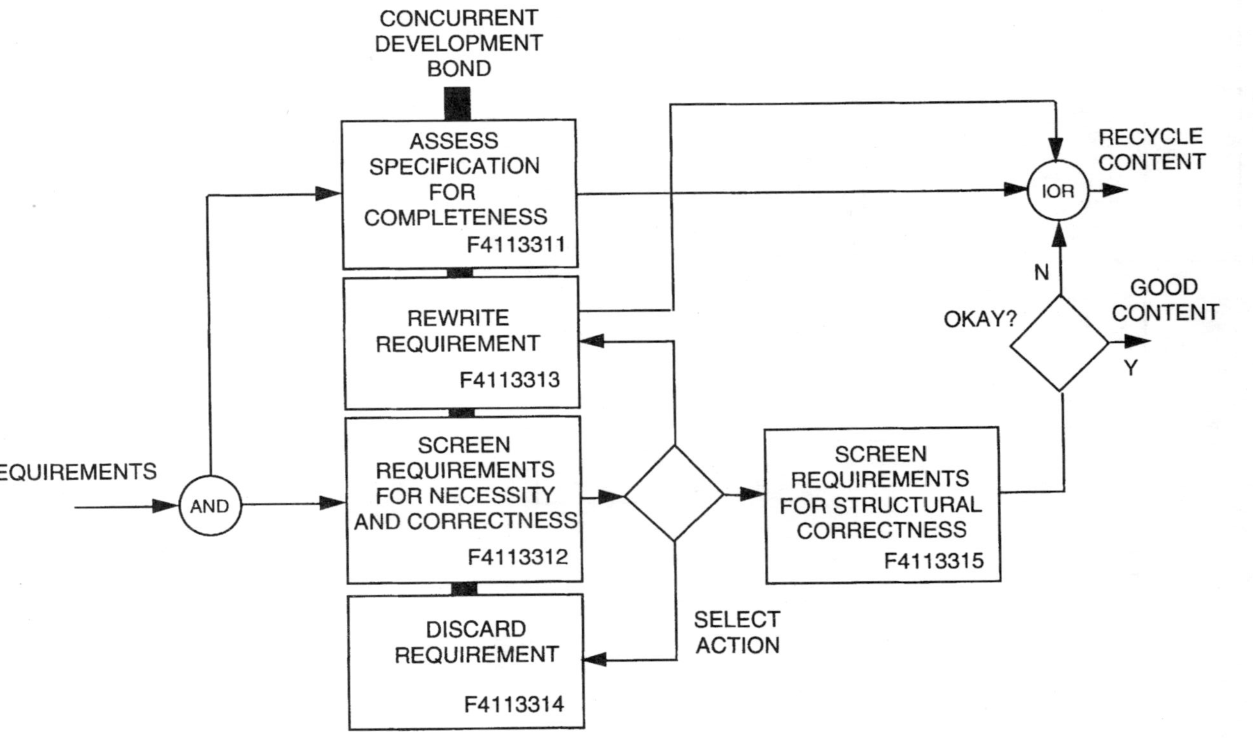

Figure A.20   Evaluate requirements, F411331.

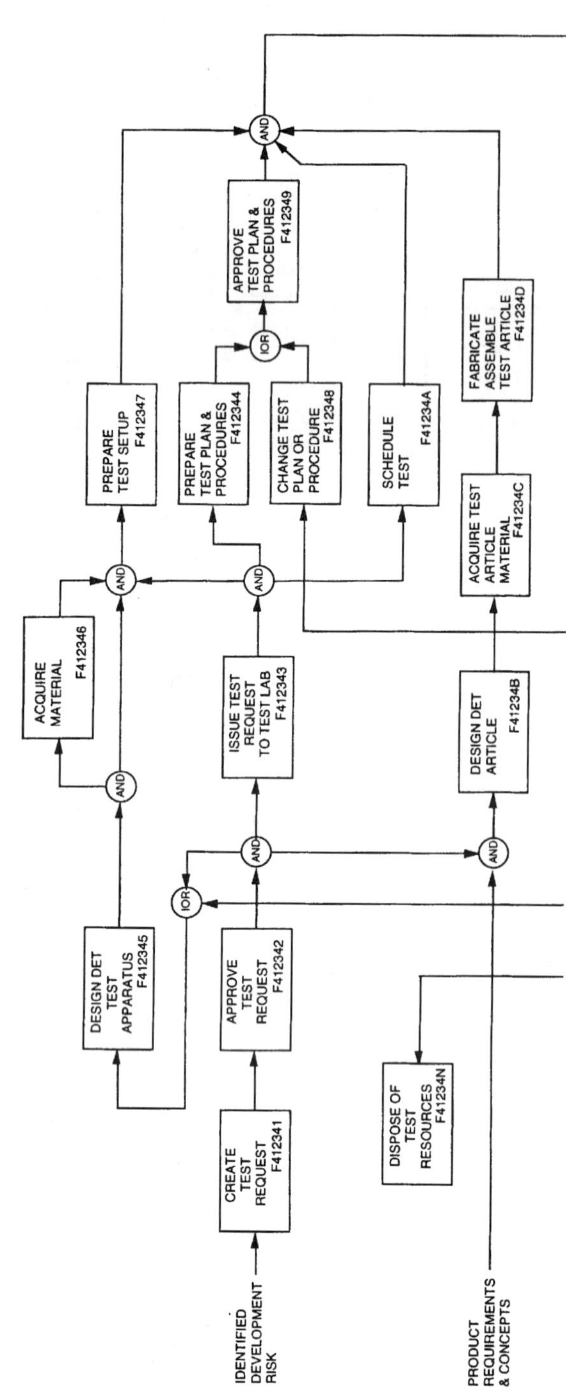

IDENTIFIED DEVELOPMENT RISK
CREATE TEST REQUEST F412341
APPROVE TEST REQUEST F412342
IOR
AND
DESIGN DET TEST APPARATUS F412345
ACQUIRE MATERIAL F412346
AND
AND
PREPARE TEST SETUP F412347
ISSUE TEST REQUEST TO TEST LAB F412343
AND
PREPARE TEST PLAN & PROCEDURES F412344
CHANGE TEST PLAN OR PROCEDURE F412348
IOR
APPROVE TEST PLAN & PROCEDURES F412349
AND
SCHEDULE TEST F41234A
DISPOSE OF TEST RESOURCES F41234N
DESIGN DET ARTICLE F41234B
AND
ACQUIRE TEST ARTICLE MATERIAL F41234C
FABRICATE ASSEMBLE TEST ARTICLE F41234D
PRODUCT REQUIREMENTS & CONCEPTS

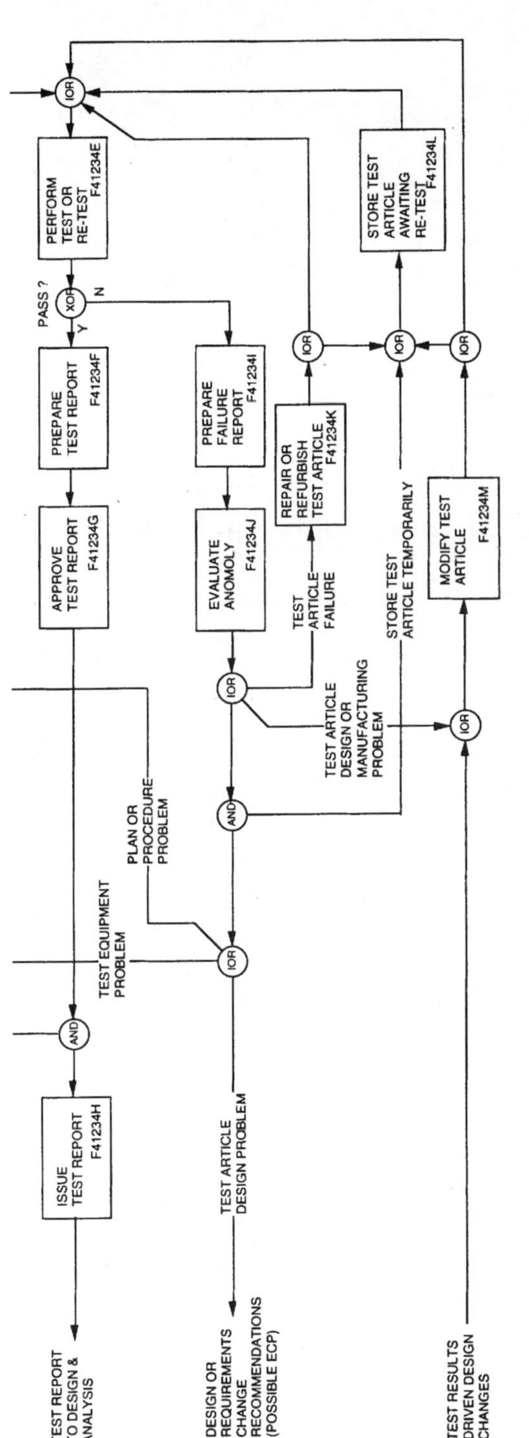

Figure A.21 Design evaluation testing (DET), F41234.

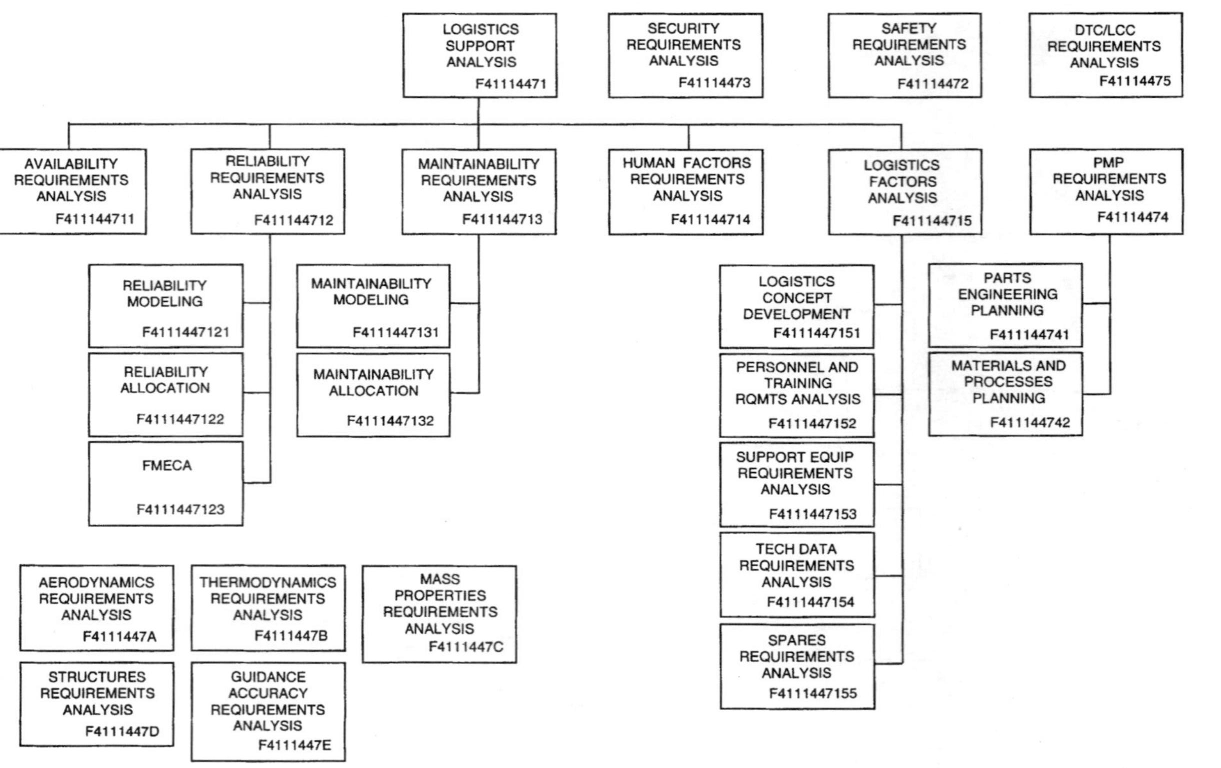

Figure A.22 Specialty engineering requirements analysis, F4111447.

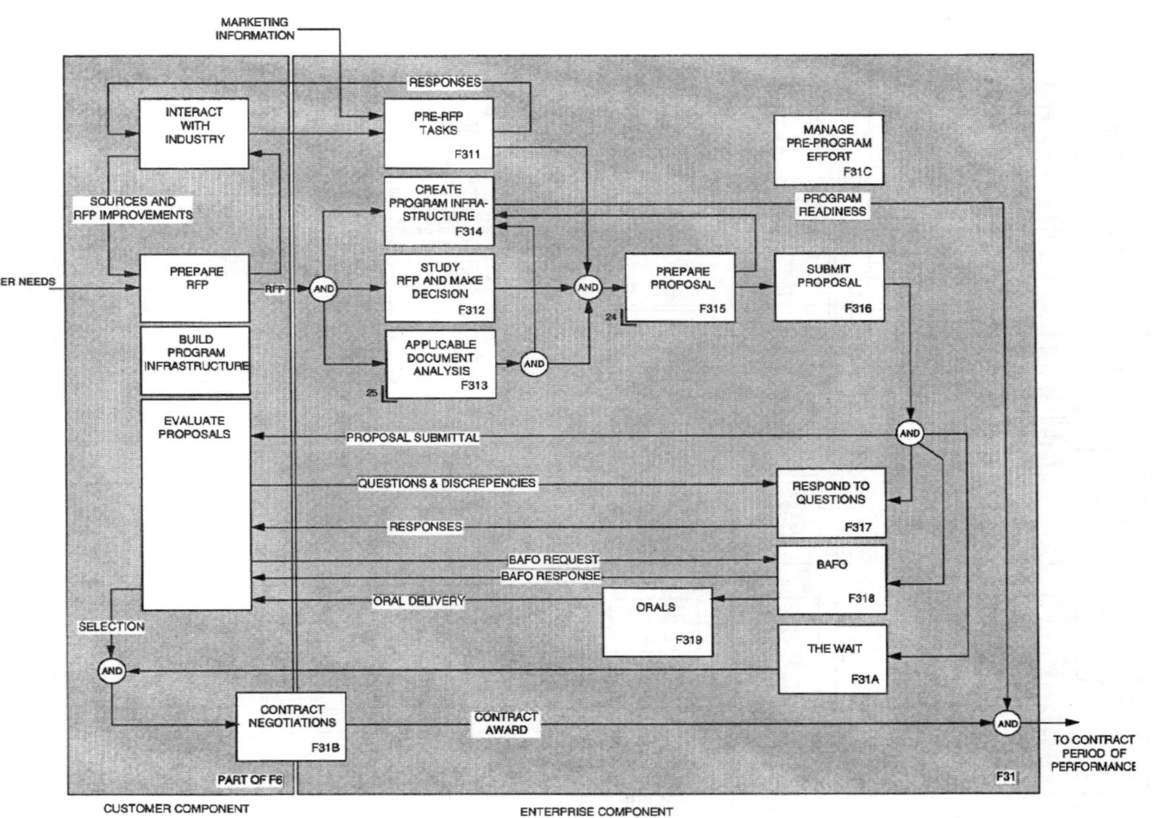

Figure A.23   RFP and proposal process, F31.

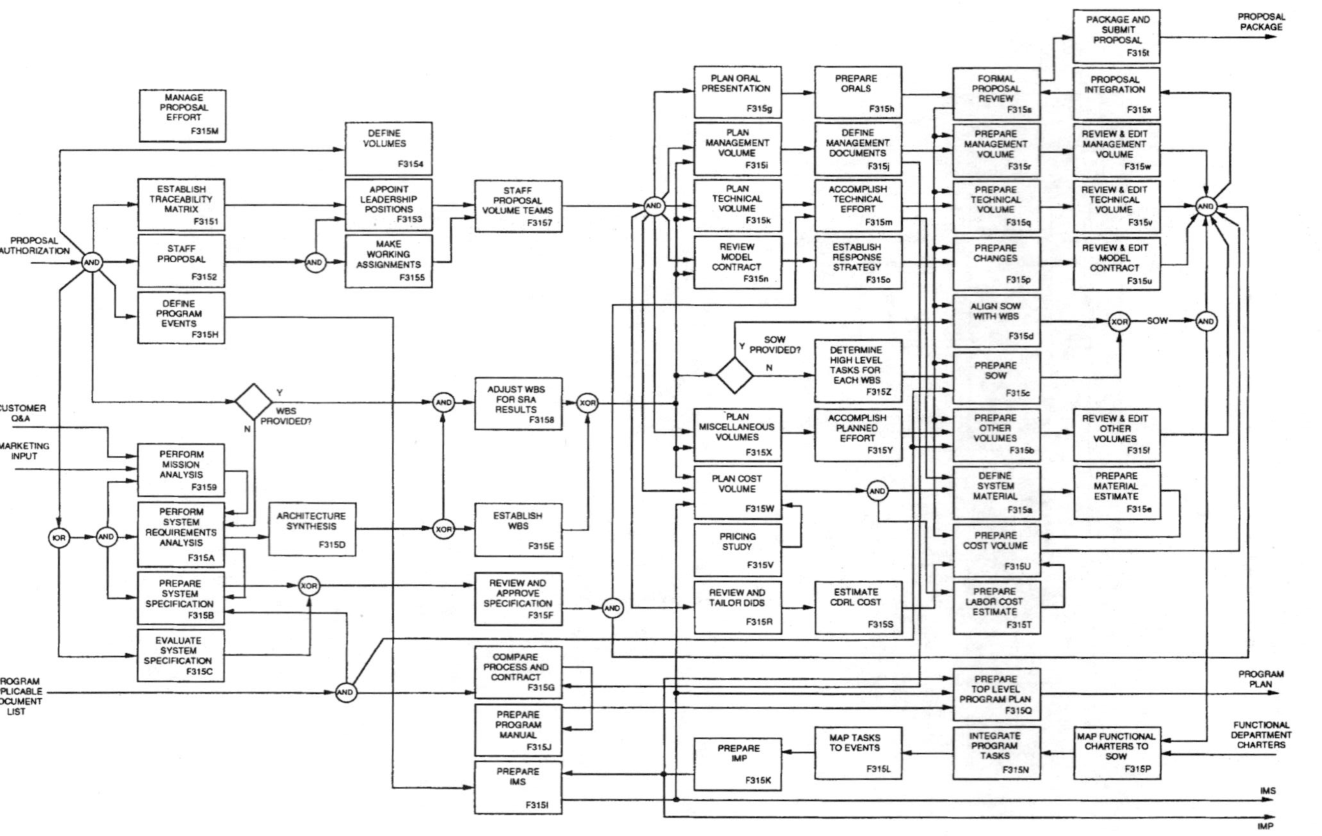

Figure A.24   Prepare proposal, F315.

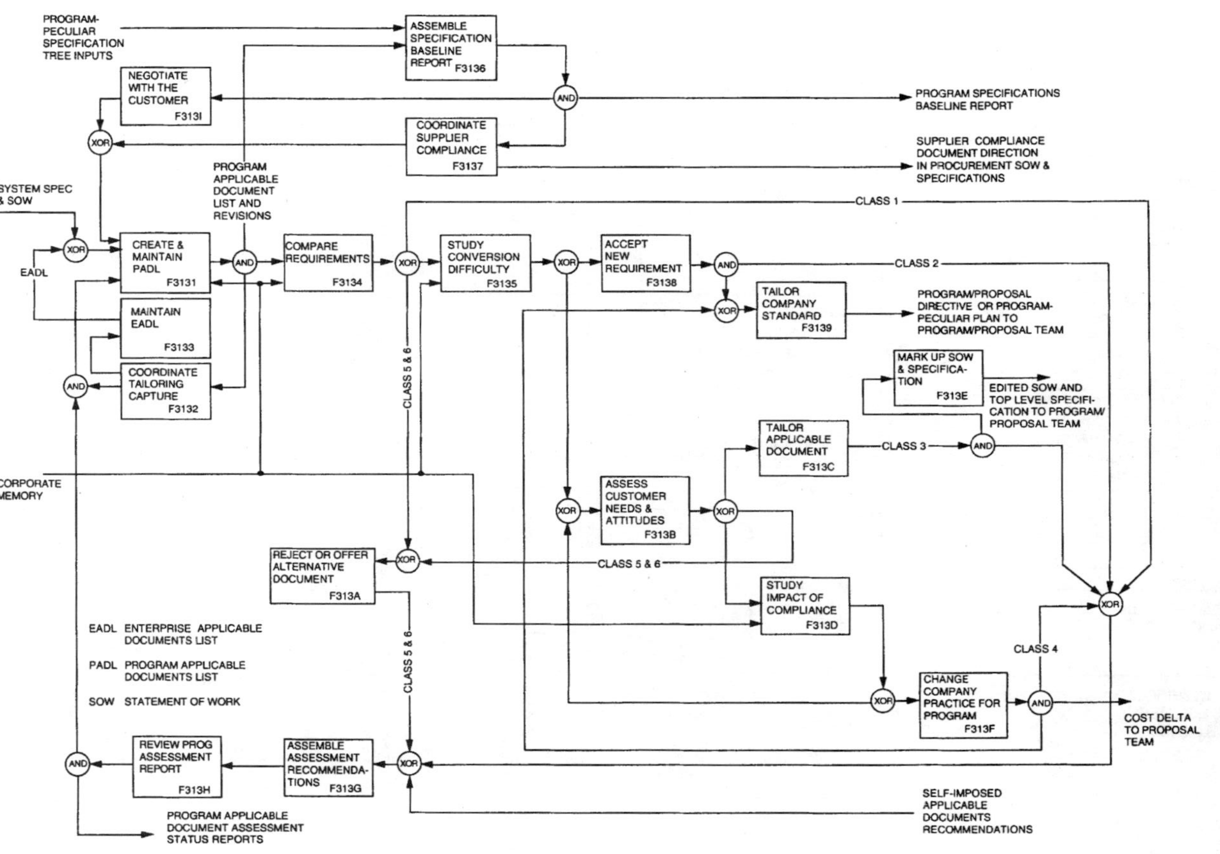

Figure A.25   Applicable documents analysis, F313.

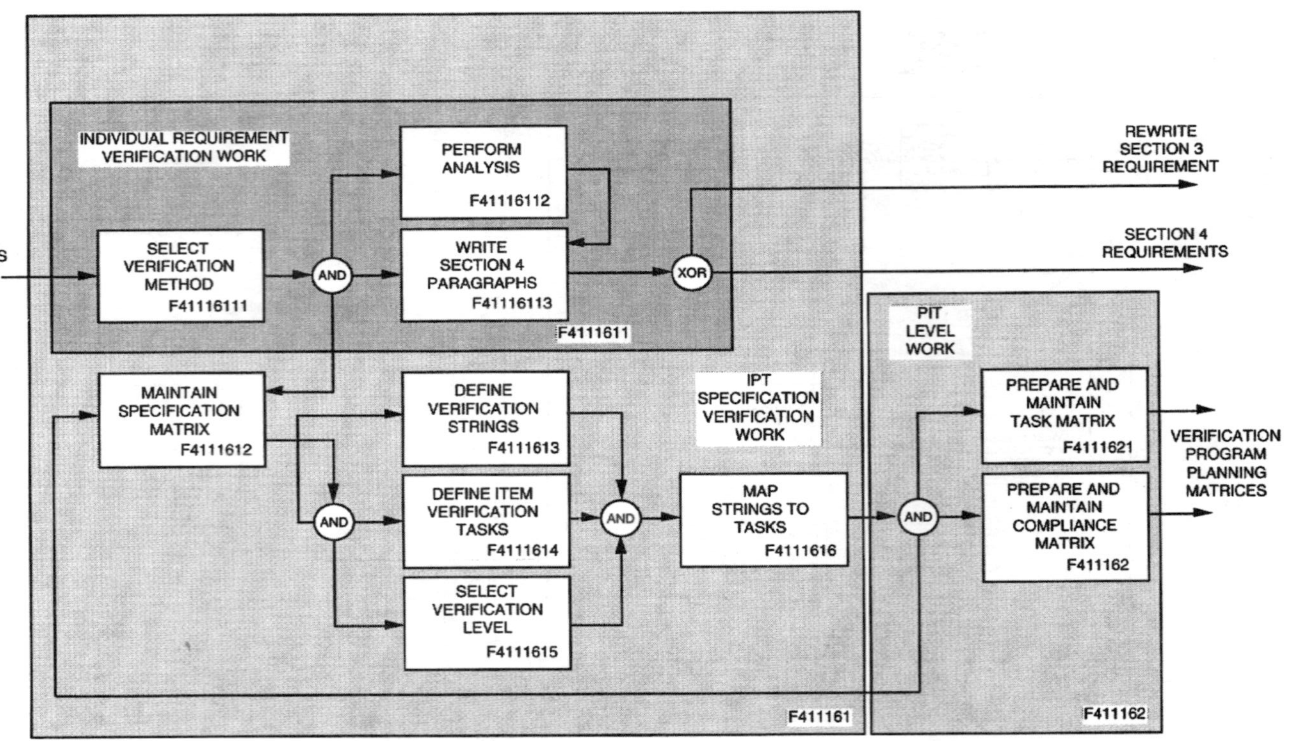

Figure A.26   Verification requirements analysis, F41116.

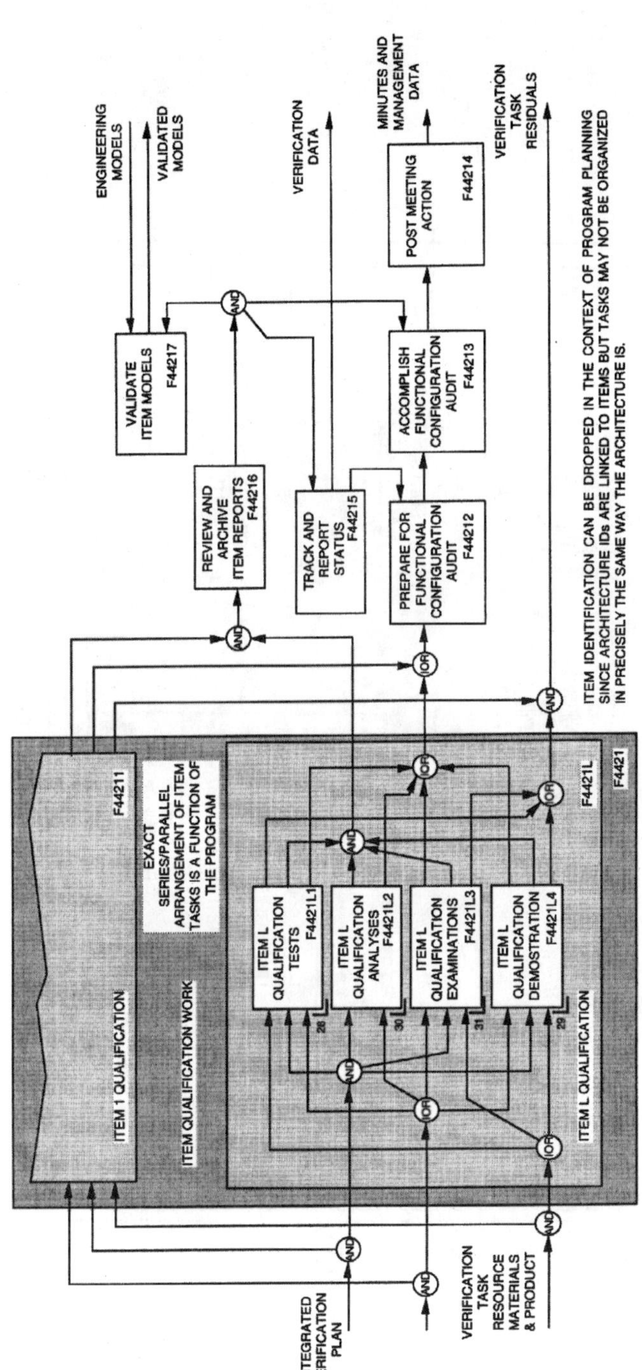

Figure A.27   Conduct item qualification, F442.

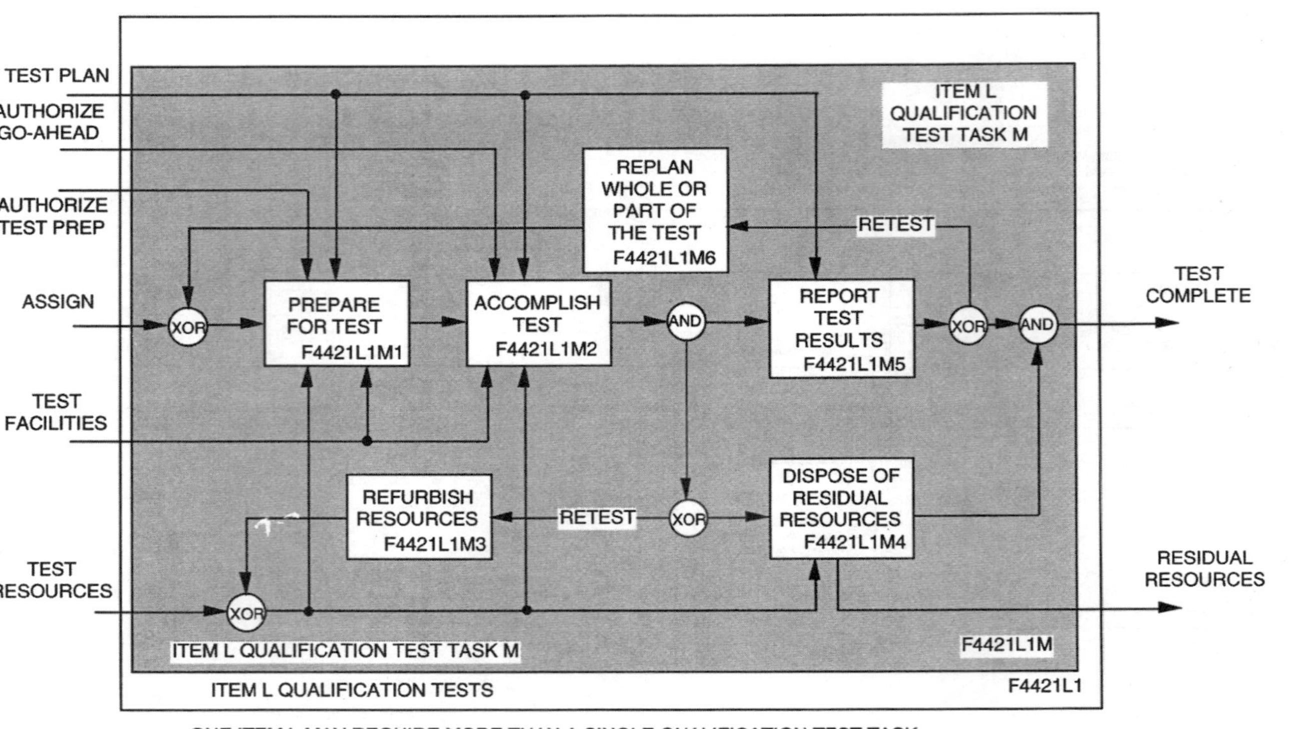

Figure A.28   Item L qualification test task M, F4421L1M.

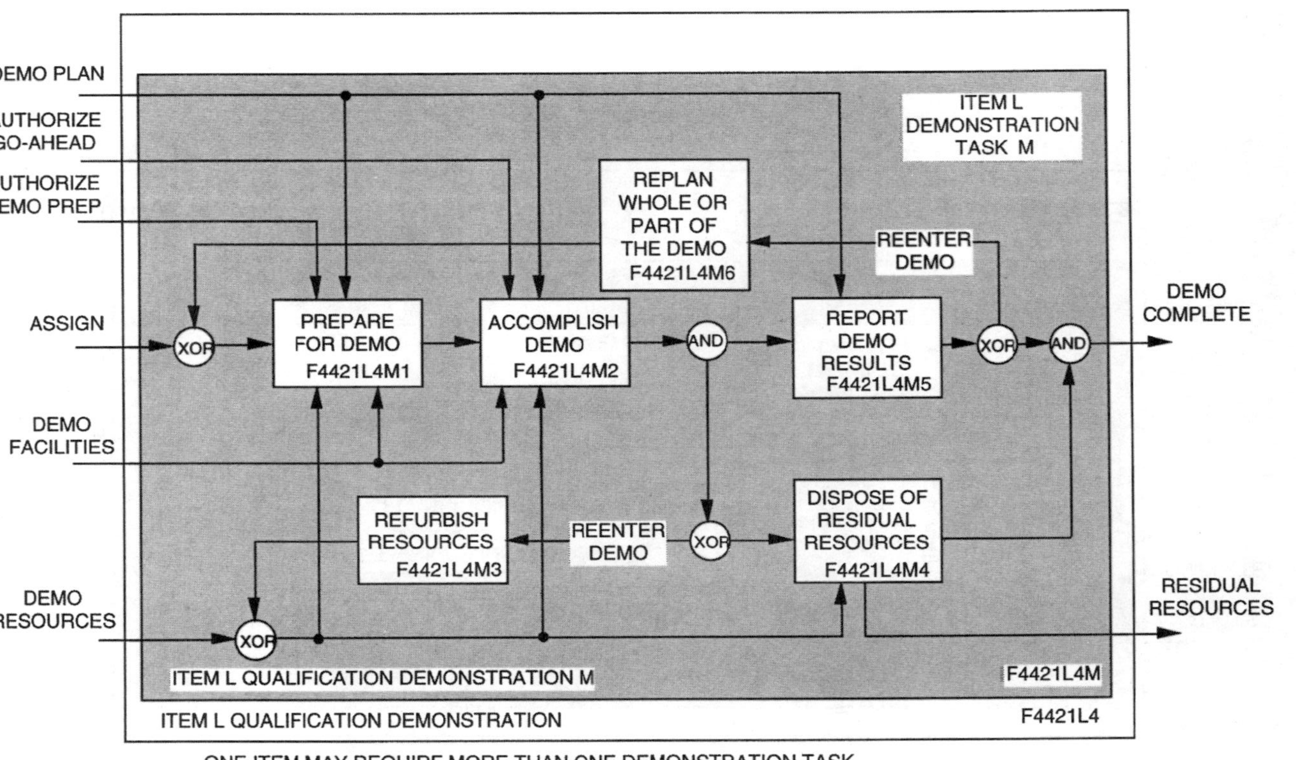

Figure A.29   Item L qualification demonstration task M, F4421L4M.

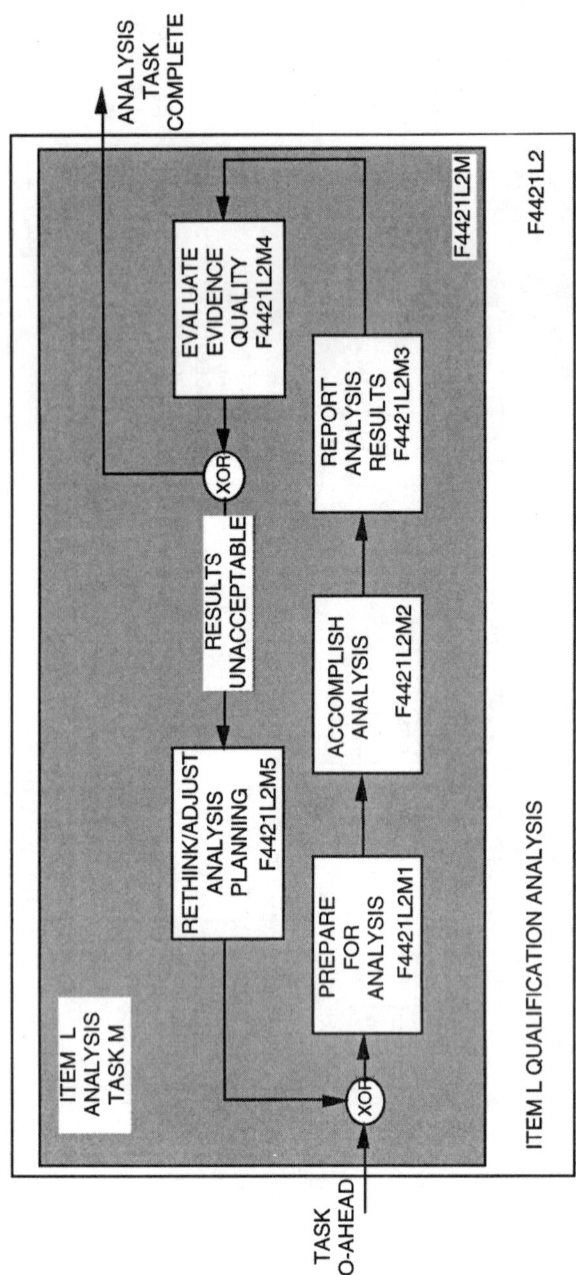

Figure A.30   Item L qualification analysis task M, F4421L2M.

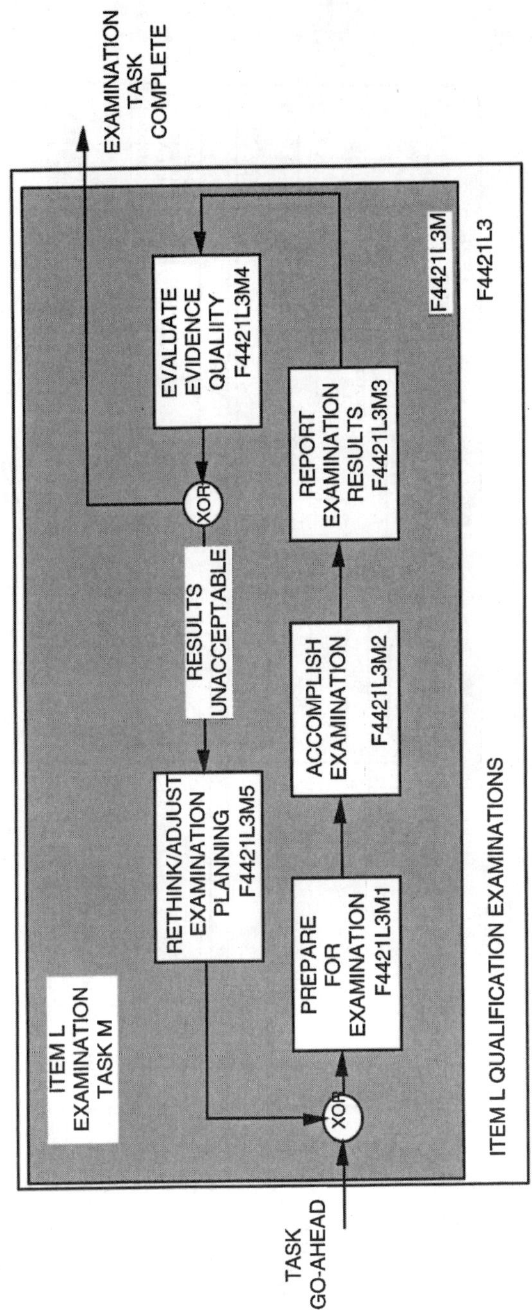

Figure A.31   Item L qualification examination task M, F4421L3M.

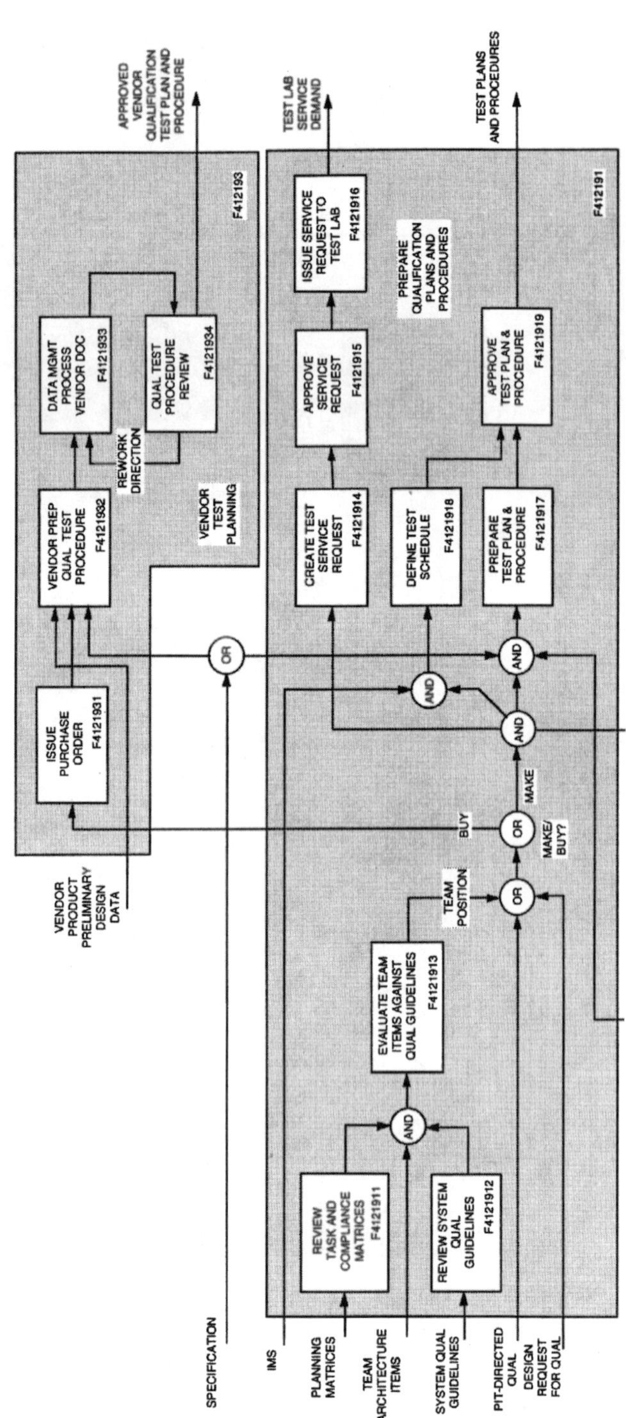

APPROVED VENDOR QUALIFICATION TEST PLAN AND PROCEDURE
TEST LAB SERVICE DEMAND
TEST PLANS AND PROCEDURES
F412193
F412191
DATA MGMT PROCESS VENDOR DOC
F4121933
QUAL TEST PROCEDURE REVIEW
F4121934
REWORK DIRECTION
VENDOR PREP QUAL TEST PROCEDURE
F4121932
VENDOR TEST PLANNING
ISSUE PURCHASE ORDER
F4121931
ISSUE SERVICE REQUEST TO TEST LAB
F4121916
APPROVE SERVICE REQUEST
F4121915
PREPARE QUALIFICATION PLANS AND PROCEDURES
APPROVE TEST PLAN & PROCEDURE
F4121919
CREATE TEST SERVICE REQUEST
F4121914
DEFINE TEST SCHEDULE
F4121918
PREPARE TEST PLAN & PROCEDURE
F4121917
OR
AND
AND
AND
MAKE
BUY
OR
MAKE/ BUY?
TEAM POSITION
OR
VENDOR PRODUCT PRELIMINARY DESIGN DATA
EVALUATE TEAM ITEMS AGAINST QUAL GUIDELINES
F4121913
AND
REVIEW TASK AND COMPLIANCE MATRICES
F4121911
REVIEW SYSTEM QUAL GUIDELINES
F4121912
SPECIFICATION
IMS
PLANNING MATRICES
TEAM ARCHITECTURE ITEMS
SYSTEM QUAL GUIDELINES
PIT-DIRECTED QUAL DESIGN REQUEST FOR QUAL

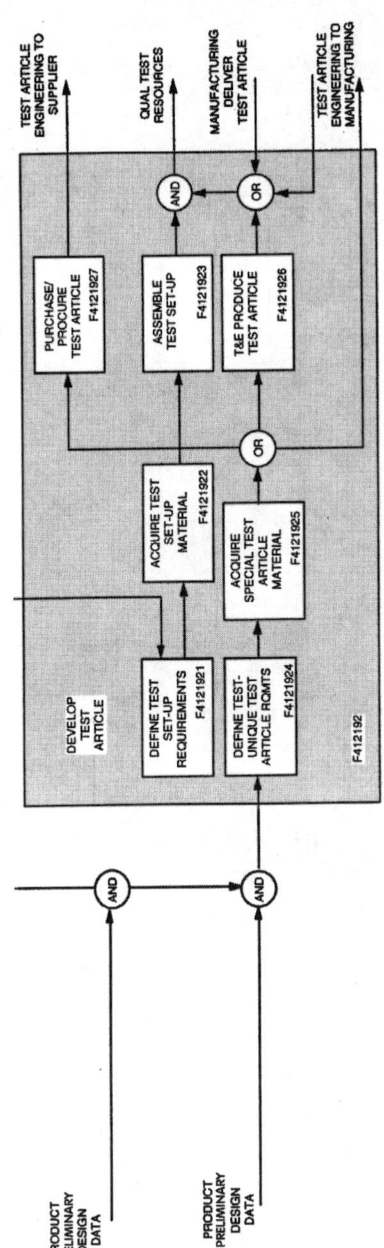

Figure A.32   Prepare qualification plans, F41219.

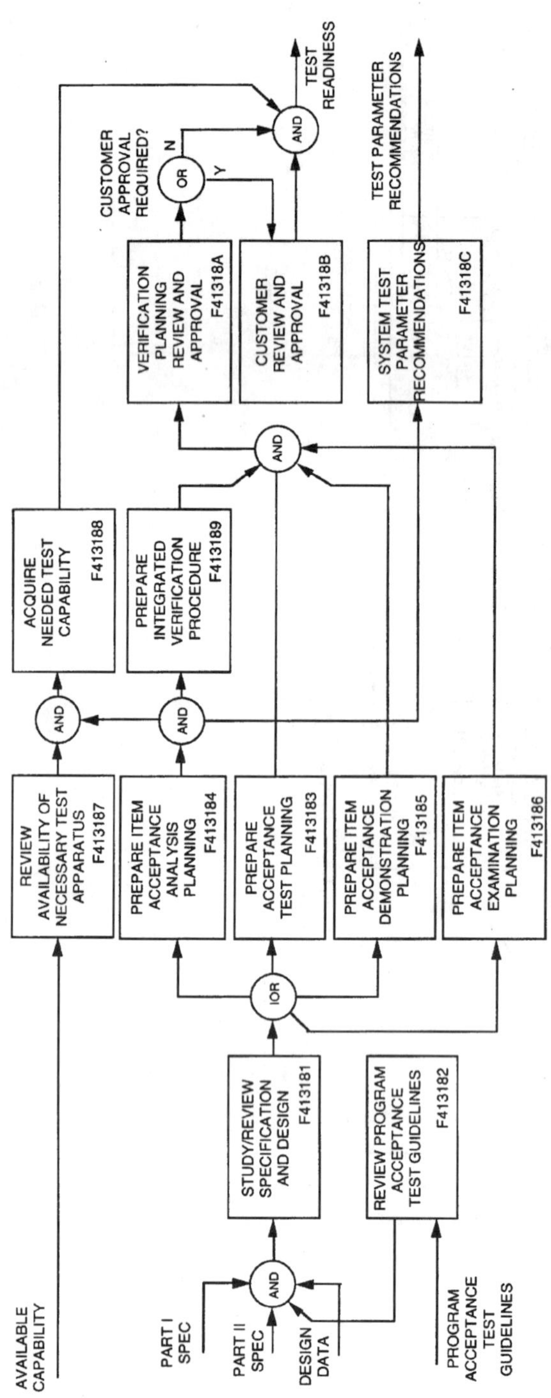

Figure A.33   Prepare acceptance verification plan, F41318.

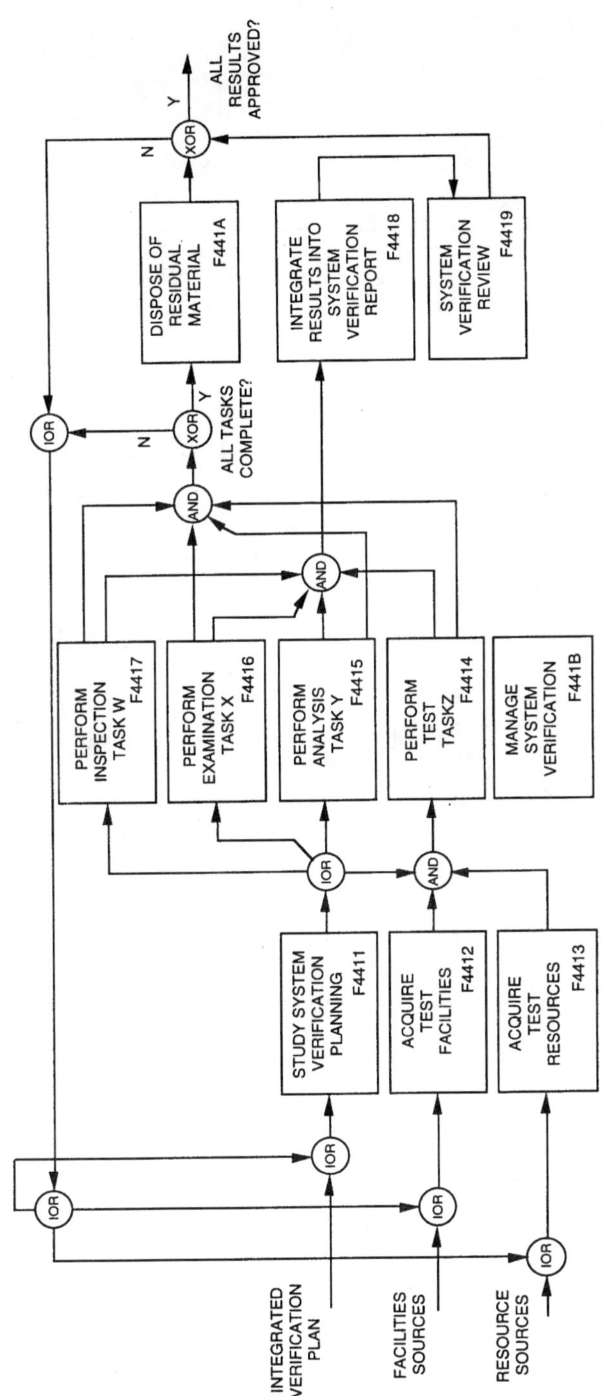

Figure A.34 Conduct system verification, F441.

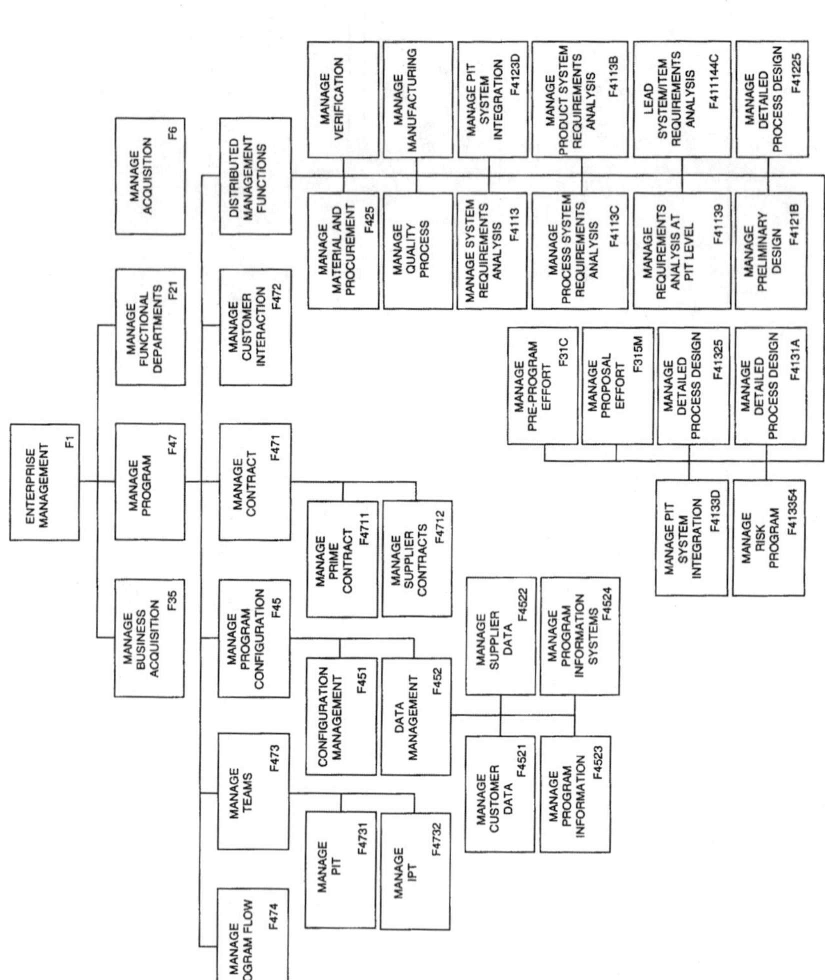

Figure A.35　Miscellaneous management functions.

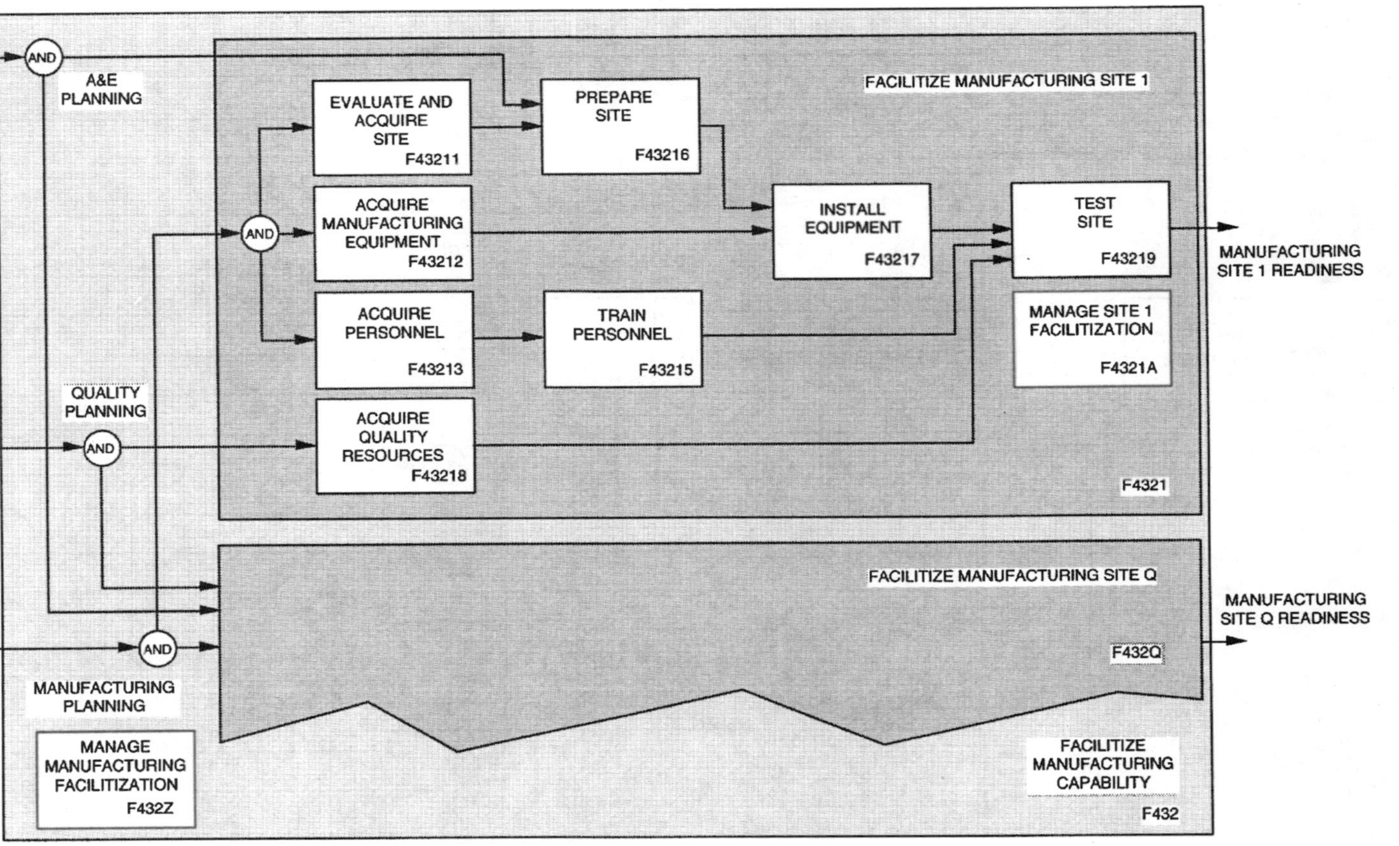

Figure A.36 Facilitize manufacturing capability, F432.

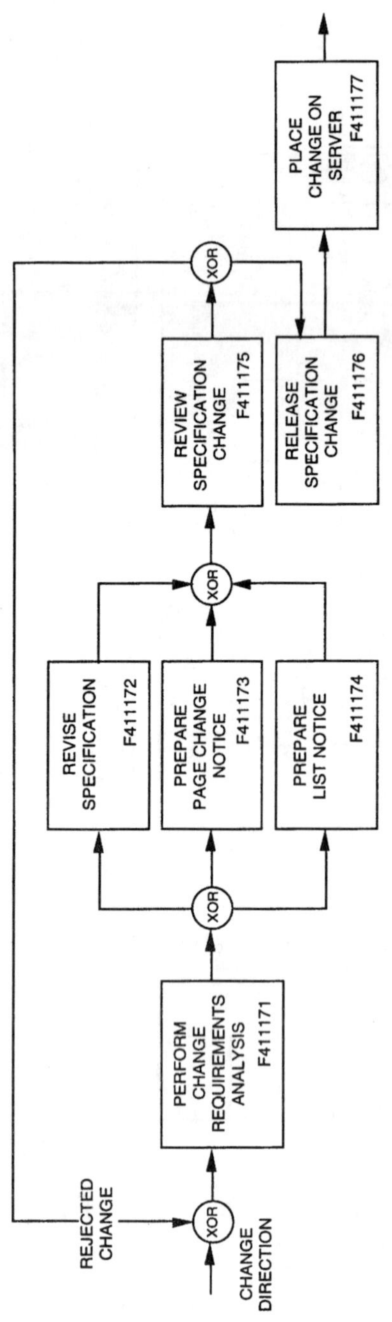

Figure A.37 Change released specifications, F41117.

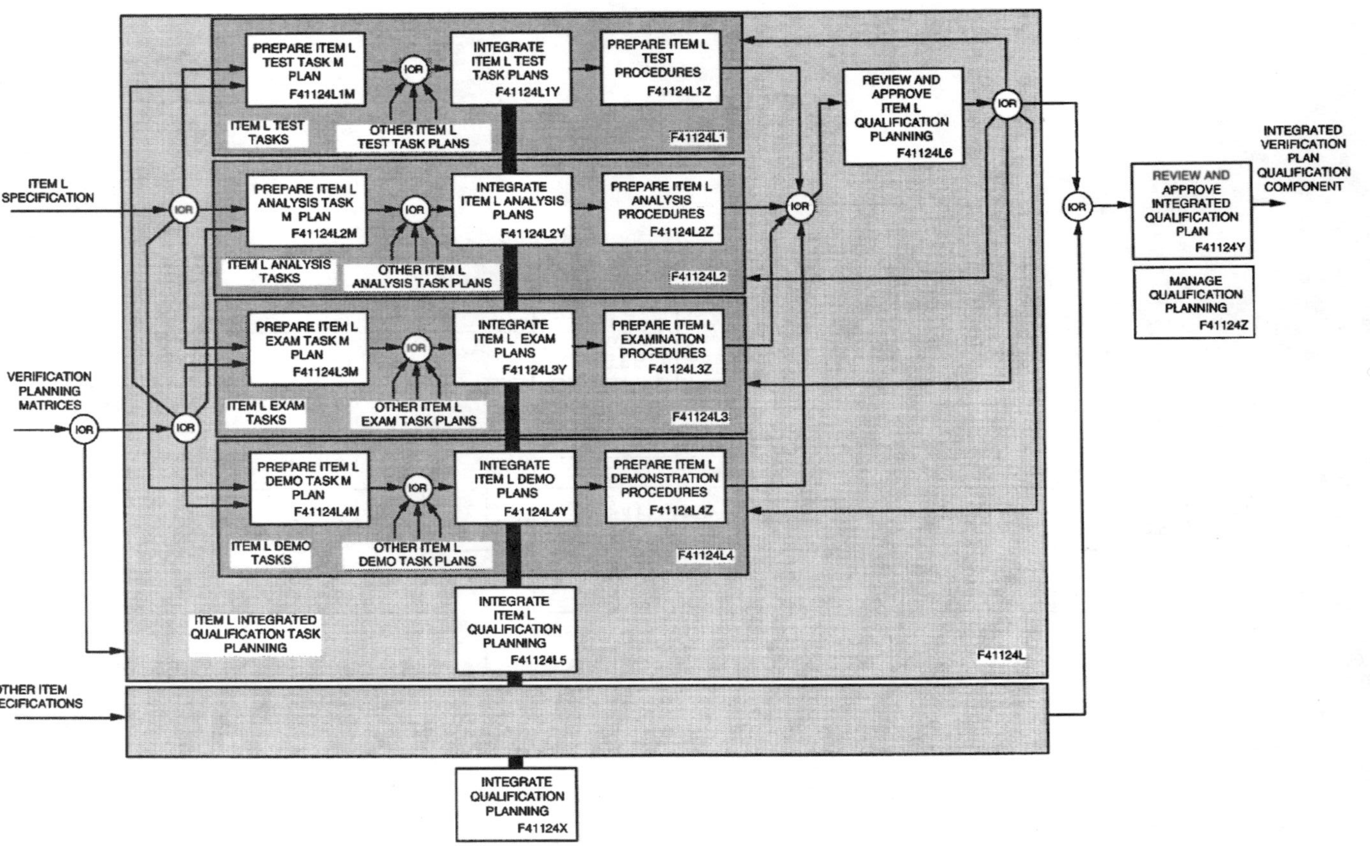

Figure A.38 Plan qualification verification process, F41124.

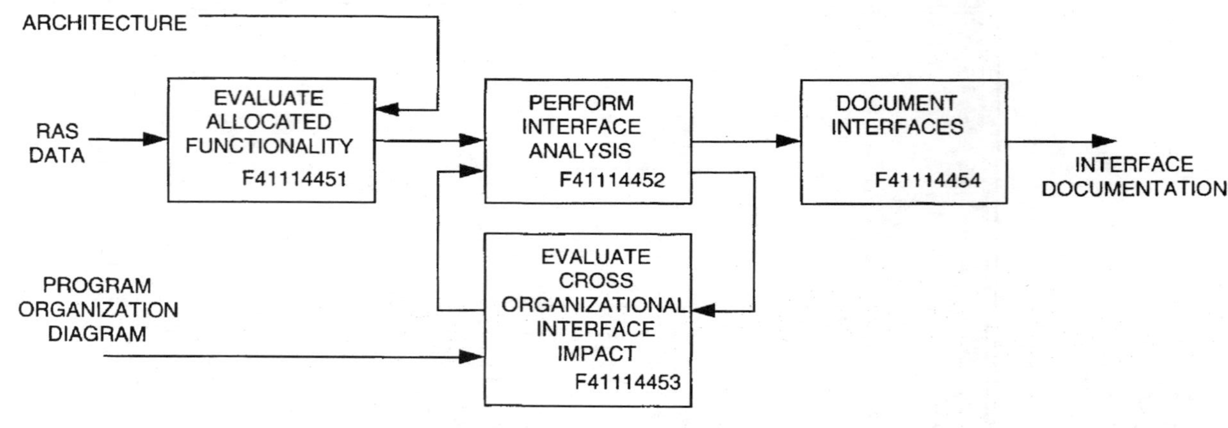

Figure A.39   Interface analysis, F4111445.

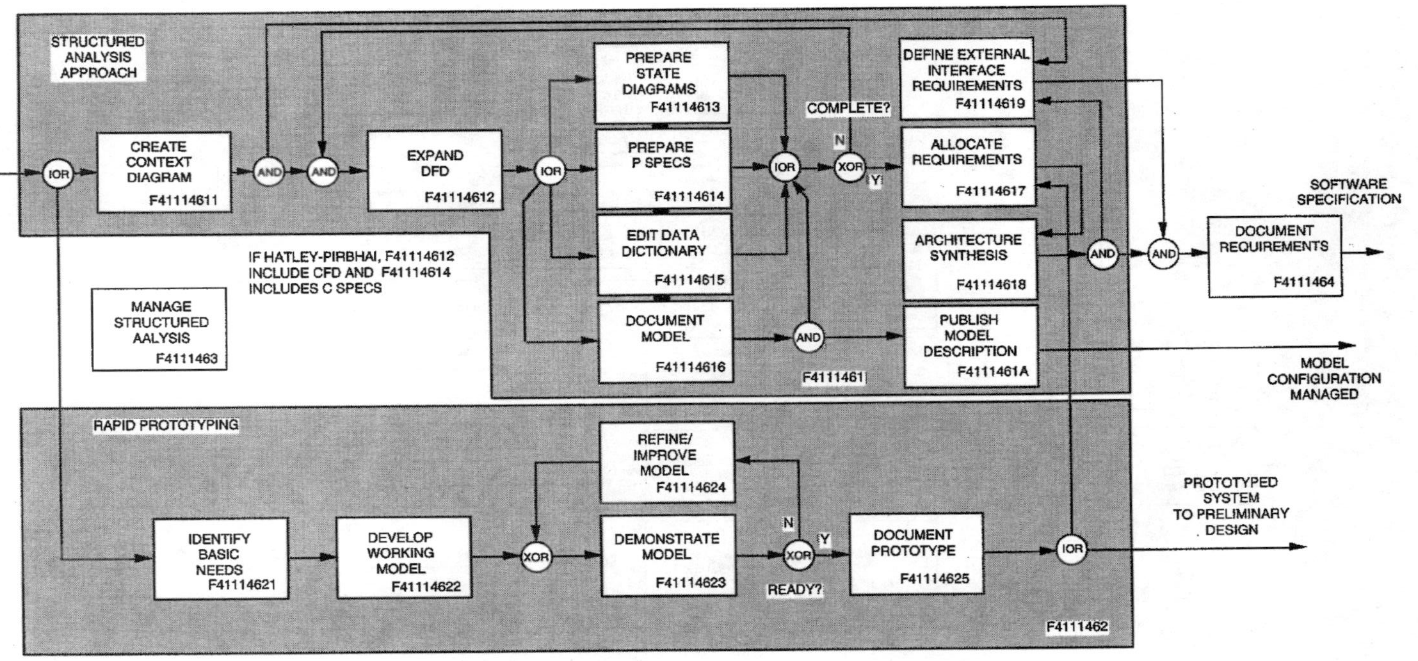

Figure A.40 Modern structured analysis, F411146.

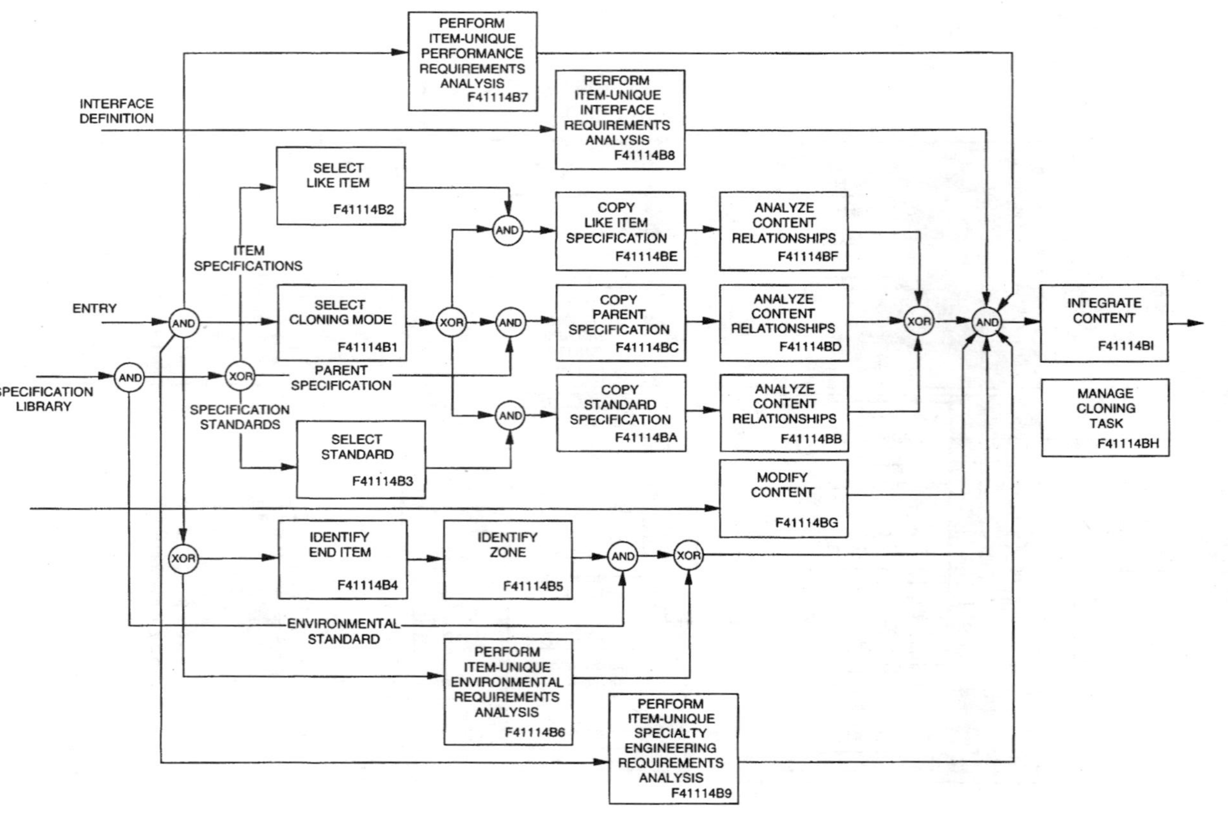

Figure A.41　Clone specification, F41114B.

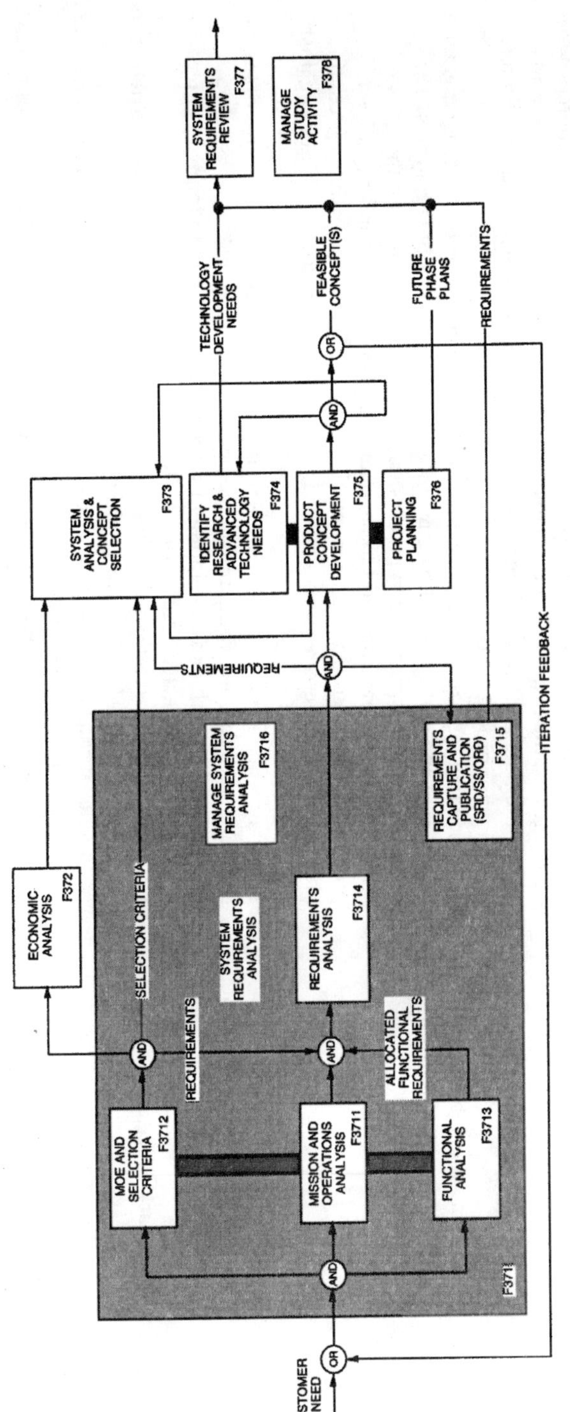

Figure A.42   Concept development, F37.

Table A.1 Enterprise Process Dictionary

Function ID	Function name	Function description
F	Enterprise Vision Statement	Develop products in our chosen product line of value to our customers that compare very favorably to those created by competing enterprises.
F1	Enterprise Management	Top-level management of the enterprise applies management techniques to establish and achieve goals. Top management depends on the heads of technical functional and business functional departments to acquire, maintain, and manage the resources needed by programs; the head of programs to manage ongoing programs; and the head of the PIT to integrate and optimize enterprise resources for the greater good of the enterprise in achieving program goals.
F2	Functional Management and Continuous Process Improvement	The functional departments are established and managed to provide programs with knowledge and skills needed by current, planned, and possible programs. These departments are owners of the process managed and integrated by the EIT and responsible for providing good tools, qualified people, and sound practices to programs. The functional departments, led by the EIT, are responsible for continuous improvement of these resources and benchmarking enterprise capabilities against industry standards.
F21	Study Lessons Learned	EIT and functional managers study the results from program work in terms of metrics, program concerns, and program status reports.
F22	Functional IRAD	Independent research and development work is accomplished to improve enterprise performance in selected areas based on lessons learned.
F23	Benchmark Process	Internal processes are compared with selected respected external processes to determine enterprise quality of performance in those tasks.
F24	Select Improvement Actions	The functional chiefs, managers, directors, and vice presidents in company with the EIT reach decisions on improvement actions based on data.
F25	Maintain Knowledge and Skills	Functional department action is taken to ensure that the enterprise continues to have access to personnel in needed skills areas.
F26	Maintain Best Practices	Functional department managers maintain their manuals as directed by EIT.

F28	Maintain Facilities and Tools	Facilities features and tools are maintained and improved to provide an effective environment within which to do system development work.
F29	Release Personnel	Personnel who have not been able to satisfy work performance criteria are separated from the enterprise.
F2A	Acquire Personnel	The enterprise constantly seeks out qualified personnel to do work defined by functional management.
F2B	Train Personnel	Enterprise personnel are provided training for process knowledge, job skills, and product knowledge through in-house courses, short courses, and university courses.
F2C	Develop/Select Training	Training courses are developed or training sources identified for specific needs.
F2D	Enterprise Integration	The aggregate process owner (EIT) coordinates improvement work across the functional departments.
F2E	Manage Functional Departments	Chiefs, managers, directors, and vice presidents manage the parts of the functional organization under their responsibility to ensure that their department continues to supply all programs with good people, good tools, and good practices.
F3	New Business Acquisition	New Business Acquisition function seeks out potential new business opportunities that fall within the enterprise vision statement, develops those potential opportunities responding to requests for information and proposal, conducts technical studies, performs market-related IRAD, and keeps current on potential customer needs and the capabilities of competitive firms.
F31	RFP and Proposal Process	The customer makes available a request for proposal to potential suppliers, and the enterprise responds with a proposal telling how it will satisfy the requirements in the RFP.
F311	Pre-RFP Tasks	
F312	Study RFP and Make Decision	
F313	Applicable Documents Analysis	
F3131	Create and Maintain Program Applicable Documents List	
F3132	Coordinate Tailoring Capture	

Table A.1 (continued)   Enterprise Process Dictionary

Function ID	Function name	Function description
F3133	Maintain Enterprise Applicable Documents List	The enterprise maintains a list of applicable documents that have been called on prior programs or that have a good chance of being called on future programs. Each document is assigned to a responsible functional department which must review it, coordinating with other functional departments as required, and determine how the content aligns with enterprise practices. To the extent that they differ, the responsible department must either draft tailoring to cause agreement or change enterprise generic work definition to agree. This information is maintained by coordinating the applicable documents work accomplished on each proposal with the EADL content. To the extent that programs wish to change this content they must get EIT approval or craft a program-peculiar instruction, which is discouraged.
F3134	Compare Requirements	
F3135	Study Conversion Difficulty	
F3136	Assemble Specification Baseline Report	
F3137	Coordinate Supplier Compliance	
F3138	Accept New Requirement	
F3139	Tailor Company Standard	
F313A	Reject or Offer Alternative Document	
F313B	Assess Customer Needs and Attitudes	
F313C	Tailor Applicable Document	
F313D	Study Impact of Compliance	
F313E	Mark-up SOW/IMP and Specification	
F313F	Change Company Practice For Program	

F313G	Review Assessment Recommendations	
F313H	Assemble and Review Program Assessment Report	
F314	Create Program Infrastructure	
F315	Prepare Proposal	
F316	Submit Proposal	
F317	Respond to Questions	
F318	BAFO	
F319	Orals	
F31A	The Wait	
F31B	Contract Negotiations	
F32	Marketing Work	New business opportunities are constantly searched for and contact is maintained with current and potential clients.
F321	Advertisement, Promotion, and Lobbying	The marketing function targets potential customers, determining how best to reach them, and the message to impress upon them. Government agencies are lobbied.
F322	Customer Prospecting	Conscious efforts are made to identify new customers, establish contact, and determine their needs.
F323	Company-Funded Feasibility Studies	Where the profit potential is adequate to justify, special studies are conducted in preparation for proposal and contract work.
F324	Define Product Base	The enterprise product base is clearly defined as a basis for marketing efforts.
F325	Study/Define Customer Base	A clear definition of the customer base is made and approved by top management.
F326	Potential Customer Interaction	Conversations take place between marketing people and customer people to understand customer needs and to expose the customer to enterprise strengths in these areas.
F32M	Manage Marketing Activities	Management skill is applied to all marketing work.
F33	Precontract Technical Studies	Feasibility studies and predesign work is performed in preparation for proposals and future contract work.
F34	Market Related IRAD	IRAD work is performed in support of or driven by marketing needs.
F35	Manage Business Acquisition	Management skills are brought to bear on the new business function to identify goals, create plans to achieve them, and monitor progress toward those goals, giving direction as needed.

Table A.1 (continued) Enterprise Process Dictionary

Function ID	Function name	Function description
F36	RFI Process	Potential customers float requests for information. The marketing function determines those that are within the scope of the enterprise vision statement and pursues them.
F4	Create Product System	Translate customer needs into a set of specifications and a compliant preferred design, manufacture the product, and verify that the product satisfies predetermined requirements.
F41	Develop Grand System	The customer need is transformed into a clear documentation of the preferred design solution coordinated with needed manufacturing, quality, logistic support, verification, and disposal process planning.
F411	Grand System Requirements Analysis	System and item product requirements, as well as program process requirements, are defined in specifications and plans.
F4111	Define Product System Requirements and Develop Concepts	Product requirements are defined in specifications.
F41111	Product System Concept Development	Mission analysis is performed once at system level though iteration may be necessary as the system concept develops.
F41112	Product System Operational Environment Definition	System level environmental analysis is performed, including identification of requirements for the following environmental elements: natural, hostile, cooperative, noncooperative, and self-induced.
F41113	Perform Applicable Documents Analysis	Applicable documents are formally selected for system level, and tailoring is accomplished as appropriate.
F411141	Study System/Item Concept	Individual system engineer or team become aware of prior concept development work to ensure that new structured analysis work will be in context with prior work.
F411142	Select Specification Template	The template (boilerplate) for a particular specification is selected by the responsible team.
F411143	Item Applicable Documents Analysis	Program and system level applicable documents work accomplished in the proposal work or prior program work is considered, looking for applicable documents that are or should be related to this item. In any case, the applicable documents called must be referenced in requirements appearing in the specification.

F411144	Traditional Structured Analysis	Functional flow diagramming or one of four other models is applied to understanding the problem space from a performance requirements and architecture needs perspective. Structured models are also applied to gain insight into design constraints.
F4111441	Functional Analysis	Five different modeling techniques are offered to identify needed functionality. Identified functions are allocated to things in the system architecture.
F41114411	Function Diagramming Steps	Subtier product functions are derived, starting with the customer need and relations established using one of several modeling techniques. Functional flow diagramming is assumed, but four other models are available.
F411144111	Identify Functions	Higher-tier functionality is expanded into lower-level details by determining one or more ways one could achieve the indicated functionality. The preferred representation of needed functionality is selected through good engineering judgment or a trade study.
F411144112	Functional Flow Analysis	Identified functions are studied for sequence possibilities and strung together into a flow diagram with directed line segments and combinatorial symbols.
F411144113	Process Analysis	Physical process analysis is applied on programs where there is an existing system being modified and the changes will not influence the basic operation and maintenance of the product system.
F411144114	IDEF 0 Functional Analysis	IDEF 0 is substituted for functional flow diagramming as the principal modeling structure.
F411144115	Hierarchical Functional Analysis	An outlining technique is applied rather than a flow-related technique.
F41114412	Functional Requirements Analysis	System measures of effectiveness are determined as a basis for trade studies, and an option is provided to accomplish performance requirements analysis prior to allocation of functionality to things in the system architecture.
F411144121	Define/Refine Measures of Effectiveness	Key parameters are identified that will be used in trade studies for a basis of selection.
F411144122	Unprecedented Performance Requirements Analysis	Offers the analyst the alternative of completing the performance requirements analysis work prior to allocation of functionality. This technique is often useful on largely unprecedented systems.
F41114413	Function Allocation Steps	Identified functions are allocated to things in the system architecture approach where several engineers are involved in a team effort.
F411144131	Form and Brief Team	Where a team is used, it is brought together with the right facilities and tools resources and prepared for the allocation work.

Table A.1 (*continued*)   Enterprise Process Dictionary

Function ID	Function name	Function description
F411144132	Review Prior Allocations	The engineer or team becomes familiar with prior work before beginning to respond to their responsibilities.
F411144133	Apply Brainstorming Technique	A team collects inputs from team members in a nonthreatening environment, organizes the inputs, and then treats the material as team data.
F411144134	Consolidate Allocation Statements	Collect inputs onto RAS.
F411144135	Make Changes	Team or engineer changes the analysis to reflect team or team leader direction.
F411144136	Engineering Review	The RAS data is reviewed by team management and approved, or direction for change is given.
F411144137	Perform Trade Study	If the allocations cannot be made through sound engineering judgment, a trade study is conducted to reach a best solution from alternatives.
F411144138	Consider Direction Impact	The individual engineer or team considers how the direction given at the review meeting will effect prior work and how they can respond to that direction.
F411144139	Capture/Document Allocations	Route the RAS data to those responsible for SDD development.
F41114414	Prepare/Maintain SDD Inputs	Function diagrams, timelines, function dictionary, and RAS are organized for inclusion in the SDD.
F411144141	Capture/Document Diagrams	Collect and organize the structured analysis diagrams and function dictionary for inclusion in SDD.
F411144142	Prepare Descriptive Function Text	A function dictionary is created, and this information is used selectively in preparing system specification content related to needed functionality.
F411144143	Format SDD Content	Inputs from functional analysis, timeline analysis, architecture synthesis, interface analysis, and functional allocation are combined into the SDD to capture the structured analysis baseline for the system.
F4111442	Performance Requirements Analysis	Allocated functions are transformed into quantified performance requirements in the final form for inclusion in the item specification.
F4111443	Timeline Analysis	Selectively, functions are evaluated for time criticality exposed in graphical or tabular (time dictionary) format. Times may be defined deterministically or probabilistically.

F4111444	Architecture Synthesis	Allocated functionality via RAS is hooked up into a hierarchical structure of the physical product system.
F4111445	Interface Analysis	Relations between the things in the architecture are identified in schematic block or n-square diagrams. An interface dictionary database is also created if automated methods are to be used for interface management. Interface responsibilities are clearly defined, and cross-organizational interfaces identified for special attention of the PIT.
F41114451	Evaluate Allocated Functionality	
F41114452	Perform Interface Analysis	
F41114453	Evaluate Cross Organizational Interface Impact	
F41114454	Document Interfaces	
F4111446	Process Analysis	A physical model of system operation with full knowledge of the architecture is crafted in support of environmental requirements analysis, system safety analysis, and logistics support analysis.
F4111447	Specialty Engineering Requirements Analysis	The many specialty engineering disciplines apply special models to identify requirements in their areas of responsibility.
F41114471	Logistics Support Analysis	
F411144711	Availability Requirements Analysis	
F411144712	Reliability Requirements Analysis	
F4111447121	Reliability Modeling	
F4111447122	Reliability Allocation	
F4111447123	FMECA	
F411144713	Maintainability Requirements Analysis	
F4111447131	Maintainability Modeling	
F4111447132	Maintainability Allocation	
F411144714	Human Factors Requirements Analysis	
F411144715	Logistics Factors Analysis	
F4111447151	Logistics Concept Development	

Table A.1 (continued)   Enterprise Process Dictionary

Function ID	Function name	Function description
F4111447152	Personnel and Training Requirements Analysis	
F4111447153	Support Equipment Requirements Analysis	
F4111447154	Technical Data Requirements Analysis	
F4111447155	Spares Requirements Analysis	
F41114472	Safety Requirements Analysis	
F41114473	Security Requirements Analysis	
F41114474	PMP Requirements Analysis	
F411144741	Parts Engineering Planning	
F411144742	Materials and Processes Planning	
F41114475	DTC/LCC Requirements Analysis	
F4111447A	Aerodynamics Requirements Analysis	
F4111447B	Thermodynamics Requirements Analysis	
F4111447C	Mass Properties Requirements Analysis	
F4111447D	Structures Requirements Analysis	
F4111447E	Guidance Accuracy Requirements Analysis	
F4111448	Service Use Profile	An architecture-process and environmental-process are combined to form a three-dimensional space defining the environmental characteristics for prime items.
F4111449	End-Item Zoning	End item environmental requirements are applied to physical zones, and component architecture items mapped to the zones defining environmental requirements for the components.

F411144A	Interface Requirements Analysis	Each line on the schematic block diagrams and/or each intersection in the n-square diagrams is translated into one or more interface requirements in the technology associated with the media.
F411144B	Write Requirements Paragraphs	Primitive requirements statements are translated into full specification text.
F411144C	Lead System/Item Requirements Analysis	A person is selected to lead the requirements analysis effort (maybe the team leader during this activity), and that person applies management skills to complete the specification in accordance with program resources allocated to this task.
F411144D	Edit/Update Current Content	Changes to the specification are made prior to review and approval.
F411144E	Integrate Item Requirements	Requirements are searched for mutual inconsistencies and conditions of suboptimization.
F411145	RDD	The requirements driven design (RDD) approach is applied to uncover information about the problem space.
F411146	Modern Structured Analysis	The methods encouraged by Yourdon, DeMarco, and Constantine are applied to develop an understanding of the problem space where the product item is clearly software. Alternatively, the Hatley-Pirbhai model could be used if the problem has real-time aspects.
F411147	Database Modeling	Data table normalization or IDEF 1X is applied to developing the requirements for a database.
F411148	Object Oriented Analysis	The Rumbaugh model is applied to gaining insight into the requirements for an item that is clearly a software entity.
F411149	Strings and States Analysis	
F41114A	Publish Specification	The content of the specification developed through one of several model bases is collected into a document in context with the template selected. Figures and tables are coordinated with the text material, and the document format established.
F41114B	Clone Specification	One of three cloning methods (parent item, like item, or document standard) is applied to develop content for the specification.
F41114B1	Select Cloning Mode	
F41114B2	Select Like Item	
F41114B3	Select Standard	
F41114B4	Identify End Item	
F41114B5	Identify Zone	

Table A.1 (continued) Enterprise Process Dictionary

Function ID	Function name	Function description
F41114B6	Perform Item-Unique Environmental Requirements Analysis	
F41114B7	Perform Item-Unique Performance Requirements Analysis	
F41114B8	Perform Item-Unique Interface Requirements Analysis	
F41114B9	Perform Item-Unique Specialty Engineering Rqmts Analysis	
F41114BA	Copy Standard Specification	
F41114BB	Analyze Content Relationships	
F41114BC	Copy Parent Specification	
F41114BD	Analyze Content Relationships	
F41114BE	Copy Like Item Specification	
F41114BF	Analyze Content Relationships	
F41114BG	Modify Content	
F41114BH	Manage Cloning Task	
F41114BI	Integrate Content	
F41115	Item Concept Development and Review	After the requirements are defined for an item, concepts are developed for each of the lower-tier items in the architecture previously defined through structured analysis in this task. This activity is performed for each specification created.
F41116	Verification Requirements Analysis	Each person with a requirements analysis responsibility must prepare one or more verification requirements for each requirement he/she writes telling any appropriate requirements for proving that product design satisfies those requirements.
F411161	IPT Specification Verification Work	

F4111611	Individual Requirement Verification Work	
F41116111	Select Verification Method	
F41116112	Perform Analysis	
F41116113	Write Section 4 Paragraph	
F4111612	Maintain Specification Matrix	
F4111613	Define Verification Strings	
F4111614	Define Item Verification Tasks	
F4111615	Select Verification Level	
F4111616	Map Strings To Tasks	
F411162	PIT Level Work	
F4111621	Prepare and Maintain Task Matrix	
F4111622	Prepare and Maintain Compliance Matrix	
F41117	Change Released Specifications	When the work has been approved by management, previously released specifications will be changed by the responsible team and reviewed and released by PIT.
F411171	Perform Change Requirements Analysis	
F411172	Revise Specification	
F411173	Prepare Page Change Notice	
F411174	Prepare List Notice	
F411175	Review Specification Change	
F411176	Release Specification Change	
F411177	Place Change On Server	
F4112	Define Process System Requirements and Develop Concepts	Process requirements are identified for seven process areas: qualification verification, quality, manufacturing, logistics, material and procurement, deployment, and schedule. Results are captured in plans and other documents. All of this work is accomplished in concert with product requirements development.
F41121	Plan Quality Process	The requirements for the quality process are defined and captured in a quality plan or an existing quality plan crafted during the proposal updated.

Table A.1 (continued) Enterprise Process Dictionary

Function ID	Function name	Function description
F41122	Plan Material and Procurement Process	Plans are created for identifying needed material, acquisition of that material, receiving and processing the material to the correct initial manufacturing site.
F41123	Manufacturing Requirements Analysis	Manufacturing process requirements are developed and documented in manufacturing plans.
F41124	Plan Qualification Verification Process	The results of Verification Requirements Analysis are reviewed and used as a basis for developing appropriate content for an evolving integrated verification plan. General process requirements are also developed and included in the plan. This activity must be accomplished for each item requirements analysis cycle.
F41124L	Item L Integrated Qualification Task Planning	
F41124L1	Item L Test Tasks	
F41124L1M	Prepare Item L Test Task Plan	
F41124L1Y	Integrate Item L Test Task Plans	
F41124L1Z	Prepare Item L Test Procedures	
F41124L2	Item L Analysis Tasks	
F41124L2M	Prepare Item L Analysis Task M Plan	
F41124L2Y	Integrate Item L Analysis Plans	
F41124L2Z	Prepare Item L Analysis Procedures	
F41124L3	Item L Exam Tasks	
F41124L3M	Prepare Item L Exam Task M Plan	
F41124L3Y	Integrate Item L Exam Plans	
F41124L3Z	Prepare Item L Examination Procedures	
F41124L4	Item L Demo Tasks	
F41124L4M	Prepare Item L Demo Task M Plan	
F41124L4Y	Integrate Item L Demo Plans	

F41124L4Z	Prepare Item L Demonstration Procedures	
F41124L5	Integrate Item L Qualification Planning	
F41124L6	Review and Approve Item L Qualification Planning	
F41124X	Integrate Qualification Planning	
F41124Y	Review and Approve Integrated Qualification Plan	
F41124Z	Manage Qualification Planning	
F41125	Plan Logistics Process	Logistics process requirements are identified and captured in an integrated logistics support plan.
F41126	Maintain Schedule	The program integrated master schedule (IMS), created during proposal work, is expanded in detail as the architecture is expanded.
F41127	Deployment Planning Analysis	Deployment or distribution requirements are identified and captured in a deployment plan.
F41131	Review and Release Specifications	Specifications are reviewed by PIT, and approved specifications are released and loaded into the program on-line specification library.
F411311	Prepare For Requirements Review	
F411312	Review Specification	
F411313	Peer and Informal Reviews	
F411314	Perform Post Review Actions	
F411315	Acquire Approval Signatures	
F411316	Release Specification	
F41132	Configuration Manage Specifications	The master copies of the specifications in electronic media are protected from unauthorized influences and records maintained to clearly identify the status of all released specifications.
F41133	Validate Product Requirements	All system and item requirements are screened for risk, and those that are risky are subjected to some form of validation to establish that it will be possible to create a responsive design within the resources made available to the responsible design team. Validation actions include development evaluation testing, technology demonstration, technology search, technical performance measurement tracking, and special studies.

Table A.1 (continued)   Enterprise Process Dictionary

Function ID	Function name	Function description
F411331	Evaluate Requirements	
F4113311	Assess Specification For Completeness	
F4113312	Screen Requirements For Necessity and Correctness	
F4113313	Rewrite Requirement	
F4113314	Discard Requirement	
F4113315	Screen Requirements For Structural Correctness	
F411332	Change Requirement	
F411333	Design Evaluation Testing	
F411334	Demo or Inspection	
F411335	Analysis	
F411336	Lead Requirements Validation	
F411337	Technical Performance Measurement	
F4113371	Track and Manage TPM	
F4113372	Terminate TPM Track	
F4113373	Design Evaluation Test	
F4113374	Demo or Inspection	
F4113375	Analysis	
F41134	Establish and Maintain Product Traceability	All requirements are reviewed for traceability to source, rationale, and parent requirements. The results are captured in the requirements database. Programs must use a requirements database in order for traceability capture to make economic sense.
F41135	Perform and Maintain Margin and Budget Analysis	The PIT must identify what requirements shall have margin or budget analysis applied and assign an engineer to manage the parameter. Product design performance is tracked against related goals and management decisions reached to encourage design compliance. Margin may be consumed through a formal design review process as one of the solutions to design difficulty.

F41136	Develop and Maintain Specification Tree	PIT develops a hierarchical organization of all program specifications and plans, determines development and maintenance responsibilities, defines types and styles, and publishes the tree on the program server. The tree may use graphical or listing technique or both in some mix.
F41137	Select Specification Templates	Standards for specification content and structure available to the program are defined and coordinated with the specifications listed in the specification tree.
F41138	Schedule Specifications	Each specification appearing on the specification tree must have a start and release date established prior to the start of work on the specification.
F4113A	Review and Approve Process Requirements	All program plans are reviewed and approved. Approved plans are loaded into the program specification library and configuration controlled by PIT.
F4113B	Manage Product System Requirements Analysis	In the development of the system specification and perhaps other, the PIT manages its own analysis activities to achieve planned goals. As the work proceeds into areas under the responsibility of the assigned IPT, PIT oversees the work of the teams from the perspectives of budget and schedule compliance, evolving traceability, margin and budget work, and requirements validation work.
F4113C	Manage Process System Requirements Analysis	PIT system level planning work and oversees lower-tier planning work accomplished by IPT.
F4114	Form PIT	The PIT leader acquires the needed team resources and initiates team action in accordance with the program plan previously crafted and maintained. Team members are acquired and trained as needed.
F41142	Identify Core Team Personnel Needs	
F41143	Acquire Core Team Members	The team leader works with functional department managers to staff the team.
F41144	Study Team Assignment	
F41145	Identify Needed Team Resources	The core team members and team leader identify equipment, facilities, utilities, and other materials needed.
F41146	Acquire Team Resources	
F41147	Select/Endorse Team Leader	Optional team endorsement of the program manager selected team leader. The program must define how team leaders will be selected.
F41148	Define Team Objectives, Scope, and Metrics	Everyone on the team must have the same understanding of the team goals. Agreements are reached about how the team will measure performance, recognizing that functional managers and the program require certain metrics.

Table A.1 (continued)   Enterprise Process Dictionary

Function ID	Function name	Function description
F41149	Team/Management Consensus	Team arrives at a common understanding with program management about the work that must be accomplished and the resources that will be available to do that work.
F4114A	Train Team	If there are areas of knowledge that team personnel do not have, training is acquired and applied.
F4114B	Co-Locate or Acquire War Room	All team personnel are brought together to encourage cross-function teaming and conversation.
F4114C	Define Team Meeting Style, and Timing	
F4115	Form IPT	The IPT leader acquires the needed team resources and initiates team action in accordance with the program plan previously crafted and maintained. Team members are acquired and trained as needed.
F41151	Identify Interim Team Leader	Program manager identifies the person to lead the team temporarily or as a permanent assignment.
F41152	Identify Core Team Personnel Needs	
F41153	Acquire Core Team Members	The team leader works with functional department managers to staff the team.
F41154	Study Team Assignment	
F41155	Identify Needed Team Resources	The core team members and team leader identify equipment, facilities, utilities, and other materials needed.
F41156	Acquire Team Resources	
F41157	Select/Endorse Team Leader	Optional team endorsement of the program manager selected team leader. The program must define how team leaders will be selected.
F41158	Define Team Objectives, Scope, and Metrics	Everyone on the team must have the same understanding of the team goals. Agreements are reached about how the team will measure performance, recognizing that functional managers and the program require certain metrics.
F41159	Team/Management Consensus	Team arrives at a common understanding with program management about the work that must be accomplished and the resources that will be available to do that work.

F4115A	Train Team	If there are areas of knowledge that team personnel do not have, training is acquired and applied.
F4115B	Co-Locate or Acquire War Room	All team personnel are brought together to encourage cross-function teaming and conversation.
F4115C	Define Team Meeting Style, and Timing	
F412	Preliminary Design	Item requirements are synthesized into preliminary designs and identified risks mitigated. Preliminary process plans are developed.
F4121	Create Product System Preliminary Design	An IPT develops the preliminary product design for the assigned item(s).
F41211	Study Prior Concepts	
F41212	Study Requirements	
F41213	Finalize Make-Buy Decision	
F41214	Support Procurement and Supplier Development	
F41215	Select Preferred Design	
F412151	Create Alternative Designs	
F412152	Define Selection Criteria and Weighting Criteria	
F412153	Define Utility Curves	
F412154	Enter Matrix Data and Get Results	
F412155	Develop Criteria Values	
F412156	Perform Sensitivity Studies	
F412157	Lead Trade Study	
F412158	Select Preferred Design	
F412159	Report Trade Study Results	
F41216	Specialty Engineering Trade Study Support	
F41217	Document Preliminary Design	
F41218	Specialty Engineering Preliminary Design Support	

Table A.1 (continued)   Enterprise Process Dictionary

Function ID	Function name	Function description
F412181	Study/Review Specification and Design	
F412182	Review Program Acceptance Test Guidelines	
F412183	Prepare Acceptance Test Planning	
F412184	Prepare Item Acceptance Analysis Planning	
F412185	Prepare Item Acceptance Demonstration Planning	
F412186	Prepare Acceptance Examination Planning	
F412187	Review Availability of Necessary Test Apparatus	
F412188	Acquire Needed Test Capability	
F412189	Prepare Integrated Verification Procedure	
F41218A	Verification Planning Review and Approval	
F41218B	Customer Review and Approval	
F41218C	System Test Parameter Recommendations	
F41219	Prepare Qualification Plans	
F412191	Prepare Qualification Plans and Procedures	
F4121911	Review Task and Compliance Matrices	
F4121912	Review System Qualification Guidelines	

F4121913	Evaluate Team Items Against Qualification Guidelines
F4121914	Create Test Service Request
F4121915	Approve Service Request
F4121916	Issue Service Request to Test Lab
F4121917	Prepare Test Plan and Procedure
F4121918	Define Test Schedule
F4121919	Approve Test Plan and Procedure
F412192	Develop Test Article
F4121921	Define Test Set-up Requirements
F4121922	Acquire Test Set-up Material
F4121923	Assemble Test Set-up
F4121924	Define Test-Unique Test Article Requirements
F4121925	Acquire Special Test Article Material
F4121926	T&E Produce Test Article
F4121927	Purchase/Procure Test Article
F412193	Vendor Test Planning
F4121931	Issue Purchase Order
F4121932	Vendor Prepare Qualification Test Procedure
F4121933	Data Management Process Vendor Documentation
F4121934	Qualification Test Procedure Review
F4121B	Manage Preliminary Design
F4121C	Evaluate/Mitigate Item Risks
F4121D	IPT Preliminary Audit/Review
F4122	Create Process Preliminary Design

The PIT creates the preliminary design for the processes associated with the assigned product entity.

Table A.1 (continued) Enterprise Process Dictionary

Function ID	Function name	Function description
F41221	Manufacturing Trade Study Support	
F41222	Quality Trade Study Support	
F41223	Logistics Trade Study Support	
F41224	Material and Procurement Trade Study Support	
F41225	Manage Process Design	
F41226	Manufacturing Process Design	
F41227	Quality Process Design	
F41228	Logistics Process Design	
F41229	Material and Procurement Process Design	
F4122A	Maintain Program Plan	
F4123	Integrate and Optimize Preliminary Design at PIT Level	PIT interacts with IPT to optimize evolving preliminary design solution at the system level.
F41231	System Optimization and Interface Integration	The ongoing preliminary design work being accomplished by the several IPT is constantly evaluated for inconsistencies and conditions of suboptimization. Cross-organizational interfaces are reviewed in the form of terminal pairs to ensure compatibility.
F41232	Systems Analysis and Specialty Engineering Assessment	The specialized analytical disciplines evaluate preliminary design concepts and interact with the designers in team meetings and joint working sessions to ensure that preliminary designs have a good chance of satisfying specialty requirements.
F41233	Integrated Verification Planning	IPT integrates cross-team verification work to ensure that a least cost, most effective plan is identified.
F41234	Design Evaluation Test	Where necessary to reduce risk, a special design evaluation test is performed that proves it will be possible to satisfy requirements.

F412341	Create Test Request	
F412342	Approve Test Request	
F412343	Issue Test Request to Test Lab	
F412344	Prepare Test Plan and Procedure	
F412345	Design DET Test Apparatus	
F412346	Acquire Material	
F412347	Prepare Test Setup	
F412348	Change Test Plan or Procedure	
F412349	Approve Test Plan and Procedures	
F41234A	Schedule Test	
F41234B	Design Test Article	
F41234C	Acquire Test Article Material	
F41234D	Fabricate/Assemble Test Article	
F41234E	Perform Test or Re-test	
F41234F	Prepare Test Report	
F41234G	Approve Test Report	
F41234H	Issue Test Report	
F41234I	Prepare Failure Report	
F41234J	Evaluate Anomaly	
F41234K	Repair or Refurbish Test Article	
F41234L	Store Test Article Awaiting Re-test	
F41234M	Modify Test Article	
F41234N	Dispose of Test Resources	
F41235	Risk Management Program	PIT constantly evaluates the evolving preliminary design to identify potential risks, assess their probability of occurrence and seriousness of the resulting problem. Risks are mitigated and tracked. Periodically the program risk signature is taken.
F412351	Technical Performance Measurement	Selected requirements values are tracked against required values in time, and actions are taken to cause closure between them.
F412352	Technology Management	All technologies available to the enterprise are cataloged and described. These are on-the-shelf technologies that can be used on programs.

Table A.1 (continued)   Enterprise Process Dictionary

Function ID	Function name	Function description
F412353	Technology Availability Assessment	New program technology needs are compared with available ones, driving conclusions about new technology needs.
F412354	Risk Analysis	All potential program risks are evaluated and assigned to specific persons to bring the risk under control in accordance with a specific work plan.
F412355	Risk Mitigation Evaluation	All risks are evaluated for probability of occurrence and degree of seriousness if they occur. Decisions are reached on what to do about these risks.
F412356	Risk Tracking and Documentation	All program risks are listed, assigned to responsible persons, and status tracked over time toward reduction in the degree of risk.
F412357	Technology Demo	Where new technologies are needed and there is some hope of developing it in-house, the technology is demonstrated to prove it is a viable technology or it is not.
F412358	Technology Search	A search is made in the literature, with other companies including suppliers, and through recognized experts in the field to find and capture that technology by some means.
F412359	Manage Risk Program	Management skills are brought to bear on risk activities.
F41236	Program Schedule Development	IPT and program schedules are compared and inconsistencies identified and worked.
F41237	DTC/LCC Integration	Design to cost and life cycle cost figures are negotiated across team boundaries.
F41238	Deployment and Operations Plans and Procedures	Deployment and operational process plans are created, evaluated, and maintained.
F41239	Quality Planning Integration	PIT and IPT quality planning is integrated and inconsistencies looked for.
F4123A	Material and Procurement Integration	PIT interacts with teams and ensures no inconsistencies exist in preliminary design of materials and processes.
F4123B	Manufacturing Process Integration	PIT interacts with IPT to ensure program product and process preliminary designs are optimum.
F4123C	Logistics Integration	PIT interacts with teams to ensure that logistics preliminary design work is consistent.
F4123D	Manage PIT System Integration	
F4124	Preliminary Design Review	The IPT stages a preliminary design review for the PIT which either approves the review or directs that work be done in product design area, process design area, or both. In change cases, the changes must be subjected to a follow-up review.
F41241	Organize Design Review Agenda	

F41242	Accomplish Additional Work	
F41243	Establish Exit Criteria	
F41244	Customer/Sponsor Review and Approve Plan	
F41245	Assign Responsibility, Schedule Review Prep, & Brief Team	
F41246	Prepare Review Materials	
F41247	Integrate Review Materials	
F41248	Review Administration	
F41249	Publish Materials	
F4124A	Present Review	
F4124B	Action Item Processing	
F4124F	Prepare Review Minutes	
F4124G	Customer/Sponsor Acceptance of Review	
F4124H	Customer/Sponsor Closure Action	
F413	Detailed Design	Engineering drawings are created for items and process designs crafted.
F413	Manage Requirements Analysis at PIT Level	The PIT oversees product and process requirements analysis at the system and item levels, reviews and releases all specifications, controls configuration of all specifications, maintains program specification tree, schedules all specification work, identifies and selects available specification templates and makes them available to teams, and reviews and approves all process requirements documents.
F4131	Create Product System Detailed Design	
F41311	Item Product Detailed Design	
F41312	Evaluate and mitigate Item Risk	
F41313	Team Drawing Package Release	
F41314	Specialty Engineering Assessment	

Table A.1 (continued)   Enterprise Process Dictionary

Function ID	Function name	Function description
F41315	Update Item Qualification Planning	
F41316	Prepare Product (Part II) Specification	
F41317	Supervise Supplier Technical Development	
F41318	Prepare Acceptance Verification Plan	
F41319	Maintain engineering	
F4131A	Manage Detailed Design	
F4131D	IPT Audit/Review	
F4132	Create Process System Detailed Design	
F41321	Prepare Manufacturing Planning	
F41322	Prepare Quality Planning	
F41323	Design Logistics Process	
F41324	Design Material and Procurement Process	
F41325	Manage Detailed Process Design	
F4133	Integrate and Optimize Detailed Design at PIT Level	PIT interacts with IPT to optimize evolving detailed design solution at the system level.
F41331	System Optimization and Interface Integration	The ongoing detailed design work being accomplished by the several IPT is constantly evaluated for inconsistencies and conditions of suboptimization. Cross-organizational interfaces are reviewed in the form of terminal pairs to ensure compatibility.
F41332	Systems Analysis and Specialty Engineering Assessment	The specialized analytical disciplines evaluate detailed design concepts and interact with the designers in team meetings and joint working sessions to ensure that detailed designs have a good chance of satisfying specialty requirements.

F41333	Integrated Verification Planning	IPT integrates cross-team verification work to ensure that a least cost, most effective plan is identified.
F41335	Risk Management Program	PIT constantly evaluates the evolving detailed design to identify potential risks, assess their probability of occurrence and seriousness of the resulting problem. Risks are mitigated and tracked. Periodically the program risk signature is taken.
F413351	Technical Performance Measurement	Selected requirements values are tracked against required values in time, and actions are taken to cause closure between them.
F413353	Technology Availability Assessment	New program technology needs are compared with available ones driving conclusions about new technology needs.
F413354	Risk Analysis	All potential program risks are evaluated and assigned to specific persons to bring the risk under control in accordance with a specific work plan.
F413355	Risk Mitigation Evaluation	All risks are evaluated for probability of occurrence and degree of seriousness if they occur. Decisions are reached on what to do about these risks.
F413356	Risk Tracking and Documentation	All program risks are listed, assigned to responsible persons, and status tracked over time toward reduction in the degree of risk.
F413359	Manage Risk Program	Management skills are brought to bear on risk activities.
F41336	Program Schedule Development	IPT and program schedules are compared and inconsistencies identified and worked.
F41337	DTC/LCC Integration	Design to cost and life cycle cost figures are negotiated across team boundaries.
F41338	Deployment and Operations Plans and Procedures	Deployment and operational process plans are created, evaluated, and maintained.
F41339	Quality Planning Integration	PIT and IPT quality planning is integrated and inconsistencies looked for.
F4133A	Material and Procurement Integration	PIT interacts with teams and ensures no inconsistencies exist in detailed design of materials and processes.
F4133B	Manufacturing Process Integration	PIT interacts with IPT to ensure program product and process detailed designs are optimum.
F4133C	Logistics Integration	PIT interacts with teams to ensure that logistics detailed design work is consistent.
F4133D	Manage PIT System Integration	
F4134	Critical Design Review	PIT reviews IPT detailed design of product and process.
F42	Material and Procurement	Material, software, and equipment defined in engineering documentation is acquired through a procurement process, received at the appropriate manufacturing facility, and placed under manufacturing control.

Table A.1 (continued)   Enterprise Process Dictionary

Function ID	Function name	Function description
F43	Manufacture Product System	The product is manufactured in accordance with engineering drawings, manufacturing and quality planning, and accepted for delivery.
F432Q	Facilitize Manufacturing Site Q	
F432Q1	Evaluate and Acquire Site	
F432Q2	Acquire Manufacturing Equipment	
F432Q3	Acquire Personnel	
F432Q5	Train Personnel	
F432Q6	Prepare Site	
F432Q7	Install Equipment	
F432Q8	Acquire Quality Resources	
F432Q9	Test Site	
F432QA	Manage Site Q Facilitization	
F432Z	Manage Manufacturing Facilitization	
F44	Verify Product System	Three layers of verification work are applied to the product. New product items are subjected to qualification to ensure their design is adequate for the application and in compliance with the specification for the item. Each manufactured item is evaluated in accordance with an acceptance process. New systems are subjected to system verification.
F4411	Study System Verification Planning	
F4412	Acquire Test Facilities	
F4413	Acquire Test Articles	
F4414	Perform Test Task Z	
F4415	Perform Analysis Task Y	
F4416	Perform Examination Task X	
F4417	Perform Inspection Task W	

F4418	Integrate Results Into System Verification Report
F4419	System Verification Review
F441A	Dispose of Residual Material
F441B	Manage System Verification
F4421	Item Qualification Work
F44212	Prepare For Functional Configuration Audit
F44213	Accomplish Functional Configuration Audit
F44214	Post Meeting Action
F44215	Track and Report Status
F44216	Review and Archive Reports
F44217	Validate Item Models
F4421L	Item L Qualification
F4421L1	Item L Qualification Tests
F4421L1M	Item L Qualification Test Task M
F4421L1M1	Prepare For Test
F4421L1M2	Accomplish Test
F4421L1M3	Refurbish Resources
F4421L1M4	Dispose of Residual Resources
F4421L1M5	Report Test Results
F4421L1M6	Replan Whole or Part of Test
F4421L2	Item L Qualification Analyses
F4421L2M	Item L Analysis Task M
F4421L2M1	Prepare For Analysis
F4421L2M2	Accomplish Analysis
F4421L2M3	Report Analysis Results
F4421L2M4	Evaluate Evidence Quality
F4421L2M5	Rethink/Adjust Analysis Planning

Table A.1 (continued)   Enterprise Process Dictionary

Function ID	Function name	Function description
F4421L3	Item L Qualification Examinations	
F4421L3M	Item L Examination Task M	
F4421L3M1	Prepare For Examination	
F4421L3M2	Accomplish Examination	
F4421L3M3	Report Examination Results	
F4421L3M4	Evaluate Evidence Quality	
F4421L3M5	Re-Think/Adjust Examination Planning	
F4421L4	Item L Qualification Demonstrations	
F4421L4M	Item L Qualification Demonstration Task M	
F4421L4M1	Prepare For Demo	
F4421L4M2	Accomplish Demo	
F4421L4M3	Refurbish Resources	
F4421L4M4	Dispose of Residual Resources	
F4421L4M5	Report Demo Results	
F4421L4M6	Replan Whole or Part of the Demo	
F45	Logistically Support Product System	Spares are stocked and routed to customer locations as requested or covered by contract. Customer training is accomplished or materials required to perform are delivered. Field engineering support is provided. If required, depot level maintenance is provided.
F46	Deliver/Deploy Product System	The product is delivered as covered by contract or plan and placed in initial operation or distribution.
F47	Manage Program	The enterprise applies management skills to the overall program from the enterprise perspective.
F5	Modify Product System	Modify product previously delivered or in production to satisfy new requirements.
F6	Manage Acquisition	For a commercial application this may be an enterprise function where an enterprise agent must play the role of the customer. Where the customer is highly organized, like a government agency, this is a customer function.

| F7 | Use Product System | The product is used as intended to satisfy customer needs. The operator is assumed to be the customer, but in some cases a customer could contract with the enterprise to use the system. This function includes maintenance and logistic support of the product as well as operation. |
| F8 | Dispose of Product System | When the product can no longer satisfy needs economically, it may be removed from service, which requires disposal of all of the elements of the system. Some residual elements may be applied on new programs as-is or after modification, while others must be disposed of. Special attention must be taken with those elements disposed which have significant health and safety consequences. |

Table A.2   Enterprise Requirements Analysis Sheet

Function ID	Dept.	Role	Comment
F1	D000	L	
F2	D100	L	Department D100 provides leadership for the business-oriented functional departments. Manager must coordinate with D200 and D300 managers under the guidance of the enterprise executive (D000).
F2	D200	L	
F2	D300	L	
F3	D900	L	
F4	D200	L	
F41	D210	L	
F4111	D216-2	L	
F41111	D216-2	C	
F41112	D216-2	L	
F41113	D216-2	L	
F41114	D216-2	L	
F411141	D216-1	L	
F411141	D216-2	C	
F411142	D216-2	L	
F411143	D216-2	L	
F411144	D212-3	C	Define necessary product structural properties.
F411144	D212-4	C	Define required weight limits, center of mass, and space and form factor needs.
F411144	D213-4	C	Conduct analysis of the thermal environment for electronic parts.
F411144	D213-5	C	
F411144	D213-6	C	Determine allowable external and internal EMI figures.
F411144	D214-3	C	
F411144	D214-4	C	
F411144	D215-1	C	In cases where it has not yet been determined to what kind of item the functionality has been allocated, this department supports the structured analysis effort. Where the functionality has been allocated to computer software, this department leads the continuing efforts.
F411144	D216-1	C	
F411144	D216-2	L	
F411144	D216-4	C	Define reliability allocations.
F411144	D216-5	C	
F411144	D216-6	C	
F411144	D216-7	C	
F411144	D216-8	C	
F411144	D221-1	C	
F411144	D231-1	C	
F411144	D241-1	C	
F411144	D242-1	C	

Table A.2 (continued)   Enterprise Requirements Analysis Sheet

Function ID	Dept.	Role	Comment
F41114411	D216-2	L	
F411144111	D216-2	L	
F411144112	D216-2	L	
F411144113	D216-2	L	
F411144114	D216-2	L	
F41114412	D216-2	L	
F411144121	D216-2	L	
F411144122	D216-2	L	
F41114413	D216-2	L	
F411144131	D216-2	L	
F411144132	D216-2	L	
F411144133	D216-2	L	
F411144134	D216-2	L	
F411144135	D216-2	L	
F411144136	D216-2	L	
F411144137	D216-2	L	
F411144138	D216-2	L	
F411144139	D216-2	L	
F41114414	D216-2	L	
F411144141	D216-2	L	
F411144142	D216-2	L	
F411144143	D216-2	L	
F4111443	D216-2	L	
F4111444	D216-2	L	
F41114441	D216-2	L	
F4111444B	D216-2	L	
F4111444C	D216-2	L	
F4111445	D216-2	L	
F41114451	D216-2	L	
F41114452	D216-2	L	
F41114453	D216-2	L	
F41114454	D216-2	L	
F4111446	D216-2	L	
F4111447	D216-2	L	
F4111448	D216-2	L	
F4111449	D216-2	L	
F411144A	D216-2	L	
F411144B	D216-2	L	
F411144C	D216-2	L	
F411144D	D216-2	L	
F411144E	D216-2	L	
F411145	D216-2	L	
F411146	D216-2	L	
F411147	D216-2	Rol	
F411148	D216-2	L	
F411149	D216-2	L	
F41114A	D216-2	L	

Table A.2 (continued) Enterprise Requirements Analysis Sheet

Function ID	Dept.	Role	Comment
F41114B	D216-2	L	
F41116	D216-2	L	
F41117	D216-2	L	
F411171	D216-2	L	
F411172	D216-2	L	
F411173	D216-2	L	
F411174	D216-2	L	
F411175	D216-2	L	
F4112	D216-2	C	
F4113	D216-2	L	
F41131	D216-2	L	
F411311	D216-2	L	
F411312	D216-2	L	
F411313	D216-2	L	
F411314	D216-2	L	
F411315	D216-2	L	
F41132	D216-2	C	
F41133	D216-2	L	
F411331	D216-2	L	
F4113311	D216-2	L	
F4113312	D216-2	L	
F4113313	D216-2	L	
F4113314	D216-2	L	
F4113315	D216-2	L	
F411332	D216-2	L	
F411336	D216-2	L	
F411337	D216-2	L	
F4113371	D216-2	L	
F4113372	D216-2	L	
F41134	D216-2	L	
F41135	D216-2	L	
F41136	D216-2	L	
F41137	D216-2	L	
F41138	D216-2	L	
F4113A	D216-2	L	
F4113B	D216-2	L	
F4113C	D216-2	L	
F41212	D216-2	L	
F41219	D216-2	L	
F4124	D216-2	C	
F4134	D216-2	C	
F5	D400	L	
F6	CUST	L	
F7	CUST	Rol	
F8	CUST	L	

Enterprise organizational structure

System engineering deployment

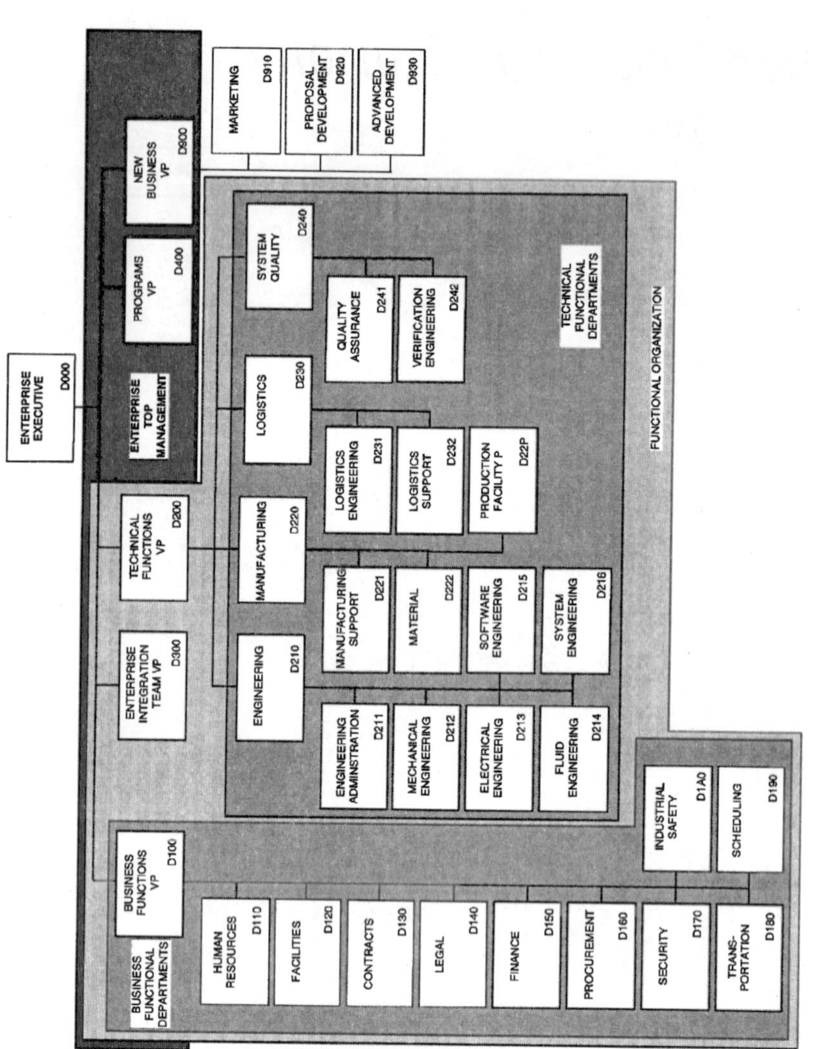

Figure B.1 Enterprise organization diagram, sheet 1.

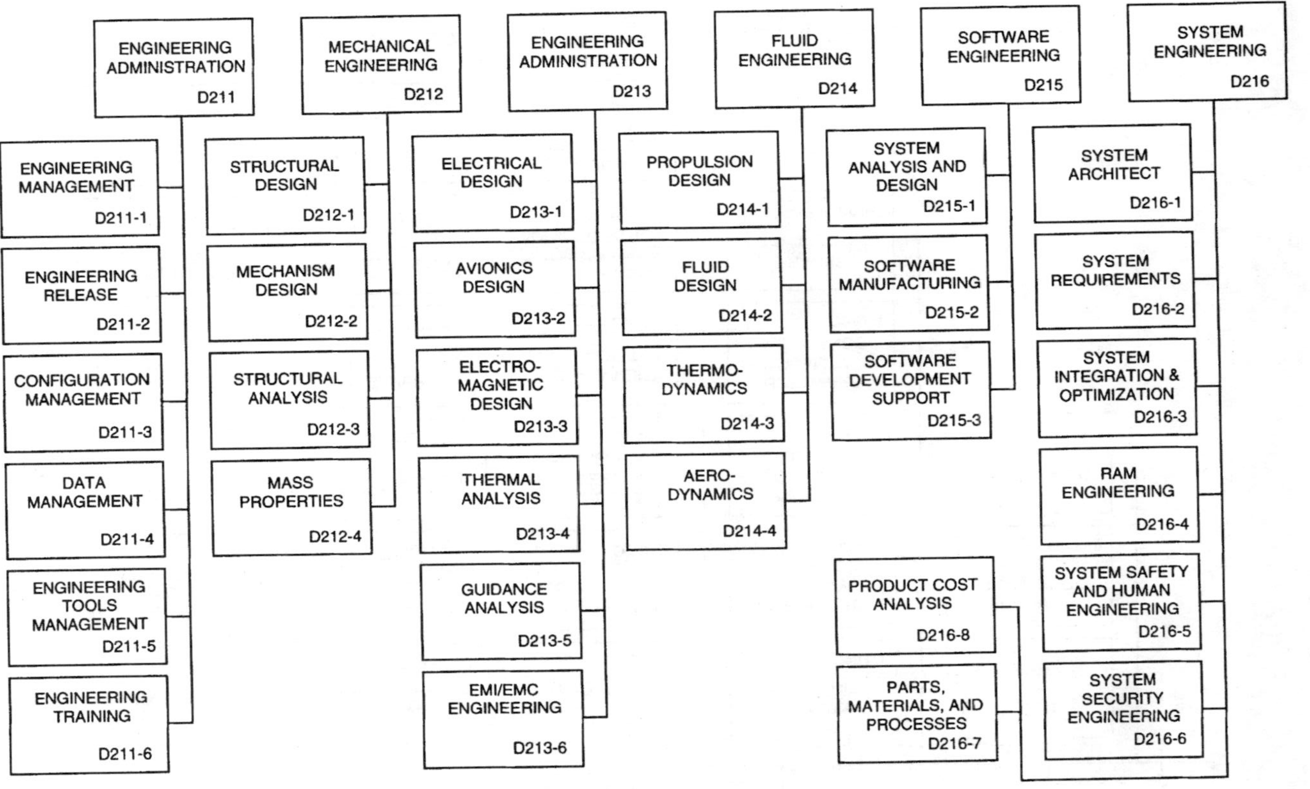

Figure B.1 Enterprise organization diagram, sheet 2.

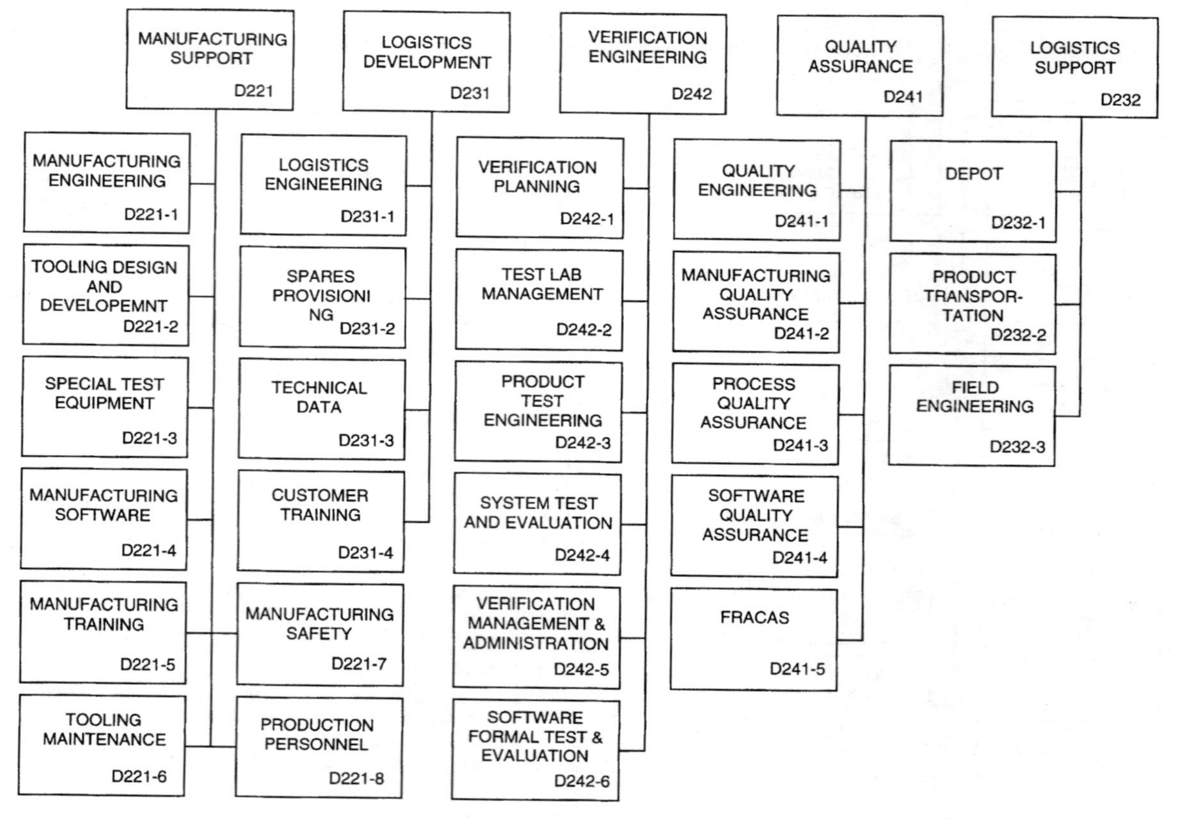

Figure B.1 Enterprise organization diagram, sheet 3.

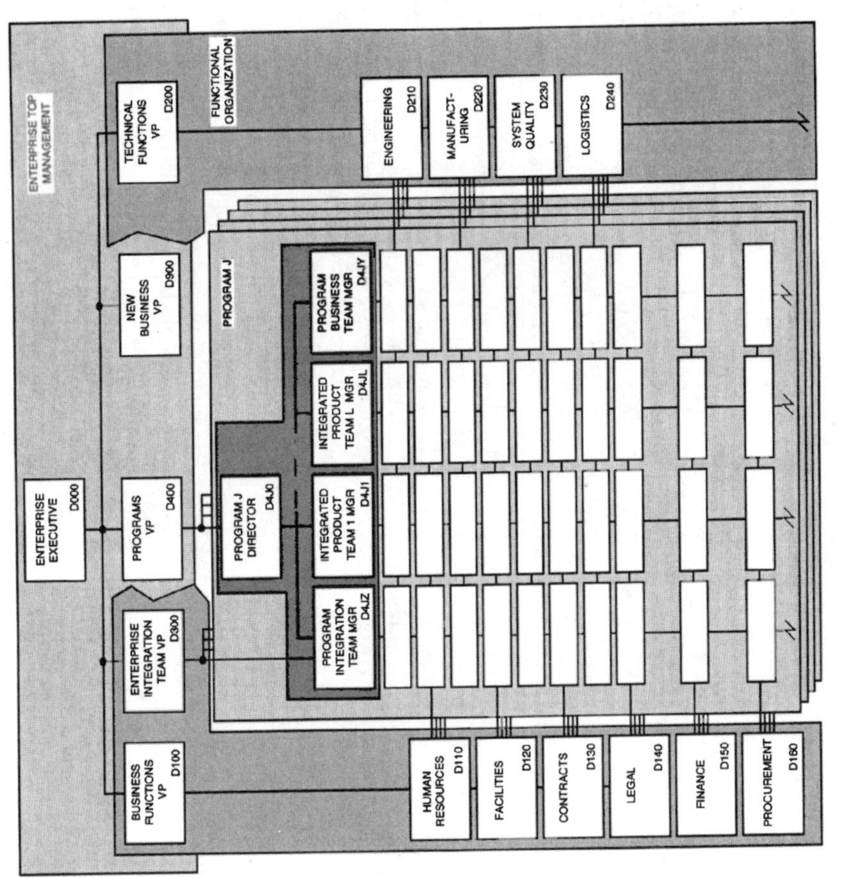

Figure B.1   Enterprise organization diagram, sheet 4.

Table B.1   Functional Department Charter Matrix

Dept	Department name	Mgmt level	Charter
D000	ENTERPRISE EXECUTIVE	President	Ultimate responsibility for enterprise operation. Includes enterprise executive, board of directors, advisory boards, and any other top-level activities sanctioned by the board of directors or executive.
D010	EXECUTIVE OFFICES	Vice President	Support the executive as required.
D100	BUSINESS FUNCTIONS	Vice President	Responsible for business functions at the enterprise and program level, the latter through the assigned program manager.
D110	HUMAN RESOURCES	Director	Responsible for acquisition, training, and separation of personnel. Keeps required records of personnel.
D111	PERSONNEL	Manager	Personnel records.
D112	RECRUITING AND SEPARATION	Manager	The conduit through which personnel pass in either direction.
D113	TRAINING AND DEVELOPMENT	Manager	Develops and maintains contact with sources of training correlated with functional departments and process steps. Facilitates training as required by departments and funded by the executive.
D120	FACILITIES	Director	Maintains, acquires, and disposes of facilities for the enterprise.
D130	CONTRACTS	Director	Represents the enterprise in all formal contractual relationships with customers and suppliers. Drafts, reviews, maintains, and changes contracts for program manager or executive approval.
D140	LEGAL	Director	Provides the enterprise with sound legal representation for contract support and such other matters that arise.
D150	FINANCE AND ACCOUNTING	Director	Satisfies all requirements defined in law and regulation regarding external financial reporting, contractual reporting, and defined in internal procedures.
D160	PROCUREMENT	Director	Responsible for acquisition of material and services needed by the enterprise in performance of enterprise and program activities. Procurement executes the formal definition of requirements for required goods and services and either implements a purchase order or arranges for contracts support and manages the delivery of covered goods and services.

Code	Name	Title	Description
D170	SECURITY	Director	Provides enterprise physical, communications, and data security in accordance with enterprise procedures and contractual provisions.
D180	TRANSPORTATION	Director	Satisfies all material and personnel transport requirements either through owned resources or by using commercial sources.
D190	SCHEDULING	Director	Performs all program scheduling work. Creates program IMS and expands it into consistent team schedules.
D1A0	INDUSTRIAL SAFETY	Director	Personnel safety and health in the workplace.
D200	TECHNICAL FUNCTIONS	Vice President	Responsible for providing programs with needed resources.
D210	ENGINEERING	Director	Manages the department. Responsible for providing programs with engineering talent needed to translate customer needs into compliant product designs providing customer value.
D211	ENGINEERING ADMINISTRATION	Manager	Manages the department.
D2111	Engineering Management	Chief	Responsible for development and improvement of engineering department practices. Responsible for development of management talent within the engineering department.
D2112	Engineering Release	Chief	Provides the machinery to formally publish program information of lasting value. This includes engineering drawings crafted on CAD and reports composed of text and graphics created by any member of the program staff. The only document that does not have to pass through release is a memo.
D2113	Configuration Management	Chief	Manages and controls the configuration of all program documentation and products.
D2114	Data Management	Chief	Provides a program information librarian for the program library consisting of all data developed on the program. Access to this data is implemented through a computer intranet connecting the databases developed by functional departments and loaded on programs by teams and their members.
D2115	Engineering Tools Management	Chief	Coordinates tool development, documentation, and maintenance within the engineering department. Maintains a map of tools vs. processes linked to tool responsibilities and users. Provides support for tool development, purchase, and modification.

Table B.1 (continued)   Functional Department Charter Matrix

Dept	Department name	Mgmt level	Charter
D2116	Engineering Training	Chief	Responsible for maintaining a map between engineering skills needed in each task and the functional department responsible for providing that skill on programs. Maintains data telling current staff capabilities relative to knowledge needs and coordinates training access to bridge that gap. Maintains the engineering map of how-to knowledge coordinated with engineering manual content.
D212	MECHANICAL ENGINEERING	Manager	Responsible for all structural and mechanical design.
D2121	Structural Design	Chief	Responsible for design of all support structures.
D2122	Mechanism Engineering	Chief	Responsible for design of mechanical fixtures and devices involving moving parts, including release and latching mechanisms, linkages, and mechanical drive trains.
D2123	Structural Analysis	Chief	Establishes structural requirements and evaluates evolving designs for compliance in support of the structural design department.
D2124	Mass Properties	Chief	Establishes center of gravity, mass, and space requirements and goals and accomplishes analysis of evolving designs for compliance.
D213	ELECTRICAL ENGINEERING	Manager	Responsible for all electrical, electronic, and avionics design.
D2131	Electrical Design	Chief	Responsible for electrical power and wiring other than RF coax, special bus wiring, waveguide, and fiber optics.
D2132	Avionics Design	Chief	Responsible for design and selection of complex electrical devices to accomplish control functions.
D2133	Electromagnetic Design	Chief	Responsible for design and selection of all RF, UV, IR, and optical devices, including the propagation media, antennas, sensors, and transducers. Communications equipment is included.
D2134	Thermal Analysis	Chief	Conducts analyses of avionics and electronic devices to determine thermal effects on circuits and advises electronic design personnel accordingly.
D2135	Guidance Analysis	Chief	Conducts analyses of the requirements for navigation and guidance, establishes error budgets and margins, and assesses designs for compliance.

D2136	EMI/EMC Engineering	Chief	Establishes EMI/EMC requirements for items, assesses designs for compliance, and plans and implements tests to detect potential EMI.
D214	FLUID ENGINEERING	Manager	Responsible for all design and selection for devices for controlling, routing, and containing gases and liquid materials.
D2141	Propulsion Design	Chief	Responsible for design and selection of energy and power-producing devices for the purpose of locomotion or as drives for other devices requiring mechanical power input.
D2142	Fluid Design	Chief	Responsible for design and selection of all devices and plumbing networks for liquid and gaseous materials.
D2143	Thermodynamics	Chief	Establishes thermal requirements and assesses designs for compliance in support of design groups.
D2144	Aerodynamics	Chief	Defines product smoothness and aerodynamic shape requirements and evaluates designs for compliance. Supports structural design by performing wind tunnel tests where appropriate.
D215	SOFTWARE ENGINEERING	Manager	Responsible for design and selection of all product computer software and firmware.
D2151	System Analysis and Design	Chief	Defines software requirements and implements a compliant software design. Works with Avionics design to select computer hardware and firmware.
D2152	Software Manufacturing (Coding)	Chief	Responsible for computer programming in accordance with requirements established for that software.
D2154	Software Development Support	Chief	Provides programs with a software development environment implementing software library, quality assurance, configuration management, and security provisions.
D216	SYSTEM ENGINEERING	Manager	Responsible for providing all programs with a systems capability involving requirements identification and specification preparation, integration and optimization across boundary conditions, and PIT/IPT team resources. Implements risk management.
D2161	System Architect	Chief	Leads or accomplishes mission analysis including identification of top-level functionality, architecture, interfaces, and environmental parameters. Participates on PIT Team.

Table B.1 (continued)   Functional Department Charter Matrix

Dept	Department name	Mgmt level	Charter
D2162	System Requirements	Chief	Responsible for providing programs with the infrastructure needed to accomplish requirements analysis. Provides principal engineer for system and other top-level specifications. Accomplishes requirements analysis at the system or IPT team level.
D2163	System Integration and Optimization	Chief	Performs integration and optimization work across boundary conditions, such as IPT teams. Responsible for internal, external, and associate interface development.
D2164	RAM	Chief	Defines reliability, maintainability, and availability requirements and assesses designs for compliance. Performs failure modes effects and criticality analysis (FMECA). Supports FRACAS.
D2165	System Safety and Human Engineering	Chief	Identifies safety and human engineering requirements and assesses designs for compliance. Identifies safety hazards and mitigates against occurrence and severity.
D2166	System Security	Chief	Defines security requirements and assesses design for compliance. Identifies security threats and mitigates against system compromise.
D2167	Parts, Materials, and Processes	Chief	Works with electrical design functions for parts selection and standardization and mechanical design functions for selection and standardization of structural materials. Works with all design functions to prepare and standardize processes called out on engineering drawings.
D2168	Product Cost Analysis	Chief	Performs DTC/LCC analysis on programs.
D220	MANUFACTURING	Director	Responsible for all enterprise manufacturing.
D221	MANUFACTURING SUPPORT	Manager	Supports all manufacturing facilities and programs with manufacturing engineering planning and development.
D2211	Manufacturing Engineering	Chief	Participates in program IPT and PIT teams to develop manufacturing requirements and compliant manufacturing designs consistent with product requirements and designs. Accomplishes manufacturing planning.

D2212	Tooling Design and Development	Chief	Defines requirements for tooling, designs and selects compliant designs, and manufactures simple manufacturing aids.
D2213	Special Test Equipment	Chief	Defines requirements for special test equipment, designs and selects STE, and manufactures simple equipment and adapters that are difficult to purchase.
D2214	Manufacturing Software	Chief	Defines manufacturing software requirements and acquire or develop compliant software.
D2215	Manufacturing Training	Chief	Responsible for planning and procurement or presentation of production training for operation of tooling and handling materials in all facilities. Responsible for tracking certification status in accordance with applicable laws and regulations.
D2216	Tooling Maintenance	Chief	Responsible for maintenance of all machine tools in all facilities.
D2217	Manufacturing Safety	Chief	Responsible for industrial safety at all manufacturing facilities.
D2218	Production Personnel	Chief	Manages all production personnel from an enterprise perspective. Personnel are actually assigned to manufacturing facilities and managed through those facilities.
D222	MATERIAL	Manager	Receives all supplier materials, stores and protects the integrity of these materials, kits materials in accordance with manufacturing instructions, and provides the kits at the time and place required in manufacturing planning. Surrenders control of material to manufacturing at the manufacturing station where it will be used.
D22P	PRODUCTION FACILITY P	Manager	Responsible for production at a typical manufacturing facility P.
D230	LOGISTICS	Director	Responsible for support of the product in its intended operational environment. Where the product is operated by the enterprise, Logistics provides the crews. Where the customer operates the product, Logistics provides spares, training, field service, depot repair, and warranty services.
D231	LOGISTICS ENGINEERING	Manager	Participates in the development process by defining logistics requirements and concepts and evaluates product requirements and designs compatible with logistics plans.

Table B.1 (continued) Functional Department Charter Matrix

Dept	Department name	Mgmt level	Charter
D2311	Logistics Engineering	Chief	Defines logistics requirements, prepares logistics plans, and monitors design progress for logistics requirements compliance.
D2312	Spares Provisioning	Chief	Selects parts for spares provisioning, prepares provisioning documentation, and manages spares development.
D2313	Technical Data	Chief	Develops and publishes maintenance, operations, and other technical data covering customer use of the product.
D2314	Customer Training	Chief	Develops, delivers, and presents training courses to the customer covering maintenance and operation of the product.
D232	LOGISTICS SUPPORT	Manager	Supports delivered product.
D2321	Depot	Chief	Provides depot maintenance for product.
D2322	Product Transportation	Chief	Moves product from manufacturing facilities to locations and at times covered in the contract. Handles product from the customer and routes to Depot Maintenance or other activity.
D2323	Field Engineering	Chief	Accomplishes such support tasks as required by the customer and provides liaison with the enterprise.
D240	SYSTEM QUALITY	Director	Overall responsibility for enterprise quality.
D241	QUALITY ASSURANCE	Manager	Responsible for all product and process quality assessment on programs.
D2411	Quality Engineering	Chief	Develops quality requirements and plans quality work on programs.
D2412	Manufacturing Quality Assurance	Chief	Responsible for defining quality requirements, assessing evolving manufacturing planning for compliance, and monitoring production process for compliance.
D2413	Process Quality Assurance	Chief	Assesses implementation of all nonmanufacturing and all nonsoftware development processes for compliance with generic practices and program plans.
D2414	Software Quality Assurance	Chief	Assesses software development for compliance with written practices and program plans.

Code	Organization	Title	Responsibilities
D2415	FRACAS	Chief	Collects and analyzes failure data. Supports program Material Review Boards to make disposition decisions. Maintains failure data records.
D242	VERIFICATION ENGINEERING	Manager	Responsible for all validation and verification planning, implementation, administration, reporting, and related audits.
D2421	Verification Planning	Chief	Prepares item verification plans and procedures. Initially prepare verification compliance and task matrices for use by Verification Management and Admin personnel.
D2422	Test Lab Management	Chief	Manages all test laboratories and simulation facilities. Facilities may be used by program PIT and IPT teams when approved by the EIT.
D2423	Product Test Engineering	Chief	Develops test capability based on test plans and procedures for development, qualification, and acceptance test purposes.
D2424	System Test and Evaluation	Chief	Develops system test plans and procedures. Staff system test programs and accomplish planned work. Prepares test and evaluation reports and evaluate the content of reports.
D2424	Verification Management and Admin.	Chief	Provides expertise to set up and run a V&V process on a program. Provides tools, audits ongoing work, and facilitates FCA and PCA.
D2425	Software Formal Test and Evaluation	Chief	Responsible for laboratory and operation of that laboratory for testing computer software
D300	ENTERPRISE INTEGRATION TEAM	Vice President	Principal enterprise integration agent. Responsible for integration across functional disciplines where conflicts arise or are threatened. This includes establishing interprogram priorities for facilities and personnel. The EIT is the ultimate process owner for the enterprise, beginning with the vision statement and running down through functional department documentation. The EIT assigns functional departments to maintain this documentation.
D400	PROGRAMS MANAGEMENT	Vice President	Responsible for providing programs with qualified program managers. Some small programs and IRAD activities may report to this office, but the program managers for all large programs report directly to the enterprise executive as a member of a Programs Management Board chaired by the enterprise executive. Identifies and develops program management talent within the enterprise.

Table B.1 (continued) Functional Department Charter Matrix

Dept	Department name	Mgmt level	Charter
D4J0	PROGRAM J MANAGER	Director	Responsible for all program activities, funds, schedule, and product performance. Manages program through cross-functional teams.
D4J1	INTEGRATED PRODUCT TEAM L	Manager	Responsible for all Team L activity.
D4JY	PROGRAM BUSINESS TEAM	Manager	Responsible for program manager support from business perspective.
D4JZ	PROGRAM INTEGRATION TEAM	Manager	Responsible for system-level work and cross-team integration and optimization.
D500	NOT ASSIGNED		
D600	NOT ASSIGNED		
D700	NOT ASSIGNED		
D800	NOT ASSIGNED		
D900	NEW BUSINESS	Vice President	Responsible for all marketing efforts and acquisition of new business through proposal development and other paths.
D910	MARKETING	Director	Maintains contact with potential and current customers and keeps enterprise informed of needs and personnel in positions to make decisions important to the enterprise.
D920	PROPOSAL DEVELOPMENT	Director	Provides the infrastructure for proposal efforts and manage proposal development efforts. Provides facilities for proposal development.
D930	ADVANCED DEVELOPMENT	Director	Department includes very experienced designers from systems, electrical, mechanical, fluids, and software domains. These people accomplish concept development and predesign work during proposals and early conceptual studies.

Appendix C

Enterprise system engineering practices manual

Contents

List of illustrations

List of tables

Introduction

1.1   Objectives

The company is determined to provide our customers excellent value in the form of machines with the highest quality, reliability, and affordability that achieve our customer's needs while maximizing shareholder value and strengthening the company's capabilities base. These ends may be achieved only through cost-efficient development of new and modified products that exceed the best the competition can supply. A systematic process for the development of products that minimizes risks while encouraging the identification of effective designs is essential. We seek a common development process across the widest possible span of the company because it will encourage a common vision of development in the different business units which will have to cooperate in the development of products that assemble into a complete machine or system.

This *System Engineering Practices Manual* (SEPM) is provided by the Enterprise Integration Team (EIT) as a standard for the performance of the systems approach in all business units. It defines and briefly describes the techniques that comprise an effective system development process. The manual does not supply the detailed how-to information needed by engineers to implement this process in the interest of creating a simple, briefly stated set of requirements for this process. A companion system engineering training course is also available from the EIT that may be completed in one, two, or three days, depending on the level of detail required by the business unit.

1.2   Glossary

Some words used in applying the systems approach can be widely interpreted and often are in the literature on the subject of this manual. The meaning of some of these words as they are intended within Enterprise business units are included below along with the meaning of any acronyms used in the manual.

1.2.1   Definitions

1.2.1.1   Architecture

Product architecture is a structure of the things that comprise the system arranged in a hierarchical block diagram. It is a model of the physical system. These are the things that should have a specification written to define the related development problem and for which design work will be accomplished responsive to the specification.

1.2.1.2   System

A system is a collection of things that interact to achieve a common purpose. The word *system* can be applied generically at any level of indenture that satisfies the basic definition. Many company products are actually systems of systems. The word *system* is reserved in the hierarchy of products to identify the functional collections of things that work together to achieve a common goal within the context of a machine level item. Subsystems may also be identified in this context. In any case, the machine level as well as the system level subordinate to it may be thought of as systems in the generic sense.

1.2.1.3   System engineering

System engineering is a process that transforms a complex technical need into a preferred solution that will surrender to detailed design and manufacture. The process entails three fundamental steps as follows: (1) define the problem in the context of a specification, (2) solve the problem through the application of sound creative design, integration, and optimization concepts and knowledge, and (3) prove that the resultant design is a valid solution to the problem upon which the design was based through test and analysis.

1.2.1.4   Validation

Requirements validation is a process for gaining confidence that it will be possible to satisfy a set of requirements within the resources made available to a program.

1.2.1.5   Verification

A process of developing evidence clearly telling whether or not a design solution satisfies the requirements that drove the design.

1.2.2   Acronyms

Each acronym used in this manual is listed below with a brief explanation of meaning.

ASIC   Application Specific Integrated Circuit — A very dense packaging of electronic logic that can be represented by VHDL code early in its development.

BIT   Business Integration Team — Program team that supports the program manager for contracts, finance, and administration.

EIT   Enterprise Integration Team — Team reporting to top management responsible for integration and optimization of the enterprise system, ownership of the enterprise process, and adjudicator of resource demand conflicts between programs.

FMEA   Failure Modes and Effects Analysis — An analytical process for determining the ways in which a product can fail and the effects of those failures.

FRACAS   Failure Reporting and Corrective Action System — The machinery for identifying failure causes during development and product manufacture and determining appropriate corrective action for those failures.

IDEF   Originally this meant integrated computer aided manufacturing (the I of ICAM) definition (DEF), a technique developed by the U.S. Air Force as a planning tool. It has since been expanded into a family of IDEF models, such as IDEF-0 for functional analysis. Originally derived from SADT, a software development model.

IPPD   Integrated Product and Process Development — The cross functional process implemented through teams to achieve program and team goals.

IPT   Integrated Product Team — This is the name of the principal teaming structure used to bring together many specialized persons to focus on a single problem or program. These teams are normally oriented toward the physical architecture of the system.

PIT   Program Integration Team — System level IPT on a program.

SEPM   *Systems Engineering Practices Manual,* this manual.

TPM   Technical Performance Measurement — A risk management technique for tracking the value of a requirement and demonstrated capability over time to show trends and coordinate actions planned to gain closure between a requirement and current capability.

VHDL   A Hardware Definition Language for defining electronic circuitry from the abstract modeling phase through the concrete circuitry phase.

1.3   *Foundations of the systems approach*

The company develops, manufactures, markets, and logistically supports products to solve very difficult and complex problems, the solution of which demands the application of a wide range of scientific and engineering knowledge and technologies as well as sound business sense and good management skills. The range of knowledge needed is far in excess of that which any one person can master because of the individual knowledge deficit of all humans relative to the tremendous expanding mass of knowledge available to mankind.

It is generally accepted that given any difficult problem, the more pertinent knowledge available to solve that problem, the better the solution will be or the more rapidly a solution can be derived and, generally, these are desirable outcomes. The fundamental system development problem thus becomes one of amassing enough knowledge to solve a very complex problem. Since no one person can master the whole knowledge base needed, it will have to be contributed to the program by many people. Man's solution to this problem is specialization. This means that many specialized persons must cooperate to master the necessary knowledge, each covering some relatively narrow slice of knowledge very deeply understood.

While individual specialization solves the knowledge mastery problem, it results in a need for very effective communications between the many people that must be brought together to cover the range of knowledge needed to solve a complex problem. These many people must also be well managed to solve the problem with which they are confronted. The process for causing these many specialists to interact efficiently to solve the same complex engineering problem is referred to in this manual as *system engineering*.

System engineering is applied on company programs through the use of cross-functional teams of specialists managed so as to solve problems which have been clearly defined in terms of their product solution requirements, required resources (money, time, facilities, and materials), and a plan of action. All of the teams identified for a program cooperate in accordance with an overall program plan to achieve three simple steps in the development of solutions to complex problems as illustrated in Figure C.1 and described as follows:

a. Define the Problem — Because of the many people who must cooperate in the development of sound designs, it is necessary for these many people to reach a shared vision of the problem they face in common. This is done by preparing a specification that contains the minimum necessary requirements that the solution must satisfy. This specification must be formally released as a precursor to spending any detailed design funding.

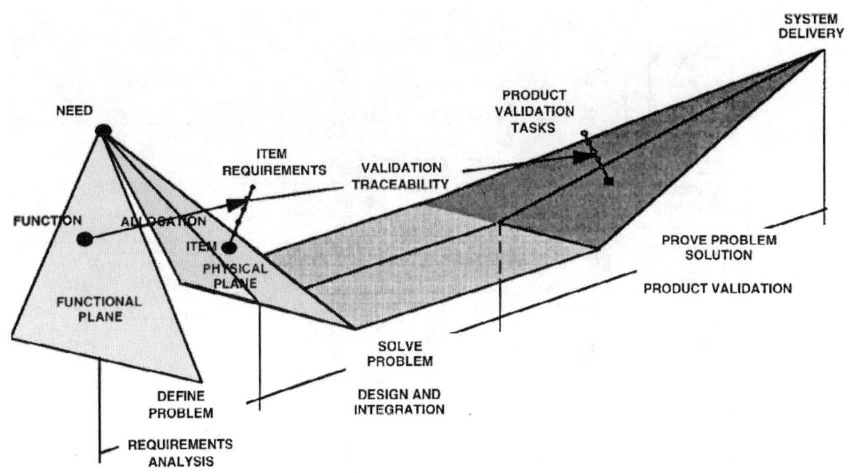

Figure C.1   N development model.

b. Solve the Problem — The problems defined in specifications are solved through the human application of creative design genius supported by many specialized persons, leading to a design solution that satisfies all of the predefined requirements.

c. Prove that the Problem is Solved by the Design — It may be possible to solve a given problem in a number of correct ways as well as many incorrect ways. We seek to solve the problems defined in specifications correctly in one continuous stream of sound reasoning, minimizing the risk and reality of expending any more than planned in terms of cost, time, and materials while achieving all technical objectives. The result of this work must be checked against the original requirements that stimulated the solution attempt. This is done by subjecting the design solution to testing and analysis focused on validating that the solution is correct.

Customer needs should drive program products and they correspond to the ultimate functionality of the system (machine) to be developed. Figure C.1 illustrates a decomposition of this need into finer functionality granularity with allocation to objects in the product physical plane, thus identifying things in the product architecture and performance requirements appropriate for them, following the sound notion of "form follows function" offered by the architect Sullivan. This process continues until the product problem has been fully described in specifications triggering an entry into detailed design, integration, and optimization. Finally, the many things designed and purchased for use on the program are produced and subjected to validation to prove that they do satisfy the requirements contained in their specifications. This process is also effective when making major changes

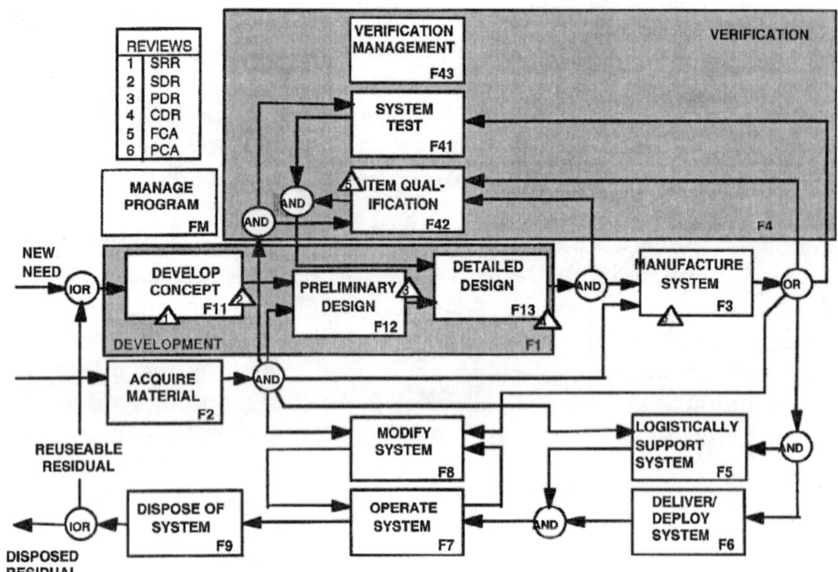

Figure C.2   System master life cycle diagram.

to existing systems wherein one compares existing system functionality with new system functionality and determines what in the existing system architecture can remain a part of the new system, what must be replaced, and what must be modified.

1.4   A guide to the manual

Section 2 provides an insight into the recommended work structure in the basic product development structure called a *program* and covers the fundamentals of the systems approach to be applied in these programs. Sections 3, 4, and 5 cover the three fundamental elements of the systems approach identified in Figure C.2 in additional detail, while Section 6 offers a management overview of the process relative to the systems engineering work to be accomplished on a program and other related program activity.

Section 3 defines the work required on programs to clearly define the requirements on a program prior to entering detailed design work on the product based on the sensible idea that we should understand a problem prior to trying to solve it. Three specification formats are offered, providing the program with different levels of control of the structure and content of its specifications. Methods are noted for identifying requirements in the several categories appropriate to company products, and these methods are discussed in detail in the companion system engineering training course.

Section 4 provides controls over the design, integration, and optimization process that will lead to a low-risk problem solution, good product reliability and durability, economical development, and good customer value. Section 5 describes a sound process for verifying that the design solution does in fact satisfy the predefined requirements.

1.5 How-to information

This manual identifies the technical and management work that should be accomplished on programs from a systems engineering perspective to encourage a clear identification of the problem the program must solve as a prerequisite to solution through design, coordinate the technical work specialized force to accomplish a unified design solution, and to develop convincing evidence that the solution satisfies the requirements developed to define the problem in the first place. This manual is not intended to provide information on how to accomplish the corresponding work on programs. That how-to information is presented in a company system engineering training course keyed to the content of this manual and available from the EIT to all company functions.

1.6 System engineering capability assessment

Section 7 offers a method for assessing systems engineering performance on programs against the content of this manual and for assessing the content of this manual against external standards for systems engineering. It is desired that company programs follow the content of this manual in performing system engineering on programs and that the content of this manual remain coordinated with widely accepted industry standards on system engineering.

Systems engineering process requirements

2.1 Program environment

All company work is accomplished on programs. Each business unit has clearly defined responsibilities relative to its product line and is responsible for developing programs to capture, maintain, and increase market share in that product line. At any one time, a business unit may have many programs in progress, some involving new development, others entailing major improvements in prior models, a manufacturing phase, product support phase, or some combination of these.

Each program makes a demand on enterprise resources in terms of qualified people, capital investment, facilities, materials, and time. These resources must be supplied by the business unit appropriate to the perceived urgency. Each program is the principal responsibility of a lead business unit that is responsible for overall management of the program. Other business units may act as suppliers to this lead unit and shall be managed relative to its contributions to the lead unit as one of the many suppliers providing material to the program.

There are several layers in the program hierarchy. The top end of most programs is the system level which entails a complete operating entity that can satisfy an end use. This system program may breed lower-tier system, segment, or end item programs at the same and other business units as appropriate to the different roles assigned to the business units. Complex systems may consist of end items and subsystems, and each of these may be operated as programs within their responsible business units. End items and subsystems consist of components which, in turn, may be procured from outside sources or run through sister business units.

The development of a system entails a network of programs within the company and connecting into other companies through contracts for supply of equipment. All programs focused on the system level must consider the interoperability of that system with other systems where multiple systems must cooperate. One component, subsystem, or system level program may serve multiple programs requiring that its product be coordinated or integrated and optimized relative to several using systems through a cross-program integration team chaired by a lead business unit designated by the enterprise integration team (EIT).

Applications may entail site and even worldwide systems within which delivered systems function, expanding the scope of the systems with which we must deal.

2.2   *Program phasing and development process definition*

Figure C.2 illustrates the standard approach to program phasing and a generic development process encouraged for all company development programs. Different customers may prefer to collect the enterprise functions depicted in Figure C.2 into phases in various ways, but if the customer does not express a preference, the program is encouraged to apply the phasing suggested by the figure. Functions or phases F1 through F5 are generally one-time activities accomplished for the purpose of creating a system. Functions F6 through F8 extend through the life cycle of the system. Function F9 is a one-time transition from system use to terminating use of the system at the end of system life. Function 2 may be involved throughout the life of the system to support logistic and modification activities as well as initial production of the system. As indicated, some residuals from a prior system may have utility in a new system application, while others are no longer required. These selections may be mandated by a customer or result from a careful study of precedent and new system relative capabilities in Function 11.

2.2.1   *Development (F1)*

During the development phase, customer needs are translated into a preferred design solution compatible with a corresponding set of process designs for manufacturing, quality assurance, operations and maintenance, logistics, procurement, and verification (qualification and system test).

2.2.1.1   *Concept phase (F11)*

The collective conclusion from people experienced in many new product introductions in many industries is that money spent well early in a program significantly reduces the development and life cycle cost because cost, schedule, and technical risks are understood and mitigated. Yet many program managers are naturally reluctant to spend money early in a development activity because money spent cannot be recalled any more than can time. The company development approach, if properly implemented by trained and experienced personnel, will produce good results and achieve the desirable effects of risk reduction and attendant success in program goals. Done right and early, the systems approach will save time, reduce cost, improve product reliability and quality, and enhance customer value. Ultimately, this means customers will be satisfied, and the company will earn a good return on investment.

During the concept phase, many critical program tasks are begun, and some of them completed. Chief among these tasks is careful discussions with customers to understand needs, emphasizing listening and searching for clear understanding. Needed product functionality is explored and allocated to a physical architecture. The elements in the architecture are associated with sources, commitments are secured, and teams formalized with clear responsibility for items and interfaces. Product and process requirements are clearly documented in technical specifications. Alternative product and process designs are brainstormed, evaluated, and preferred designs selected as a result of trade studies.

2.2.1.2 *Preliminary design phase (F12)*

The term *preliminary design* applies to a process of refining and expanding requirements definition and validation of those requirements through design sketches, special development testing, and special analyses showing that it is possible to satisfy those requirements. The exit criteria for this phase is a set of approved specifications for selected things in the product architecture, system diagrams, key installation drawings, identification of risks with mitigation plans for all medium and high risks, preliminary designs for manufacturing, quality, procurement, verification, and logistics processes.

2.2.1.3 *Detailed design phase (F13)*

Three cycles of the detailed design phase exist. First, it is necessary to develop the design for the items that will be manufactured to support the verification process. This cycle is complete when 95% of all planned engineering drawings have been originally released, and it is terminated by a preliminary design review (PDR) that authorizes manufacturing and procurement of product necessary to accomplish verification. During and subsequent to verification, signaled by functional configuration audit (FCA), it will become obvious that engineering changes are needed, and in a second cycle of the detailed design process these changes are implemented, leading to manufacturing of the first article in the manufacturing function. This article is inspected, leading to a physical configuration audit within the manufacturing function. If satisfactory, production continues through possible low-rate and high-rate production phases. If problems are uncovered in the design, another cycle of detailed design changes ensues, followed by possible qualification work, process changes, manufacturing action, and closure of prior PCA concerns, clearing the way for continued production. Additional detailed design work may be required to support future modifications to the system (F8).

2.2.2   *Acquire material (F2)*

Sources for the materials defined on engineering drawings and in manufacturing planning are identified and sources selected. Later in the program, as this material flows into the production plants, it does so in accordance with plans defined in this activity. These plans also tell how the materials are to be kited and routed to the point they are introduced into the production process.

2.2.3   *Manufacture system (F3)*

Company programs commonly use four steps in the production process. End items and components are produced and subjected to qualification tests. Findings from these tests are introduced into the design, leading to a functional configuration audit and to production of a first article that is subjected to acceptance tests followed by a physical configuration audit. Any necessary changes are made, and low-rate production is entered into. To the maximum extent possible, the manufacturing techniques applied for qualification items, first article, and low-rate production should be those planned for the articles produced at high rate to encourage realistic verification of the product. Some programs may never reach a high production rate owing to the small numbers of systems required.

2.2.3.1   *Manufacture qualification items*

All items in the system architecture should be screened for qualification for use in the application of the system in terms of operational and environmental stresses. If it can be shown that the item in question has been exposed to stresses expected in this application, it may be possible to claim qualification based on past application of a similar nature. Otherwise the item should be subjected to a qualification process to prove that its design satisfies its requirements. Where this verification process must involve demonstration, examination, or test (nonanalysis methods), one or more articles must be manufactured or procured and subjected to planned qualification activities. Where possible, these items should be acquired in exactly the same way production items will be obtained, but this is not always possible due to the current status of the production and/or procurement process relative to the item. In these cases it may be necessary to fabricate and assemble the test article in an engineering laboratory. Special care must be exercised in these cases to replicate planned manufacturing processes and to account for any planned differences in the qualification process.

2.2.3.2   *First article*

The first production article is a pathfinder producing evidence of acceptability of the production plans, procedures, and manufacturing personnel training to produce the product. This first article may be subjected to special inspections by quality assurance, with the evidence collected for use at the physical configuration audit to follow completion of first article production.

2.2.3.3 *Low-rate production*

On a program that will have to produce articles at a high rate, the production process may first have to transition through a low-rate period to prove out the process, personnel skills, FRACAS capabilities, procurement, transportation, and material handling provisions. This process is also appropriate to produce initial product resources needed for system testing. These articles may also require special instrumentation installed during production which will cause their production process to differ from later production where the articles are moving to the customer for normal use.

2.2.3.4 *High-rate production*

The only difference between low-rate and high-rate production is the speed of the manufacturing process, but the increase in speed must carry with it high-rate procurement, shipment, and internal material processing. Provisions are also necessary to move produced product out of the production facility and into the hands of the distribution or delivery process.

2.2.4 *Verify system (F4)*

Verification is a process for proving that the product design satisfies the requirements that drove the design. Company programs apply analytical, physical demonstration, examinations, and test verification activities for the purpose of proving that the product developed satisfies its development requirements.

2.2.4.1 *System test and evaluation (F41)*

When item qualification is complete and the FCA has resulted in customer acceptance of the results, entry into system tests will be authorized. These tests should be planned during the development period based on the content of the system specification. The purpose of system test and evaluation is to prove that the system performs in accordance with the content of that specification. Testing should be progressively staged. First, simple, safe steps are planned followed by progressively more stressful conditions until testing is accomplished at the limits of required capability. System testing may be carried out in two steps, with company-operated development test and evaluation followed by customer-conducted operational test and evaluation. Commonly the former will be focused on an engineering view of the product and its requirements, while the latter is oriented toward operational missions that the customer/user needs to have accomplished. These two test phases can produce very different conclusions about the readiness of the system to satisfy system requirements. The program must be very careful to review customer operational test and evaluation planning to ensure that these are based on the approved system specification used for development test and evaluation.

2.2.4.2   *Item qualification (F42)*

All items in the system must be qualified for the application defined by its operational capabilities, environmental stresses, and interface relationships. This is accomplished by accomplishing planned tests, analyses, demonstrations, and examinations covered in an integrated verification plan that defines verification tasks to be accomplished, items to be tested, schedule demands, personnel and organizational responsibilities, and other resources required. Each verification task is linked to some number of verification requirements contained in the item specification and thus to the item requirements in that specification via a method selected (test, examination, analysis, and demonstration), the level at which it will be verified (item, parent item, or child item), and a verification task within which the verification work will be accomplished in accordance with planning and procedures in the integrated verification plan. The results of each verification task must be reported in a verification report. All of the verification reports may be collected as sections in an integrated verification report or be separately released, but they must be linked to the task upon which they provide documentation of results.

Item qualification terminates through successful completion of the functional configuration audit. On a small program, it may be possible to complete the FCA in a single meeting. On a more complex program involving many items, it may require a series of FCAs followed by a system FCA to close out the whole series.

2.2.4.3   *Manage verification (F43)*

The qualification verification process stretches from the preparation of the item specifications through the FCA and subsequently into system testing. This whole process must be managed to control risk through careful definition of the work that must be performed, identification of responsibilities, and provisions for adequate resources for all tasks. Given a sound plan, the actual performance to that plan must be followed by management to ensure that planned work occurs and any discontinuities in continuing work be removed and resolved as soon as possible. This process will generate a tremendous amount of documentation, and the management of the process must provide for adequate storage and access to this information.

2.2.5   *Logistic support (F5)*

Logistics support activities include depot repair, field spares, continuing maintenance and operator training, intermediate or shop maintenance (if used), spares inventory maintenance and transport, and technical data. All of these resources must be defined and developed during the development phase so as to be consistent with the product design. On different programs, there are various ways to balance the contributions between the company and the customer. These range from the extremes of complete company operation of the product to unsupported customer operation of the product and everything in between.

2.2.6   Deliver/deploy system (F6)

Product is delivered in accordance with contractual requirements and deployed into operational condition. The delivery and/or deployment must be planned during the development process.

2.2.7   Operate system (F7)

The customer receives deployed product and applies it to their work. Operation of the system should be defined in technical data delivered to the customer as part of the program.

2.2.8   Modify system (F8)

As systems age in use, customers often find a need for additional capabilities leading to modifications of the system. These are developed and verified, where required, just like major programs.

2.2.9   Dispose of system (F9)

At some point, the product will begin to wear out such that repair is less advantageous than replacement. The product may be recycled into raw materials or refurbished for continued use as a piece of used equipment. In the later case, the product is recycled into the Deliver/Support Product function. This step may entail a new development effort focused on modifying a particular machine model, resulting in the manufacture of modification kits that are shipped to the dealers where they are installed.

2.2.10   Manage program (FM)

This function runs throughout the program providing an ongoing and thoughtful look at where the program is, where it wants to go, and how it can get there. This process starts prior to a formal program initiation to explore possibilities for a new product, analyze competitor positions and capabilities, assess how the program will affect the company's competitiveness, responsiveness, and profitability. Goals are established and all of the people and organizations that should be involved are identified. This may involve strategic partners as well as company entities. Every effort consistent with protecting the company's competitive position is made to fully inform everyone involved about the selected strategy, since it offers everyone a common vision of what it will take to succeed.

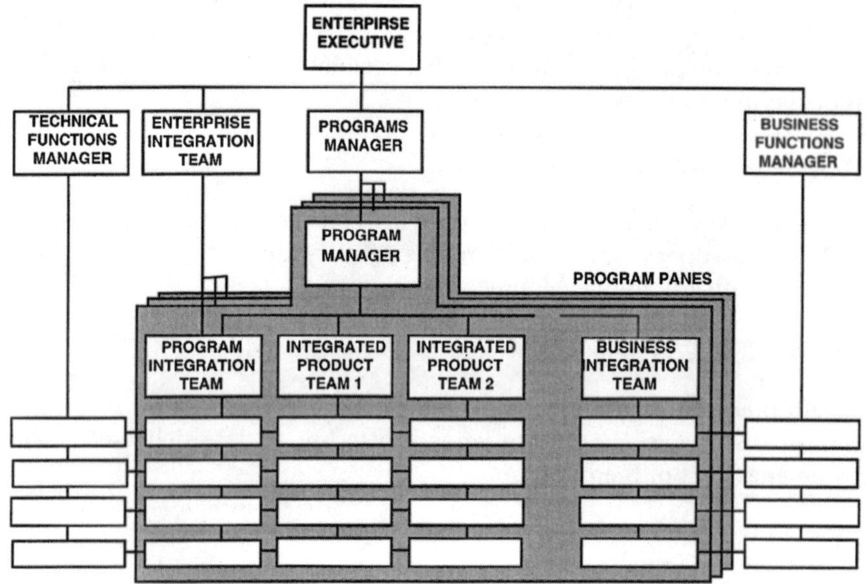

Figure C.3   Program organizational structure.

In early program phases (proposal and concept development), the program should be managed through a program integration team (PIT) and business integration team (EIT) as shown in Figure C.3. At that time it will not be necessary to staff integrated product teams (IPT), because it is during the concept phase that the product architecture is established, and it becomes the basis for team formation subsequently. Each program manager reports to the programs manager and the top-level teams on a program report to that program manager. The technical and business functions managers are charged with the responsibility for providing all programs with the resources they require to succeed (qualified personnel, tools, facilities, and sound practices). The strong management axis in Figure C.3 is the vertical one.

2.3   Major reviews and audits

A very specific sequence of reviews is prescribed for all programs in the interest of controlling risk while moving toward a production decision for a program. Figure C.2 illustrates this sequence, focusing on three phases and one overarching strategy influence interspersed with six major reviews and audits where progress toward phase goals are evaluated and direction given for subsequent phases. Figure C.2 is annotated with review identifiers in the form of numbers 1 through 6 in triangles.

2.3.1   System requirements review (SRR)

The SRR is held as early as possible in the concept phase to ensure a clear understanding of the customer need and top-level requirements. This is identified as review 1 in Figure C.2. The system specification should be reviewed and approved at this review.

2.3.2   System design review (SDR)

The SDR provides an opportunity to review the system design concept responsive to the system requirements as a prerequisite to movement to the preliminary design phase. This is identified as review 2 in Figure C.2.

2.3.3   Preliminary design review (PDR)

The PDR falls at the end of the preliminary design phase to review the requirements for the system elements and corresponding design concepts. Remaining risks and work accomplished to mitigate them is reviewed as well. Process requirements and preliminary designs are also covered. This review authorizes expenditure of funds for detailed design and procurement of long lead items needed for qualification. It is identified as review 3 in Figure C.2.

2.3.4   Critical design review (CDR)

The CDR is conducted at the end of detailed design and provides an opportunity to review the completed design for compliance with driving requirements, interface compatibility, residual risk intensity, and completeness. By definition, the design is complete when 95% of the planned drawings have been released. Acceptance of the CDR should yield a decision to build the items needed for qualification testing.

2.3.5   Functional configuration audit (FCA)

The FCA is held after completion of all verification tasks and is intended to review the content of the verification task reports, resulting in a conclusion that the item satisfies its requirements or does not and, if not, what requirements have not been satisfied.

The FCA should be conducted at the item level (configuration or end item as defined by the customer or program management). If the program involves multiple end items, there may have to be multiple FCAs followed by a system FCA.

2.3.6   Physical configuration audit (PCA)

The PCA is the second step in the verification process conducted in association with the first article product item. The PCA, like the FCA, may be done at the item level with multiple PCAs follow by a system PCA. Data must be collected during the manufacturing and acceptance processes, and this data is reviewed at the audit along with the results of other work, depending on customer requirements. Commonly the PCA falls into one of three patterns:

a. Document-Oriented — Customer personnel review the documentation related to production and acceptance.
b. Product-Oriented — The customer disassembles the product and compares it with engineering drawings and analysis report content.
c. Combined — Both a and b are employed.

2.4 Development process steps

The systems approach entails three fundamental steps: (1) define the problem, (2) solve the problem, and (3) prove that you solved the problem as described in association with Figure C.1. The company system engineering process recognizes these three sequential steps, encourages the application of concurrent engineering techniques in their performance, and simultaneously developing the product system and related processes. Subordinate paragraphs briefly describe these three fundamental steps that occur in Functions 1 and 2 of Figure C.3.

2.4.1 Define the problem

2.4.1.1 Customer requirements definition

Every effort should be made to clearly understand the customer's requirements in the most detail possible. There are two recommended approaches to this. First, ultimate users, dealers, and marketing company personnel should be approached and asked to identify any requirements they can. A checklist may be helpful in pursuing this conversation, or a list of requirements for the current model could be used as a basis for discussion. A structured approach is encouraged as a way of getting insight into questions of merit, to uncover needed functionality, and to identify needed architecture for the product. The final product of this process should be a written list of customer requirements, with corresponding values reviewed and approved by program management.

2.4.1.2 Development requirements analysis

This process leads to the paper or electronic publication of a specification for selected items in the system architecture which define the design problem the responsible design team must solve. A structured approach to the work necessary to complete these specifications is encouraged rather than a resort to an *ad hoc* approach, which generally will result in omission of needed requirements, inclusion of unnecessary requirements, and introduction of resultant program risks. Refer to Section 3 for detailed process requirements.

A system-level team should first develop the system-level requirements based on customer requirements, available technology, and sound business decisions. Then the system team should accomplish such work as necessary so as to prepare and release a subordinate specification for each next-tier item in the architecture, the design responsibility for which will be assigned to a lower-tier team. Each team thus formed about the evolving architecture should be given the responsibility for developing any lower-tier specifications under their area of responsibility or delegating that responsibility to lower-tier teams formed about items subordinate to their item. In all cases, a team should begin work with a top-level specification for its area of responsibility in hand.

Specifications should be reviewed and approved and then formally released as a prerequisite to authorizing use of program budget to accomplish detailed design work. Some preliminary design work will generally be necessary in order to encourage identification of all of the appropriate requirements for an item and its subordinate parts.

These first specifications developed should be design independent and contain the minimum requirements needed to define the problem the designers must solve in order to provide the design team with the greatest possible design solution space. These requirements drive the qualification process, leading to a functional configuration audit so they have to be coordinated with verification requirements defining the corresponding verification problem. Verification plans and procedures for the design of the verification process have to be responsive to the verification requirements. Programs are encouraged to capture the item verification requirements in the item specification. These specifications should be complete and approved as a PDR entry criteria.

2.4.1.3   Product requirements analysis

Product requirements define the requirements for acceptance of manufactured articles for delivery. They are design dependent and must focus on the product design features. These requirements must be complete as a CDR entry condition. Acceptance test plans and procedures should be developed from them.

2.4.1.4   Combined requirements analysis

Programs may create two-part specifications as described under the two previous paragraphs or create a single part specification progressively by expanding upon the development specification to add the product peculiar requirements, perhaps in a tabular fashion along with the development requirements.

2.4.1.5   Use of computer tools

Small programs with very few specifications may be well served by crafting the specifications on a word processor. Large programs with many specifications and a need to employ requirements traceability and traceability to verification work should use a requirements database to capture and publish program specifications. Computer database use is encouraged on all programs.

2.4.2   *Solve the problem*

Programs should define design problems in specifications as a prerequisite to solving them through design. Design is the application of creative ideas motivated by a clear understanding of requirements toward the solution of the problem defined by those requirements. These problems will commonly be so complex that no one person will be able to act independently to synthesize all of the requirements into a design solution. Thus, the company encourages the use of integrated product teams staffed by the right disciplines for the problems assigned. This design process has three stages: concept development, preliminary design, and detailed design.

During the concept development process, programs are encouraged to evaluate alternative solutions to customer needs and to carefully weigh the relative advantages and disadvantages of those alternatives using a trade study approach. The trade approach calls for identification of a set of key requirements that all candidates must satisfy, a set of key requirements that will be allowed to vary and will be used as a selection criteria and program value system, and a small number of alternative or candidate solutions. The candidate solutions should be evaluated relative to the selection criteria, and the overall better solution selected. To the extent that this decision is made based on uncertain data, it should be tested for sensitivity to variation of the values, weighting, and any other parameters employed.

Upon selection of the preferred concept, some work should be accomplished to identify and mitigate program risks and gain confidence in the program's ability to complete the program leading to a preliminary design review that approves the design approach or directs changes in the interest of cost, schedule, or performance characteristics.

The responsible team, or individual design engineer supported by others, must understand the requirements prior to accomplishing design to ensure that the results of that design will work synergistically with respect to all other system or machine elements to achieve the customer need and company business goals. This is a creative process requiring the employment of gifted designers coordinating with specialists appropriate to the technology. This work should progress from a design concept to a preliminary design solution and, finally, a detailed design solution. Peer and formal design reviews should be used at each of these points to ensure that the best possible design is evolving. Refer to Section 4 for detailed requirements.

Throughout the design process, it is necessary to coordinate the work of the several teams and engineers working on a program to ensure that the best possible overall design solution is evolving. It is a common characteristic of people and teams to integrate and optimize at their own level of responsibility and knowledge. This may not always be in the best interest of the whole. Thus, it is necessary for an integration and optimization agent on all branches and at all levels of system architecture.

2.4.3   *Proof through verification*

The third step in the systems approach is to produce evidence that the design solution is a solution to the original problem defined in the specification. Proof is offered in the form of test, analysis, demonstration, and examination reports that tell the results of verifications tasks. This process must start when the requirements are being defined. For each requirement in Section 3 or 5, one or more verification process requirements shall be identified and placed in Section 4 of the specification. These verification process requirements will be used as the basis for verification planning work.

Requirements definition

3.1 Requirements defined

A requirement is an essential characteristic of a product or process entity. A requirement is a characteristic that the design solution must respect, and it must be verified that the design solution does in fact respect it. Development requirements should be design independent and must be captured and documented in specifications written for specific items prior to initiating detail design for those items. Development specifications define the technical problem that the design team must solve. Product requirements define requirements that must be proven during acceptance of the product subsequent to manufacture and are design dependent.

3.2 Requirements identification timing

The development requirements for an item or process must be defined and formally released as a prerequisite to expenditure of detail design funding. Product requirements must be defined no later than CDR.

3.3 Requirements categories

Two fundamental kinds of requirements are recognized: performance requirements and design constraints. A performance requirement answers the question, what must the item do and how well must it do it? Design constraints provide boundary conditions within which the designer must remain while satisfying the performance requirements. Three kinds of constraints are recognized: interfaces between items, environmental requirements, and specialty engineering requirements. The companion systems engineering training course offers a set of thinking tools with which to identify requirements in each of these categories.

3.4 *Structured analysis vs. ad hoc approaches*

Structured analysis process should be employed to identify product requirements rather than relying on *ad hoc* approaches. Structured methods have been identified for each requirements category. The benefit from structured approaches is that they encourage completeness while avoiding identification of unnecessary requirements that may narrow the solution space and deprive the program of alternative good design solutions. The principal elements of the structured approach are outlined in subordinate paragraphs.

3.4.1 *The customer need*

A clear statement of customer need shall be developed by or for the customer for each program. The customer may be a single, clearly identified entity or a collective entity defined by a marketing function. In the latter case, this criterion may be provided by citing a perceived customer need based on market research.

3.4.2 *Product functionality*

An organized means shall be in place to decompose the large problem expressed by the customer's need into a more detailed understanding of the problem and assignment or allocation of the functionality thus exposed to elements in the system architecture (or physical solution space) consisting of hardware, computer software, facilities, materials, processes, and personnel. As a means of understanding what the product must achieve in use, the program should develop a functional analysis diagram of some type. Some examples of these diagrams are functional flow, hierarchical functional, state transition, IDEF-0, and process flow diagrams.

3.4.3 *Product architecture definition*

As a result of a structured analysis process, the program should identify a preferred architecture expressed in an architecture block diagram, simply a family tree of the things that comprise the product.

3.4.4 *Product interfaces definition*

As a means of identifying and agreeing upon all of the interfaces required in the product, the program should develop some combination of schematic block and N-square diagrams illustrating the interfaces that must exist between the things identified on the architecture block diagram.

3.4.5 *Environmental requirements definition*

System or machine level environmental requirements shall be identified in terms of a list of environmental characteristics with corresponding parameter value ranges drawn from an environmental standard appropriate to the spaces within which the product will be operationally employed. A life cycle environmental profile is encouraged at the machine level as a way to understand the aggregate environmental situation over the complete product life. Environmental impact parameters shall also be developed consistent with the environmental regulations common to areas in which the product will be operated.

3.4.6 Specialty engineering requirements analysis

The senior system agent for the system will identify the specialty engineering requirements categories that will be applied to all items in the system and identify those disciplines that will maintain requirements math models. For each item in a system, the senior system agent shall identify any other specialty engineering requirements that will be required.

3.5 Responsibility assignment

An organized method shall be in place for clearly and comprehensively assigning responsibility for development of the product system elements and their associated employment process to personnel and/or teams. This method must be mutually consistent with the definition of work breakdown (in support of effective cost collection) and the company development team concept applied on the program.

3.6 Team structures

Programs shall be staffed through cross-functional teams focused on the product architecture. These teams should be physically collocated to the maximum possible extent. Teams must be provided with good inter- and intrateam communication (phone, computer, fax, etc.), wall space for big picture exposure, and meeting space (separate meeting rooms are ideal but not essential).

3.7 Requirements validation

The requirements validation process involves gaining confidence that it will be possible to satisfy the requirements prior to doing the detailed design work. All requirements shall be reviewed prior to or during the specification review and approval process to determine the degree of risk in satisfying them in the design process to follow specification approval. All requirements that appear risky to satisfy should be challenged, and clear evidence established that the requirement is necessary and its value reasonable. Where necessary, special tests, analyses, examinations, and demonstrations shall be accomplished to mitigate risky requirements. Where it will require considerable time to gain control over a particular requirement, it should be considered for tracking via technical performance measurement.

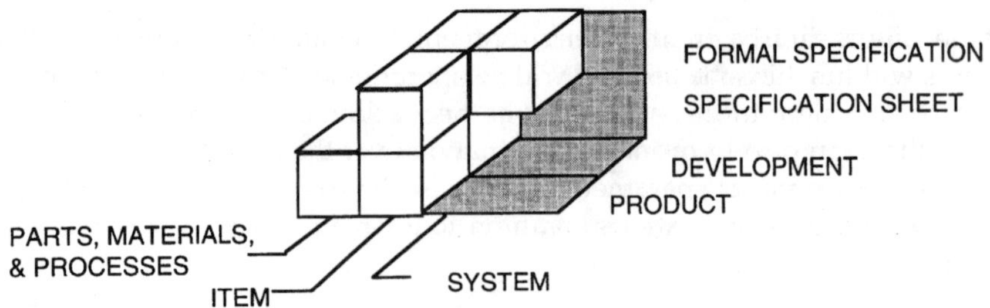

Figure C.4   Company specification types.

3.8   Requirements documentation

Specifications shall be prepared for a predetermined list of items in the system architecture. Types of specifications and the schedule for development shall be defined.

3.8.1   Specification types

Figure C.4 illustrates all of the possible kinds of specifications that may have to be prepared on a company program. Three different levels of specifications are identified including system, item, and parts, materials, and processes. Two levels of formatting control are identified from specification sheets and formal specifications. The specification sheet is simply a nonordered but numbered list of requirements in primitive format consisting of attribute–relationship–value–units following the example of weight ≤ 1820 pounds. The formal specification is a formatted document with six sections. The development specification is intended for development efforts defining design independent characteristics. The product specification adds design dependent information for the purpose of supporting acceptance test planning. There may be multiple levels of item specifications on a program focused on hardware or software.

3.8.2   Specification review and approval

All specifications of all kinds shall be reviewed and approved by the manager at the next higher level from the organizational entity that created the specification. Detailed design should not be sanctioned until this occurs for the item development specification.

3.8.3   Specification release and control

All approved specification masters shall be retained in protected (read-only) computer memory storage accessible by all program personnel through computer network connections to a central program specification library. Version control shall be maintained, and a master list of the current requirements baseline maintained.

A means shall be put in place to clearly define the current system require-ments and functional baseline during the period leading up to the SRR and SDR. Subsequent to these reviews the allocated baseline shall be captured in a set of item development specifications beginning at PDR. Between PDR and SDR the product specifications shall be developed and in combination with the engineering drawings and test planning documentation shall define the product baseline.

3.8.4   Requirements support tools and applications

Use of a requirements database tool is encouraged. When such tools are used, programs are encouraged to maintain requirements hierarchical trace-ability and validation traceability.

3.8.5   Requirements relationships

Values for requirements that appear in all item specifications in a system and are numerically stated, like reliability and weight, shall be developed by a single discipline using a mathematical model and requirements value allocation techniques. Where significant risks are identified for particular requirements categories, margin techniques are encouraged. Requirements traceability should be accomplished and maintained on all programs where there are more than five specifications and/or three or more levels in the specification tree hierarchy.

Design, integration, and optimization

4.1 Design

4.1.1 Design responsibility

An organized method shall be in place for clearly and comprehensively assigning responsibility for development of the product system elements and their associated employment process to personnel and/or teams. This method must be mutually consistent with the definition of work breakdown (in support of effective cost collection) and the development team concept applied on the program. Teams will be assigned responsibility for specific items in the system architecture. Individual engineers may be assigned responsibility for component level items with the understanding that they may draw upon the talent pool of the responsible team.

4.1.2 Design capture

An organized way shall be put in place to capture and retain the results of development team product and process concept and design work so as to be easily accessible at any time by all program personnel and available for iterative improvement by responsible engineers and analysts. The use of computer-aided design (CAD) is encouraged with linkage to computer-aided manufacture (CAM) capability, analytical and modeling tools base, and computer-aided procurement.

4.2 Integration

4.2.1 Integration and optimization responsibility

Design and integration responsibility is aggregated at increasingly higher levels. The person responsible for an item is also responsible for all items subordinate to it. The responsibility for integration between two items should be most intensely pursued at the lowest level that includes responsibility for both items but extends through all higher levels of authority.

4.2.2 System integration

An organized means shall be in place to integrate the work of separate product development teams and to optimize at the system level across design, manufacturing, quality, test, operations, and logistics support to overcome problems resulting from suboptimization at the individual team level or individual discipline. Where multiple team levels are used on a program, the integration and optimization responsibility must be taken on by any team with subteams.

4.2.3 Interface development

An organized way shall be devised and put in place to develop compatible internal and external interfaces between elements depicted on the system architecture, to minimize the cross-organizational interface, and to clearly assign responsibility for the development of each interface. Interfaces between items under the control of the company and associates will be developed in accordance with a plan approved by all parties. Interfaces with items to be provided by suppliers shall be controlled by the same company agent responsible for the item being procured.

4.2.4 Specialty engineering

Each NPI team is responsible for coordinating the work of the several specialty engineering disciplines simultaneously at work on a program within the context of their team's assigned work so that their actions come to bear jointly in the most effective way to achieve team and program goals.

4.2.4.1 Reliability and warranty provisions

Product reliability shall be assured so as to be consistent with planned product warranty provisions. The program should consider applying failure rate allocation or apportionment and prediction; failure modes and effects analysis (FMEA); a failure reporting and corrective action system (FRACAS) to positively identify causes of failure during development, test, and manufacture leading to optimum solution, fault trees; and Tuguchi methods in the design of experiments.

4.2.4.2 Maintainability, serviceability, and supportability

The design should maximize customer value by minimizing down time for maintenance, servicing, and logistics support reasons. The program should consider component remove and replace time studies using three-dimensional CAD or people with maintenance experience using two-dimensional drawings imaginatively. All items which can fail that flow from the FMEA and as a result have an adverse impact on immediate use of the machine for productive work should be evaluated for corrective maintenance actions to restore them to full functionality. This should include the impact on spares maintained by dealers, the technical data supplied to dealers for maintenance instruction purposes, training of dealer mechanics and technicians, and special test and support equipment needed by dealer organizations to diagnose and correct faults.

4.2.4.3 *Product safety and human factors*

Throughout the development of new products the development staff shall consciously pursue identification of safety hazards and mitigation of them through reduction of the probability of occurrence and reduction in the severity foreseen should the hazard occur in use. A safety hazard list is encouraged combined with active work to move all hazards listed down into the low risk categories defined by low probability of occurrence and low impact of occurrence.

The capabilities of humans in terms of physical strength, dimensional relationships, manipulative capabilities, sensory ranges, and speed and effectiveness of reasoning abilities should be consciously considered in the development of designs to satisfy performance requirements.

4.2.4.4 *Other disciplines*

Program goals, past experience, required technologies, and design considerations should be taken into account when planning the program to determine what other, if any, specialty engineering disciplines should be assigned to the program.

4.3 *Product representations*

All product representations should be identified along with the engineer responsible for maintaining configuration control of it. This extends to physical models, mock-ups, mathematical models, and simulations. The responsible person shall be held accountable for knowledge of the representation relative to the current product baseline and ensuring that the representation always reflects the desired product configuration when it is used.

4.4 *Design rationale capture*

A means shall be put in place to capture design rationale and traceability of the decision process that led to the preferred system concept. All requirements and the designs corresponding to those requirements shall be reviewed by responsible management personnel and approved. Relevant details of these reviews shall be captured for the purpose of communicating direction and for subsequent reference.

4.5   Development configuration control

A means shall be put in place to positively control the configuration of the product design subsequent to the formal approval of that design and to control the configuration of all representations of that design in the form of models, simulations, mock-ups, test articles, test results, and analyses. Design configuration management should begin no later than the A review.

4.6   Product configuration control

The configuration of physical products produced through procurement, assembly, and manufacturing shall be clearly defined and tracked in time as a function of changes. Each product entity (prototype, pilot model, or production article) shall be clearly linked to its design and requirements configuration definition.

4.7   Risk management

A means shall be in place to systematically identify potential cost, schedule, and performance risks and to mitigate these risks over the life of the program such that their effects can be positively responded to should they materialize during the period the program is in force. This activity must deal with all potential risks and include provisions to: encourage and accept identification of risks from all parties to the program; evaluate candidate risks relative to the probability of occurrence and the seriousness of impact; assign responsibility for mitigation; and review current status resulting in clear direction of any appropriate actions based on current status.

section 5

Verification

5.1 Verification defined

Verification is a process of establishing proof that a design satisfies the requirements for which it was developed. This is extended to include gaining confidence in one's ability to satisfy requirements through design as a precursor to the actual design work. These goals are achieved through planned verification tasks driven by the requirements contained in program specifications. The company uses a three-stage verification process. The first stage is called *qualification* and is driven by the development specifications. The second stage is called *acceptance* and is driven by product specifications. Finally, a collection of accepted product components, all having been previously qualified, will generally have to be subjected to system level verification to verify compliance with system specification content.

5.2 Verification methods

Four methods of verification have commonly been applied: analysis, test, demonstration, and examination. These methods apply to both qualification and acceptance.

5.2.1 Analysis

Special analyses shall be conducted to verify compliance with requirements where testing is not feasible or can be avoided without incurring risk. In the past, the quality of the evidence produced through testing has generally been preferred to that produced through analysis. Analysis has been generally preferred to testing as a function of cost. With the advent of computer-aided design and manufacturing supported by integrated models and simulation capabilities, analytical capabilities focused on these computerized models are becoming better and are encouraged. In that these models are available early in programs, they should be applied as early as possible to confirm requirements compliance of alternative and preferred design solutions. The precise mix of test and analysis for verification on a program involves a balance between evidence quality and cost of acquisition of that evidence.

Analyses for specific items shall be performed by the responsible IPT. Analyses that span the boundaries established for one or more IPT shall be performed by the lead plant IPT, which is also responsible for managing the overall program analysis process for the purpose of verifying requirements. On programs with more than two or more levels of teams, each team shall be responsible for all subordinate analytical work and should review the results.

5.2.2   Test

A test involves manipulation, commonly in some fairly complex way, of a test article by some combination of human test engineers and a test apparatus. A test apparatus causes the product entity to be stimulated in some controlled way and record the response compared to a standard. Testing shall be applied to product development for one of three purposes when it is determined to be the most cost-effective way of accomplishing the corresponding ends. Subordinate paragraphs describe each of these three purposes and provide guidance in each case.

Every test shall be characterized by a goal, test requirements, a test plan, and/or procedure. The goal defines the overall objectives. The test requirements provide detailed definition of what must be accomplished and should be covered in the corresponding product specification or result from grouping together requirements from several sources in the interest of economy. Test plans offer guidance on the use of company resources to accomplish desired test results defined in test goals and requirements. Test procedures define in detail what has to be done to satisfy test goals and requirements. All test goals, requirements, plans, and procedures may be grouped together into an integrated program test plan.

Software, firmware, application-specific integrated circuit (ASIC), and programmable gate array code should be developed using computer development tools and in that context shall be tested throughout its development process.

5.2.3   Other verification methods

Requirements may also be verified by demonstration or examination. Validation by demonstration is appropriate where human activities are involved. An activity is accomplished, and in the process it is shown that the requirement has been satisfied. Examination is a process of observation of a reality and comparison of results with a standard without the use of complex instruments.

Where items have been used in previous system applications with similar environmental influences and requirements, it may be possible through analysis to show that further testing is not required for the new application because the product has been qualified for the application based on similarity with the prior application. All items falling into this category shall be evaluated in a design equivalency review. Those that pass this review shall not require new validation testing or analysis for the application. Those that fail to pass this review shall be subjected to appropriate testing or analysis actions in accordance with the new program requirements.

5.3 Qualification verification

New and modified products must be proven adequate for their application through qualification at the component and/or end item level. This process entails establishing that the item design satisfies the development requirements contained in the item specification. Qualification evidence is acquired throughout the verification methods discussed above in the form of reports containing the evidence of compliance with the requirements.

It is preferred that programs prepare and maintain a record of traceability between specification requirements and the verification evidence using the same requirements computer system used to capture and maintain the specification content. As a minimum, a compliance traceability matrix should be maintained, showing the relationship between individual requirements, the verification tasks where they will be verified. A separate task matrix, ideally coordinated from the same data used to generate the compliance matrix, is encouraged, showing the relationship between verification tasks and reports for task results, task responsibilities, and giving the status of those tasks.

5.4 Functional configuration audit

The compliance evidence should be reviewed internally and a decision made as to whether or not that evidence supports the conclusion that the product design satisfies specification content. The evidence in the form of reports should also be made available to the customer, independent agent, certifying agent, or other party with legitimate interest in the qualification evidence for independent review. The results of these reviews should be consummated at a major program review where a formal decision is reached and agreed upon by all parties that the system is qualified. On a large program entailing many items, this review may be segmented into reviews for each item, followed by a system-level review to account for any discrepancies identified in segment reviews.

5.5 Acceptance verification

Every product item produced or some periodic sample thereof should be examined to verify that it is acceptable to the customer base as defined in a product or detail specification. Any acceptance testing should be structured to avoid reducing product life while verifying compliance. Records should be captured of the results of acceptance verification.

5.6   *Physical configuration audit*

The results of acceptance work applied to the first production article should be examined carefully for compliance with the appropriate specification by all parties to the program, including any customers covered through a contractual relationship with the company. The results of this review should be consummated in a major program review where a conclusion is formally arrived at regarding the acceptability of the product. As in qualification verification, the acceptance audit may be segmented followed by a system audit.

5.7   *System verification*

When dealing with a complex product system, it may be necessary to follow up the qualification process with a series of system level tests that combines qualified items into a complete or partial system subjected to tests crafted to expose the product capabilities relative to system specification content.

Systems engineering management requirements

6.1 Organization and facilitization

Programs are encouraged to organize using integrated product teams (IPT) consisting of people from the specialized disciplines that relate to the technologies to be applied, implementing a concurrent product and process development approach.

Physical collocation of the team should be used to the maximum practical extent because the effectiveness of human communication drops off very rapidly as the distance between persons increases. Since it is through human communications that the adverse effects of specialization are overcome, physical collocation is very important. Where this is not practical, the use of virtual office techniques can be partially effective in minimizing the communications loss. This may include the use of computer networks through which a program or team common database can be maintained, good telephone communications, and videoconferencing.

6.2 Team structures

The IPT approach should be applied throughout program organizational structures. These teams are formed from personnel assigned to them based on the technologies (structural, electronic, hydraulic, or propulsion for example) that will have to be applied to solve the team design problem. A machine-level team shall be led by the overall program manager. Subordinate teams shall be formed at supplier divisions, and all teams, including non-company suppliers, shall be managed relative to their cost, schedule, and performance results relative to plan.

6.3   Program management support

An organized method shall be put in place for gaining management insight into the results of development team work, coupled with a means to compare these results with planned work and to direct the work of teams based on these insights and program goals and requirements. Control measures should be established to monitor and influence program cost and schedule as defined in program planning and product performance as defined in product specifications.

6.3.1   Cost control

A means shall be in place to allocate and collect development and production cost to/for specific product entities. Cost performance shall be tracked as a means of assuring that cost targets will be satisfied at some level.

6.3.2   Schedule control

A means shall be in place to define planned work scheduling and to track work performance and results to that schedule at some level.

6.3.3   Product performance control

A means shall be in place to identify and track product performance in selected parameters. Select requirements that history has shown difficult to achieve on similar products, requirements that are critical to program success, and higher-level requirements that drive the design. Track the values of these requirements over time along with demonstrated capability and future trend predictions. Take actions necessary to cause closure between required values and demonstrated values. This process is called technical performance measurement (TPM).

6.4   Program phasing and reviews

Program management shall establish the program schedule keyed to the Generic company phasing milestones and interspersed by major reviews, as described in Section 2. All reviews will examine current program status relative to phase goals established prior to entry into the phase. A decision shall be made relative to future program actions: program termination, adjustment, or continuation as planned.

6.5   Risk

An effective program shall be in place to identify risks, assign those risks to specific persons charged with mitigating them in accordance with a plan, and periodically review the program risk status. Risk status should be reviewed at every team meeting, program meeting, and program major review.

6.6   Supplier technical management

Programs shall manage the technical work of their suppliers by requiring contractual compliance with appropriate program and process requirements.

Program system engineering maturity assessment

7.1   Assessment process

The company employs a two-dimensional assessment system to monitor system engineering performance and assure that all programs are applying an effective systems approach based on a specific criteria defined in this appendix that is traceable to the contents of this manual. The first dimension is for program compliance with internal practices defined in this document. The second dimension is for benchmarking internal practices against recognized standards for performance of systems engineering. The former metric is focused on individual programs, while the latter is focused on the company as a whole, since the enterprise employs a common systems approach on all programs. Paragraph 7.2 covers the metrics used. Paragraph 7.3 provides the internal assessment criteria and questions related to that criteria, upon which the internal metric is based.

Figure C.5 illustrates the general process used to determine overall process metrics. For lower-tier practices, the corresponding task is selected on a specific program and performance audited against internal practices for that task. In the case of the overall systems engineering process that this document is focused on, the process definition is contained in this manual. A person is selected by the process owner, the enterprise integration team, to accomplish the audit for a specific program. That person conducts an audit based on a series of questions included in the paragraph 7.3.2 checklist. The result is a number in the range of 0 to 100. A program that faithfully implements this manual will receive a high score. A program that does not will suffer a low score.

Other specific metrics are also used to assess performance as well as the overall systems engineering metric discussed in this appendix. Those are defined by other means.

Programs are called upon to make improvements in implementation where low scores are detected. This will entail program study of the problem

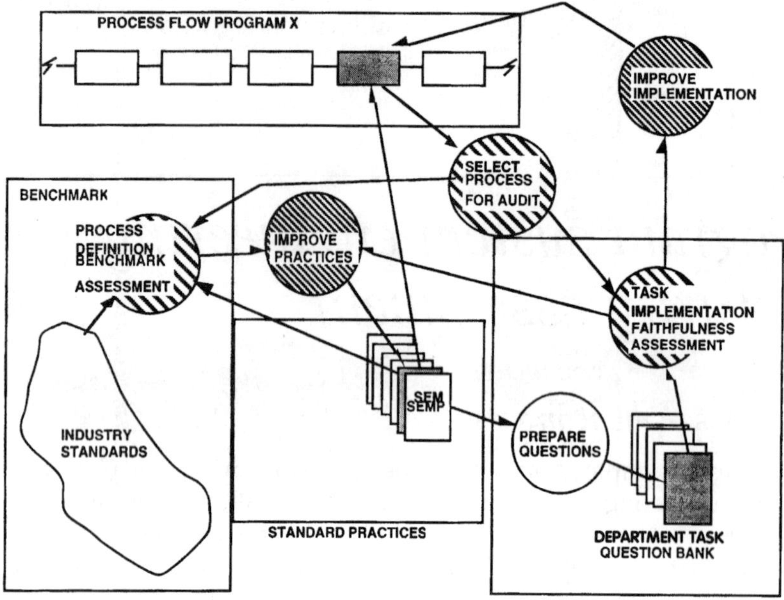

Figure C.5   System engineering audit process.

followed by program determination of a corrective course of action reviewed by the process owner. The process owner tracks program performance on all audit corrective action plans. Programs with low scores may also be selected for audit more frequently than those that exhibit higher scores.

At the same time or at some other time, this manual is checked against a recognized benchmark for compliance by the process owner or under their guidance. The most recent benchmarking score is used in establishing the combined system engineering metric for a program. This metric takes the form of X,Y, where X is the internal practices assessment figure and Y is the most recent benchmarking score. This combination tells how well the program is performing the system engineering process with respect to our internal practices and how those internal practices stack up against a particular standard of system engineering performance. The process owner is responsible for improvements suggested through benchmark audits to this *Systems Engineering Practices Manual* (SEPM).

7.2   *Assessment metrics and tracking*

The company uses two metrics to determine the adequacy of its systems engineering performance and to measure progress over time toward a goal of world-class systems engineering. This goal is a moving target because there is continuing progress in industry in improving the process. The company must avoid institutionalizing any of its practices. Rather, it must constantly be seeking ways to improve them. The company maintains contact with customers,

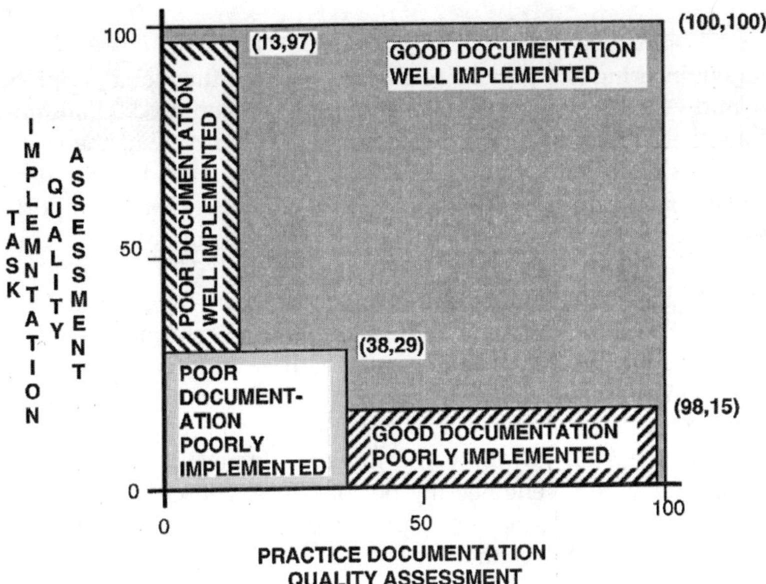

Figure C.6　Two-dimensional assessment scoring example.

other companies, and engineering societies to ensure access to the latest process methods and standards. Periodically the company conducts an assessment of this manual with respect to an accepted industry standard to ensure internal identification of the most effective techniques available.

Figure C.6 illustrates the two-dimensional nature of the overall systems engineering metric. The vertical axis corresponds to the internal practices metric covered in paragraph 7.3 below. The horizontal dimension is for the benchmarking figure determined by comparison of this document with an external source reference as described in paragraph 7.5.

The two metrics computed are M_y, the internal practices compliance metric (vertical axis of Figure C.6) and M_x, the practices quality metric (horizontal axis of Figure C.6). These should be charted in time in the Figure C.6 format, with each x,y data point annotated for date and program. Commonly the M_x metric will remain fixed for a period of time subsequent to an internal manual update and benchmarking exercise, while the M_y metric for a given program more frequently adjusts for improvements centered about the current practices. Over time, the metric M_x,M_y should move up and to the right toward the top right corner of the chart. If it remains stationary, it indicates a lack of process improvement.

An aggregate M_y metric for the whole company may be calculated and tracked by averaging the individual program M_y metrics, with or without scaling factors for relative program importance.

It is the responsibility of the EIT to respond to low M_x metrics uncovered through audits, and it is the responsibility of a program PIT to respond to a low M_y metric with corrective action.

7.3 Minimum systems engineering criteria

System engineering is an interdisciplinary, or cross-functional, approach to evolve and verify an integrated balanced set of system product and process solutions that satisfy customer needs across the life cycle of the system. It encompasses the scientific and engineering efforts related to the development, manufacturing, verification, deployment, operations, support, and disposal of system products and processes. Twenty minimum system engineering process requirements are listed in paragraph 7.3.1 below. A process that satisfies these minimum requirements and where the indicated work is executed well and in a timely way is by definition performing an adequate system engineering process. In paragraph 7.4 these requirements are mapped to the content of this document by paragraph numbers to demonstrate document completeness with respect to this criteria and provide assurance that programs which seek to follow the process defined in this manual will receive accurate system engineering performance metrics when assessed in accordance with the criteria. Programs must satisfy these minimum standards in planning and execution.

Paragraph 7.3.2 provides a set of questions in checklist form that the person accomplishing a specific program systems engineering audit will use to derive the metric for the aggregate effect of the criteria listed in paragraph 7.3.1.

7.3.1 The criteria

a. *The Customer Need.* A clear statement of system need shall be developed by or agreed to by the customer. The customer may be a single, clearly identified entity as in a collective entity defined by a marketing function on a commercial program. In the latter case, this criterion may be provided by citing a perceived customer need based on market research that must be satisfied.

b. *System Requirements Capture.* A set of system requirements shall be derived from the need to define the product mission, system performance goals/requirements/figures of merit, external interface definition, system environment definition, and top-level specialty engineering constraints in the simplest format acceptable to the customer.

c. *Problem Space Decomposition.* An organized means shall be in place to decompose the large problem expressed by the customer's need into a more detailed understanding of the problem space and assignment or allocation of the functionality thus exposed to elements in the system architecture (or physical solution space) consisting of hardware, computer software, facilities, materials, and personnel.

d. *Cost Definition.* A means shall be in place to allocate and collect recurring and nonrecurring cost to/for specific product entities and capture them within the company's, and possibly the customer's, cost accounting system.

Cost performance shall be tracked as a means of ensuring that cost target will be satisfied.

e. *Schedule Definition.* A means shall be in place to define planned work scheduling and to track work performance to that schedule.

f. *Development Responsibility.* An organized method shall be in place for clearly and comprehensively assigning responsibility for development of the product system elements and their associated employment process to personnel and/or teams. Where the customer requires financial reporting, this method must be mutually consistent with customer's definition of work breakdown (in support of effective cost collection) and the company development team concept applied on the program.

g. *Requirements Before Design.* A means shall be in place to define and enforce the definition of requirements for items prior to design work being permitted. It may be necessary to define an item concept prior to completion of some of the design constraints, but these must be in place and approved by appropriate authority before a commitment to detailed design is initiated.

h. *Requirements Capture and Publication.* A means shall be in place to capture all requirements, correlated with their architecture item, and to make them available for reference, discussion, and critical review by all program personnel. If required by the customer, specifications shall be published and maintained in a format compatible with customer documentation requirements. Programs may use requirements database content directly without paper publication if desired and if it is not inconsistent with customer requirements.

i. *Traceability and Flowdown.* Traceability of requirements to sources, other requirements, and the development process shall be accomplished. Ideally, all requirements should trace to the ultimate requirement, the customer's need. This process shall ensure that higher-tier requirements have properly influenced lower-tier requirements, especially for procured items and that customer requirements have been appropriately flowed down to lower-tier elements. This requirement encourages the use of a computer database tool for the purpose of capturing requirements and providing for traceability.

j. *Requirements Verification.* An organized means shall be in place for verifying that the final product design satisfies the requirements prescribed. This means shall provide for associating specific compliance evidence in the form of analysis and test reports to specific requirements. This requirement encourages the use of a computer database tool linked to the requirements database suggested in item "i" above.

k. *Interface Development.* An organized way shall be devised and put in place to develop compatible internal and external interfaces between elements depicted on the system architecture, to minimize the cross-organizational interface, and to clearly assign responsibility for the development of each interface. Interfaces between items under the control of the company and associates will be developed in accordance with a plan approved by all parties including the customer. Interfaces with items to be provided by suppliers shall be controlled by the same company agent responsible for the item being procured.

l. *Design Capture.* An organized way shall be put in place to capture and retain the results of development team product and process concept and design work so as to be easily accessible at any time by all program personnel and available for iterative improvement by responsible engineers and analysts.

m. *Management Oversight.* An organized method shall be put in place for gaining management insight into the results of development team work, coupled with a means to compare these results with planned work and to direct the work of teams based on these insights and program goals and requirements.

n. *System Integration.* An organized means shall be in place to integrate the work of separate product development teams and to optimize at the system level across design, manufacturing, quality, test, operations, and logistics support to overcome problems resulting from suboptimization at the individual team level or individual discipline.

o. *Specialty Engineering Integration.* A means shall be put in place to perform product and process development work concurrently. The concurrent development process shall encourage cooperative development work between all relevant specialists so as to apply the right disciplines at the best possible time and period to achieve the most desirable results in the following areas unless in conflict with program requirements:

1. This shall include a means to define what the system and its components must do in the customer's environment to maintain needed readiness and operational capability and to attain mission success.
2. This method must provide for development and evaluation of the system maintenance and logistic support concepts and coordination of those concepts with the evolving system design concept and a mechanism for gaining insight into needed procedural and human task-oriented requirements.
3. The means shall also provide for evaluation of the system and its components for reliable and safe operation.
4. The concurrent development mechanism shall result in development of requirements and plans for testing, inspection, and production processes concurrently with product design requirements and processes.

p. *Design Rationale Capture.* A means shall be put in place to capture design rationale and traceability of the decision process that led to the preferred system concept. All requirements and the designs corresponding to those requirements shall be reviewed by responsible management personnel and approved. Relevant details of these reviews shall be captured for the purpose of communicating direction and for subsequent reference.

q. *Requirements Configuration Control.* A means shall be put in place to clearly define the current system requirements and concept baseline during the period leading up to the P review. Subsequent to the P review the requirements baseline shall be captured in a set of specifications and reports, or computer data equivalents.

r. *Design Configuration Control.* A means shall be put in place to positively control the configuration of the product design subsequent to the formal approval of that design and to control the configuration of all representations of that design in the form of models, simulations, mock-ups, test articles, test results, and analyses. On some programs design configuration management can be delayed until the P review but should begin no later.

s. *Product Configuration Control.* The configuration of physical products produced through procurement, assembly, and manufacturing shall be clearly defined and tracked in time as a function of changes. Each product entity shall be clearly linked to its design and requirements configuration definition.

t. *Continuity Restitution.* There shall be in place a means to detect that a discontinuity has occurred in planned program work and to reach sound decisions that will return the development process to a condition of control.

u. *Risk Management.* A means shall be in place to systematically identify potential cost, schedule, and performance risks and to mitigate these risks over the life of the program such that their effects can be positively responded to should they materialize during the period the program is in force.

7.3.2   *Internal practices checklist*

The final three pages of this appendix provide a systems engineering internal practices checklist, a copy of which may be used directly for program audit purposes. It includes a series of questions for each item in the paragraph 7.3.1 criteria list. To use this checklist note the weighting factor (WT) and place that number in the SCORE column for that question if the answer is yes. Make no entry if the answer is no. Upon completion of the questions, add up the numbers in the SCORE column to yield S_t (score total), divide by 50 (ΣW_t, weighting factor total), and multiply the quotient by 100 to give the final metric value in the range 0 to 100. This metric M_v corresponds to the vertical axis of Figure C.6.

In the case of each question a simple answer by a program person should not be the only source of information. The assessor should gain access to tangible evidence of corresponding program performance in the form of a document, entered computer information, or observed behavior.

Depending on the program phase at the time of the assessment, the questions may have a slightly different flavor. In an early phase, there will be no evidence of accomplishment of later phase work, but there should be evidence of corresponding planning. In later program phases, there will be no current activity on earlier work, but there should be evidence of having completed the related work in the earlier phase.

7.4 System engineering process criteria traceability

Paragraph 7.3.1 lists a minimum system engineering process criteria and Tables C.1 and C.2 map the criteria to the content of this document for the purpose of demonstrating process coverage completeness with respect to the company's own system engineering criteria. Refer to paragraph 7.3.1 for process criteria names. Table C.1 covers criteria a through k, and Table C.2 picks up the remainder, l through u. Table C.3 maps this same criteria to the generic company program phases illustrated in Figure C.1 and described in related text.

Table C.1   Content to Process Criteria Map, Criteria Group 1

SEPM Para	Title	a	b	c	d	e	f	g	h	i	j	k
2.1	Program Environment											
2.2	Program Phasing and Development Process											
2.2.1	Development (F1)											
2.2.1.1	Concept Phase (F11)											
2.2.1.2	Preliminary Design Phase (F12)											
2.2.1.3	Detailed Design Phase (F13)											
2.2.2	Acquire Material (F2)											
2.2.3	Manufacture System (F3)											
2.2.3.1	Manufacture Qualification Items											
2.2.3.2	First Article											
2.2.3.3	Low Rate Production											
2.2.3.4	High Rate Production											
2.2.4	Verify System (F4)										X	
2.2.4.1	System Test and Evaluation (F41)											
2.2.4.2	Item Qualification (F42)											
2.2.4.3	Manage Verification (F43)											
2.2.5	Logistics Support (F5)											
2.2.6	Deliver/Deploy System (F6)											
2.2.7	Operate System (F7)											
2.2.8	Modify System (F8)											
2.2.9	Dispose of System (F9)											
2.2.10	Manage Program (FM)											
2.3	Major Reviews and Audits											
2.3.1	System Requirements Review (SRR)											

Table C.1 (*continued*)   Content to Process Criteria Map, Criteria Group 1

SEPM Para	Title	a	b	c	d	e	f	g	h	i	j	k
2.3.2	System Design Review (SDR)											
2.3.3	Preliminary Design Review (PDR)											
2.3.4	Critical Design Review (CDR)											
2.3.5	Functional Configuration Audit (FCA)											
2.3.6	Physical Configuration Audit (PCA)											
2.4	Development Process Steps											
2.4.1	Develop Concept, F1											
2.4.1.1	Customer Requirements Definition	X	X									
2.4.1.2	Development Requirements Analysis							X				
2.4.1.3	Product Requirements Analysis											
2.4.1.4	Combined Requirements Analysis											
2.4.1.5	Use of Computer Tools											
2.4.2	Solve the Problem											
2.4.3	Proof Through Verification											
3.1	Requirements Defined											
3.2	Requirements Identification Timing							X				
3.3	Requirements Categories											
3.4	Structured vs. *ad hoc* Approach											
3.4.1	The Customer Need	X										
3.4.2	Product Functionality		X	X								
3.4.3	Product Architecture Definition						X					
3.4.4	Product Interfaces Definition											X
3.4.5	Environmental Requirements Definition											
3.4.6	Specialty Engineering Requirements Analysis											
3.5	Responsibility Assignment						X					
3.6	Team Structures						X					
3.7	Requirements Validation											
3.8	Requirements Documentation								X			
3.8.1	Specification Types											
3.8.2	Specification Review and Approval								X			
3.8.3	Specification Release and Control								X			
3.8.4	Requirements Support Tools and Applications											
3.8.5	Requirements Relationships									X		
4.1	Design											
4.1.1	Design Responsibility						X					
4.1.2	Design Capture											
4.2	Integration											
4.2.1	Integration and Optimization Responsibility											
4.2.2	System Integration											
4.2.3	Interface Development											X
4.2.4	Specialty Engineering											
4.2.4.1	Reliability and Warranty Provisions											

Table C.1 (continued) Content to Process Criteria Map, Criteria Group 1

SEPM		Generic process criteria										
Para	Title	a	b	c	d	e	f	g	h	i	j	k
4.2.4.2	Maintainability, Serviceability, and Supportability											
4.2.4.3	Product Safety and Human Factors											
4.2.4.4	Other Disciplines											
4.3	Product Representations											
4.4	Design Rationale Capture											
4.5	Development Configuration Control											
4.6	Product Configuration Control											
4.7	Risk Management											
5.1	Verification Defined										X	
5.2	Verification Methods										X	
5.2.1	Analysis										X	
5.2.2	Test										X	
5.2.3	Other Verification Methods										X	
5.3	Qualification Verification										X	
5.4	Functional Configuration Audit											
5.5	Acceptance Verification											
5.6	Physical Configuration Audit											
5.7	System Verification										X	
6.1	Organization and Facilitization											
6.2	Team Structures						X					
6.3	Program Management Support											
6.3.1	Cost Control				X							
6.3.2	Schedule Control					X						
6.3.3	Product Performance Control											
6.4	Program Phasing and Reviews											
6.5	Risk											
6.6	Supplier Technical Management											
A	Section 7, Systems Engineering Maturity Assessment											

Table C.2 Content to Process Criteria Map, Criteria Group 2

SEPM		Generic process criteria									
Para	Title	l	m	n	o	p	q	r	s	t	u
2.1	Program Environment										
2.2	Program Phasing and Development Process										
2.2.1	Development (F1)										
2.2.1.1	Concept Phase (F11)										
2.2.1.2	Preliminary Design Phase (F12)	X									
2.2.1.3	Detailed Design Phase (F13)	X									

Table C.2 (continued) Content to Process Criteria Map, Criteria Group 2

SEPM		Generic process criteria								
Para	Title	l	m	n	o	p	q	r	s	t	u
2.2.2	Acquire Material (F2)										
2.2.3	Manufacture System (F3)										
2.2.3.1	Manufacture Qualification Items										
2.2.3.2	First Article										
2.2.3.3	Low-Rate Production										
2.2.3.4	High-Rate Production										
2.2.4	Verify System (F4)										
2.2.4.1	System Test and Evaluation (F41)										
2.2.4.2	Item Qualification (F42)										
2.2.4.3	Manage Verification (F43)										
2.2.5	Logistics Support (F5)										
2.2.6	Deliver/Deploy System (F6)										
2.2.7	Operate System (F7)										
2.2.8	Modify System (F8)										
2.2.9	Dispose of System (F9)										
2.2.10	Manage Program (FM)										
2.3	Major Reviews and Audits										
2.3.1	System Requirements Review (SRR)										
2.3.2	System Design Review (SDR)										
2.3.3	Preliminary Design Review (PDR)										
2.3.4	Critical Design Review (CDR)										
2.3.5	Functional Configuration Audit (FCA)										
2.3.6	Physical Configuration Audit (PCA)										
2.4	Development Process Steps										
2.4.1	Develop Concept, F1										
2.4.1.1	Customer Requirements Definition										
2.4.1.2	Development Requirements Analysis										
2.4.1.3	Product Requirements Analysis										
2.4.1.4	Combined Requirements Analysis										
2.4.1.5	Use of Computer Tools										
2.4.2	Solve the Problem										
2.4.3	Proof Through Verification										
3.1	Requirements Defined										
3.2	Requirements Identification Timing										
3.3	Requirements Categories										
3.4	Structured vs. *ad hoc* Approach										
3.4.1	The Customer Need										
3.4.2	Product Functionality										
3.4.3	Product Architecture Definition										
3.4.4	Product Interfaces Definition										
3.4.5	Environmental Requirements Definition										
3.4.6	Specialty Engineering Requirements Analysis							X			

Table C.2 (continued) Content to Process Criteria Map, Criteria Group 2

SEPM Para	Title	l	m	n	o	p	q	r	s	t	u
3.5	Responsibility Assignment										
3.6	Team Structures										
3.7	Requirements Validation										
3.8	Requirements Documentation										
3.8.1	Specification Types										
3.8.2	Specification Review and Approval						X				
3.8.3	Specification Release and Control						X				
3.8.4	Requirements Support Tools and Applications										
3.8.5	Requirements Relationships										
4.1	Design										
4.1.1	Design Responsibility										
4.1.2	Design Capture	X									
4.2	Integration			X							
4.2.1	Integration and Optimization Responsibility			X							
4.2.2	System Integration			X							
4.2.3	Interface Development										
4.2.4	Specialty Engineering				X						
4.2.4.1	Reliability and Warranty Provisions										
4.2.4.2	Maintainability, Serviceability, and Supportability										
4.2.4.3	Product Safety and Human Factors										
4.2.4.4	Other Disciplines										
4.3	Product Representations										
4.4	Design Rationale Capture	X				X					
4.5	Development Configuration Control							X			
4.6	Product Configuration Control								X		
4.7	Risk Management										
5.1	Verification Defined										
5.2	Verification Methods										
5.2.1	Analysis										
5.2.2	Test										
5.2.3	Other Verification Methods										
5.3	Qualification Verification										
5.4	Functional Configuration Audit										
5.5	Acceptance Verification										
5.6	Physical Configuration Audit										
5.7	System Verification										
6.1	Organization and Facilitization										
6.2	Team Structures										
6.3	Program Management Support										
6.3.1	Cost Control		X								
6.3.2	Schedule Control		X								
6.3.3	Product Performance Control		X								

Table C.2 (continued)   Content to Process Criteria Map, Criteria Group 2

SEPM Para	Title	l	m	n	o	p	q	r	s	t	u
6.4	Program Phasing and Reviews		X								
6.5	Risk									X	X
6.6	Supplier Technical Management										
7	Systems Engineering Maturity Assessment										

Table C.3   Generic Phase to Process Criteria Map

Program Phase	a	b	c	d	e	f	g	h	i	j	k	l	m	n	o	p	q	r	s	t	u
Concept	X	X	X	X	X	X	X	X	X		X		X		X		X			X	X
Prelim Design				X	X	X	X	X	X		X	X	X	X	X	X	X	X		X	X
Detail Design				X	X	X		X	X		X	X	X	X	X	X	X	X		X	X
Test and Eval				X	X	X				X			X				X	X		X	X
Production				X	X	X							X				X	X	X	X	X

7.5 *Practice documentation quality assessment*

The external systems engineering assessment process begins with the selection of a standard for comparison. This standard is reviewed by the process owner, and each specific requirement identified, ideally by a paragraph number. Each of these references is listed in tabular form and mapped to the content of this manual. Where the item maps to this manual, a score in the range of 1 to 10 is assigned, using the list offered below for guidance. If no corresponding practices content can be found, a score of 0 is recorded. Intermediate scores may be entered. Obviously, there is a degree of subjectivity in this scoring process and, ideally, more than one person should contribute to the scoring within the boundaries established by time and budget.

   0   There is no reference in company practices to this matter
   1
   2
   3
   4
   5   Company practices address this area, but improvements can be made
   6
   7
   8
   9
 10   Company practices are fully compliant or judged to be better than the external standard

The aggregate count of these mapping events (ΣE_i) is divided by 10 times the total number of source document requirements ($10E_t$) and multiplied by 100 to derive a metric in the range of 0 to 100. The formula for this metric, therefore, is:

$$M_x = 100\Sigma\ E_i/10E_t$$

This external audit should be accomplished or updated under any of the following circumstances:

a. Upon initial release of this manual.
b. When this manual undergoes a significant change.
c. When a different external standard has been selected.
d. If three years have passed since the most recent audit.

7.6 *Traceability to external standards*

The content of this practice may be mapped to the content of external standards such as IEEE Standard on Systems Engineering, 1220, and the results captured in a table similar to Table C.4. In the process of doing so, the content of this practice may have to be adjusted to ensure a good score as defined in paragraph 7.5. Also, the content of IEEE 1220 should be tailored to delete and/or edit any portions thought not to be consistent with the enterprise product line, customer base, or culture. In that this enterprise system engineering standard was constructed for traceability to IEEE 1220, it has by definition a score of 100 relative to the tailored standard. The enterprise integration team may choose in the future to map this practice to other external standards, such as EIA 632, as they mature and gain a degree of acceptance.

Table C.4 IEEE 1220 Traceability to This Practice			
IEEE		Practice	
Para	Title	Para	Title

SYSTEMS ENGINEERING
INTERNAL PRACTICES CHECKLIST
PROGRAM AUDIT FORM

PROGRAM ____ DATE _______________

AUDITOR _______________________________

	QUESTION	WT	SCORE
A1	Does a customer need statement exist?	1	_____
A2	If the customer need statement exists, ask three people picked at random on the program to tell you what the customer need is. If two can show you a copy of the need, point to a document which contains it or verbally give an interpretation of it that in your judgment is in agreement with the actual statement, score the question yes.	1	_____
B1	Does a system specification, or a specification for the highest level item for which the company program is responsible, exist?	2	_____
C1	Determine how the architecture of the system was determined or is being determined. If an organized decomposition process defined in this manual was used for those portions of the system that entail new development, score the question yes. Yes may be assigned if a very large percentage of the system is predetermined and an impact analysis was accomplished.	2	_____
D1	Does the program have a cost structure defined for all work with allocations made for specific cost centers?	1	_____
D2	Is cost tracked to the predetermined cost allocations and management effort effectively applied to encourage cost compliance?	1	_____
E1	Does a current program schedule exist?	1	_____
E2	Is work accomplished tracked to the schedule and management effort effectively applied to encourage schedule compliance?	1	
F1	Is all program work defined as in a statement of work?	1	_____
F2	Has all of the work been clearly allocated to specific organizational entities?	1	_____
F3	Is the work allocated to specific organizational entities traceable to the cost and schedule requirements?	1	_____
G1	Is there evidence that requirements are being or have been developed and approved before detailed design is undertaken?	3	_____
H1	Does a program specification tree exist?	1	_____
H2	Review the program specification tree. Is there a clearly defined mechanism for identification of specification types, forms, and formats?	1	_____
H3	Select one specification at random and determine if it follows the program standard for specifications. Does it comply?	1	_____
I1	Does the program maintain requirements traceability data?	1	_____
I2	Examine traceability data. Are you able to find inconsistencies in the data within a period of no more than 10 minutes?	1	_____
J1	Do specifications contain a verification methods matrix?	1	_____

SYSTEMS ENGINEERING
INTERNAL PRACTICES CHECKLIST
PROGRAM AUDIT FORM

QUESTION	WT	SCORE
J2 Does evidence exist of verification planning in the form of a verification compliance matrix, verification planning documentation, or integrated test planning?	1	_____
K1 Is there a clear definition of interface responsibilities on the program in the form of a schematic block diagram or N-square diagram with some means to correlate architecture responsibility with interface responsibility?	2	_____
L1 Are the results of product and process design work captured in an enduring way that can easily be accessed by all program personnel?	2	_____
M1 Does the program have an effective way for management to gain insight into work performance and status?	2	_____
N1 Does the program have an effective means to integrate the product work of program personnel and teams?	2	_____
N2 Does the program have an effective means to integrate the product and process work on the program to the end that manufacturing, quality, test, and logistics processes are compatible and optimum with respect to the product design?	2	_____
O1 Is there a written list defining what specialty engineering disciplines shall be applied on the program?	1	_____
O2 Are all specialty disciplines involved in the requirements definition process?	1	_____
O3 Is there evidence of designer-specialty engineer interaction to understand specialty engineering requirements?	1	_____
O4 Are the documented results of specialty engineering assessments of the design included within the verification evidence?	1	_____
P1 Can the program produce the rationale for design decisions and actions?	2	_____
Q1 Is there evidence of an ability to baseline the requirements at specific points in time and that these baselines are respected?	2	_____
R1 Does the program actively control the configuration of product representations (models, test articles, simulations, etc.)?	2	_____
S1 Does the program have a means to clearly control the configuration of specific designs and their physical product entities?	2	_____
T1 If there has been any history of discontinuity in past program work, is there evidence of having expeditiously identified and corrected the problem?	2	_____
U1 Does the program have a means to identify cost, schedule, and technical risks, a way to assess their impact, assign mitigation responsibility, and track status?	3	_____
TOTAL SCORE (ΣS_i)		_____
TOTAL WEIGHT (ΣW_t)	50	

$$M_y = 100 \Sigma \ S_i / \Sigma W_t = 100 \ (\ )/(50) = \underline{\hspace{2cm}}$$

AUDIT REMARKS

Index